AF678182

METHODS & NEW FRONTIERS IN NEUROSCIENCE

Series Editors
Sidney A. Simon, Ph.D.
Miguel A.L. Nicolelis, M.D., Ph.D.

Published Titles

Apoptosis in Neurobiology
Yusuf A. Hannun, M.D., Professor of Biomedical Research and Chairman/Department of Biochemistry and Molecular Biology, Medical University of South Carolina
Rose-Mary Boustany, M.D., tenured Associate Professor of Pediatrics and Neurobiology, Duke University Medical Center

Methods for Neural Ensemble Recordings
Miguel A.L. Nicolelis, M.D., Ph.D., Professor of Neurobiology and Biomedical Engineering, Duke University Medical Center

Methods of Behavioral Analysis in Neuroscience
Jerry J. Buccafusco, Ph.D., Alzheimer's Research Center, Professor of Pharmacology and Toxicology, Professor of Psychiatry and Health Behavior, Medical College of Georgia

Neural Prostheses for Restoration of Sensory and Motor Function
John K. Chapin, Ph.D., Professor of Physiology and Pharmacology, State University of New York Health Science Center
Karen A. Moxon, Ph.D., Assistant Professor/School of Biomedical Engineering, Science, and Health Systems, Drexel University

Computational Neuroscience: Realistic Modeling for Experimentalists
Eric DeSchutter, M.D., Ph.D., Professor/Department of Medicine, University of Antwerp

Methods in Pain Research
Lawrence Kruger, Ph.D., Professor or Neurobiology (Emeritus), UCLA School of Medicine and Brain Research Institute

Motor Neurobiology of the Spinal Cord
Timothy C. Cope, Ph.D., Professor of Physiology, Emory University School of Medicine

Nicotinic Receptors in the Nervous System
Edward D. Levin, Ph.D., Associate Professor/Department of Psychiatry and Pharmacology and Molecular Cancer Biology and Department of Psychiatry and Behavioral Sciences, Duke University School of Medicine

Methods in Genomic Neuroscience
Helmin R. Chin, Ph.D., Genetics Research Branch, NIMH, NIH
Steven O. Moldin, Ph.D, Genetics Research Branch, NIMH, NIH

Methods in Chemosensory Research
Sidney A. Simon, Ph.D., Professor of Neurobiology, Biomedical Engineering, and Anesthesiology, Duke University
Miguel A.L. Nicolelis, M.D., Ph.D., Professor of Neurobiology and Biomedical Engineering, Duke University

The Somatosensory System: Deciphering the Brain's Own Body Image,
Randall J. Nelson, Ph.D., Professor of Anatomy and Neurobiology, University of Tennessee Health Sciences Center

New Concepts in Cerebral Ischemia
Rick C. S. Lin, Ph.D., Professor of Anatomy, University of Mississippi Medical Center

DNA Arrays: Technologies and Experimental Strategies
Elena Grigorenko, Ph.D., Technology Development Group, Millennium Pharmaceuticals

Methods for Alcohol-Related Neuroscience Research
Yuan Liu, Ph.D., National Institute of Neurological Disorders and Stroke, National Institutes of Health
David M. Lovinger, Ph.D., Laboratory of Integrative Neuroscience, NIAAA

IN VIVO OPTICAL IMAGING OF BRAIN FUNCTION

Edited by
Ron D. Frostig

CRC PRESS

Boca Raton London New York Washington, D.C.

Library of Congress Cataloging-in-Publication Data

Frostig, Ron D.
In vivo optical imaging of brain function / Ron D. Frostig.
p. ; cm. — (Methods & new frontiers in neuroscience)
Includes bibliographical references and index.
ISBN 8493-2389-4
1. Brain—Imaging. I. Title. II. Series.
[DNLM: 1. Brain—physiology. 2. Diagnostic Imaging. WL 300 F939i 2002]
QP376.6 .F765 2002
612.8′2—dc21 2002019238
CIP

This book contains information obtained from authentic and highly regarded sources. Reprinted material is quoted with permission, and sources are indicated. A wide variety of references are listed. Reasonable efforts have been made to publish reliable data and information, but the author and the publisher cannot assume responsibility for the validity of all materials or for the consequences of their use.

Neither this book nor any part may be reproduced or transmitted in any form or by any means, electronic or mechanical, including photocopying, microfilming, and recording, or by any information storage or retrieval system, without prior permission in writing from the publisher.

All rights reserved. Authorization to photocopy items for internal or personal use, or the personal or internal use of specific clients, may be granted by CRC Press LLC, provided that $.50 per page photocopied is paid directly to Copyright Clearance Center, 222 Rosewood Drive, Danvers, MA 01923 USA. The fee code for users of the Transactional Reporting Service is ISBN 0-8493-2389-4/02/$0.00+$1.50. The fee is subject to change without notice. For organizations that have been granted a photocopy license by the CCC, a separate system of payment has been arranged.

The consent of CRC Press LLC does not extend to copying for general distribution, for promotion, for creating new works, or for resale. Specific permission must be obtained in writing from CRC Press LLC for such copying.

Direct all inquiries to CRC Press LLC, 2000 N.W. Corporate Blvd., Boca Raton, Florida 33431.

Trademark Notice: Product or corporate names may be trademarks or registered trademarks, and are used only for identification and explanation, without intent to infringe.

Visit the CRC Press Web site at www.crcpress.com

© 2002 by CRC Press LLC

No claim to original U.S. Government works
International Standard Book Number 0-8493-2389-4
Library of Congress Card Number 2002019238
Printed in the United States of America 1 2 3 4 5 6 7 8 9 0
Printed on acid-free paper

Methods & New Frontiers in Neuroscience

Series Editors

Sidney A. Simon, Ph.D.

Miguel A.L. Nicolelis, M.D., Ph.D.

Our goal in creating the Methods & New Frontiers in Neuroscience Series is to present the insights of experts on emerging experimental techniques and theoretical concepts that are, or will be, at the vanguard of neuroscience. Books in the series cover topics ranging from methods to investigate apoptosis to modern techniques for neural ensemble recordings in behaving animals. The series also covers new and exciting multidisciplinary areas of brain research, such as computational neuroscience and neuroengineering, and describes breakthroughs in classical fields like behavioral neuroscience. We want these to be the books every neuroscientist will use in order to get acquainted with new methodologies in brain research. These books can be given to graduate students and postdoctoral fellows when they are looking for guidance to start a new line of research.

Each book is edited by an expert and consists of chapters written by the leaders in a particular field. The books are richly illustrated and contain comprehensive bibliographies. Chapters provide substantial background material relevant to the particular subject. Hence, they are not only "methods books." They contain detailed "tricks of the trade" and information as to where these methods can be safely applied. In addition, they include information about where to buy equipment and web sites that are helpful in solving both practical and theoretical problems.

We hope that as the volumes become available, the effort put in by us, the publisher, the book editors, and individual authors will contribute to the further development of brain research. The extent to which we achieve this goal will be determined by the utility of these books.

Preface

These are exciting times for the emerging field of optical imaging of brain function. This field has advanced rapidly in recent years in its theoretical and technical domains and in its contributions to brain research, which have helped our understanding of brain function. Thus, the description of these techniques and their applications should be attractive to both basic and clinical researchers interested in brain function.

This book brings together, for the first time, a description of the wide variety of optical techniques that have recently emerged for the specific study of neuronal activity in the living brain and their application for animal and human functional imaging research. These *in vivo* techniques can vary by their level of temporal resolution (milliseconds to seconds), spatial resolution (microns to millimeters), degree of invasiveness to the brain (removal of the skull above the imaged area to complete noninvasiveness), and current application in clinical studies. Besides their usefulness for the study of brain function, these optical techniques are also appealing because they are typically based on compact, mobile, and affordable equipment; therefore, they can be more readily integrated into a typical neuroscience, cognitive science, or neurology laboratory as well as into a hospital operating room.

The chapters of this book describe theory, setup, analytical methods, and examples of experiments highlighting the advantages of the different techniques. Because the book focuses on *in vivo* imaging techniques, examples of the applications are focused on living brains rather than tissue cultures or brain slices. In addition, this book's focus on optical imaging techniques precludes nonoptical techniques such as fMRI, PET, and MEG techniques.

I have no doubt that the emerging study of brain function using optical imaging techniques, as described in this book, is just the beginning of a very promising future. Increased interest in these methods coupled with a continuous, rapid pace in relevant theoretical and technological advances should promote these techniques as major players in future brain research.

Ron Frostig

Editor

Ron Frostig, Ph.D. is an associate professor in the Department of Neurobiology and Behavior and the Center for Biomedical Engineering at the University of California, Irvine (UCI). He earned a B.Sc. in biology and M.Sc. in neurobiology from the Hebrew University, Jerusalem, Israel, and a Ph.D. in neurobiology from the University of California, Los Angeles (UCLA). Dr. Frostig did his postdoctoral training at the Rockefeller University in New York. In 1990 he joined the Department of Neurobiology and Behavior (formerly Psychobiology) as an assistant professor, and since 1997 he has been an associate professor of neurobiology. Since 2000 he has also been associate professor at the Center of Biomedical Engineering and has served as the director of graduate studies at the Department of Neurobiology and Behavior. His main research interests are the functional organization and plasticity of sensory cortex.

Dr. Frostig is a member of the American Society for Neuroscience and the American Physiological Society. He received an Excellence in Teaching Award for his teaching at UCI. For his research, he has received the Alexander Heller Prize and the Arnold and Mabel Beckman Young Investigator Award, in addition to research grants from the National Institute of Health and the National Science Foundation.

Contributors

Cynthia Chen–Bee
Department of Neurobiology and Behavior
The Center for the Neurobiology of Learning and Memory
University of California at Irvine
Irvine, California

David A. Boas
NMR Center
Massachusetts General Hospital
Harvard Medical School
Charlestown, Massachusetts

Lawrence B. Cohen
Department of Cellular and Molecular Physiology
Yale University School of Medicine
New Haven, Connecticut

Earl M. Dolnick
Department of Physics
University of California at San Diego
San Diego, California

Andy K. Dunn
NMR Center
Massachusetts General Hospital
Harvard Medical School
Charlestown, Massachusetts

Monica Fabiani
Psychology Department and Beckman Institute
University of Illinois
Champaign, Illinois

Chun X. Falk
Department of Cellular and Molecular Physiology
Yale University School of Medicine
New Haven, Connecticut

Maria Angela Franceschini
NMR Center
Massachusetts General Hospital
Harvard Medical School
Charlestown, Massachusetts
and Department of Electrical Engineering and Computer Science
Tufts University
Medford, Massachusetts

John S. George
Biophysics Group MS-D454
Los Alamos National Laboratory
Los Alamos, New Mexico

Gabriele Gratton
Psychology Department and Beckman Institute
University of Illinois
Champaign, Illinois

Ronald M. Harper
Department of Neurobiology and the Brain Research Institute
University of California at Los Angeles
Los Angeles, California

Ehud Kaplan
The Laboratory of Visual Neuroscience
Departments of Ophthalmology, Physiology, and Biophysics
Mount Sinai School of Medicine
New York, New York

David Kleinfeld
Department of Physics and Graduate Program in Neurosciences
University of California at San Diego
San Diego, California

Morten P. Kristensen
Department of Neurobiology and the Brain Research Institute
University of California at Los Angeles
Los Angeles, California

Nozomi Nishimura
Department of Physics
University of California at San Diego
San Diego, California

Gina R. Poe
Department of Neurobiology and the Brain Research Institute
University of California at Los Angeles
Los Angeles, California

Daniel B. Polley
Department of Neurobiology and Behavior
The Center for the Neurobiology of Learning and Memory
University of California at Irvine
Irvine, California

Nader Pouratian
Laboratory of Neuro Imaging
Department of Neurology
UCLA School of Medicine
Los Angeles, California

David M. Rector
Department of Neurobiology and the Brain Research Institute
University of California at Los Angeles
Los Angeles, California
and Biophysics Group MS-D454
Los Alamos National Laboratory
Los Alamos, New Mexico

Christopher A. Richard
Department of Neurobiology and the Brain Research Institute
University of California at Los Angeles
Los Angeles, California

Lawrence Sirovich
Laboratory of Applied Mathematics
Department of Biomathematical Sciences
Mount Sinai School of Medicine
New York, New York

Gary Strangman
NMR Center and Neural Systems Group
Massachusetts General Hospital
Harvard Medical School
Charlestown, Massachusetts

Arthur W. Toga
Laboratory of Neuro Imaging
Department of Neurology
UCLA School of Medicine
Los Angeles, California

Philbert S. Tsai
Department of Physics
University of California at San Diego
San Diego, California

Matt Wachowiak
Department of Cellular and Molecular Physiology
Yale University School of Medicine
New Haven, Connecticut

G. Allen White
Department of Physics
University of California at San Diego
San Diego, California

Elizabeth J. Yoder
Department of Radiology
University of California at San
Diego School of Medicine
San Diego, California

Michal R. Zochowski
University of Michigan
Ann Arbor, Michigan

Table of Contents

1 Voltage-Sensitive and Calcium-Sensitive Dye Imaging of Activity: Examples from the Olfactory Bulb

Michal R. Zochowski, Lawrence B. Cohen, Chun X. Falk, and Matt Wachowiak

CONTENTS

1.1 INTRODUCTION

An optical measurement using voltage-sensitive or calcium-sensitive dyes as indicators of activity can be beneficial in several circumstances. In this chapter we

0-8493-2389-4/02/$0.00+$1.50
© 2002 by CRC Press LLC

describe two examples in the study of olfactory processing. In both measurements a large number of neurons or processes are imaged onto each pixel of the camera. Thus, the signals report the average of the changes that occur in this population. For each example we provide some details of the methods used to make the measurements. In addition, because the optical signals are small (fractional intensity changes, $\Delta I/I$, of between 10^{-4} and 5×10^{-2}), optimizing the signal-to-noise ratio in the measurements is important. We will discuss the choice of dyes, sources of noise, light sources, optics, and cameras. The general approach to improving the signal-to-noise ratio is three-pronged. First, test dyes to find the dye with the largest signal-to-noise ratio. Second, reduce the extraneous sources of noise (vibrations, line frequency noise, etc.). Third, maximize the number of photons measured to reduce the relative shot noise (noise arising from the statistical nature of photon emission and detection).

An important advantage of an optical measurement is the ability to make simultaneous measurements from many locations. The two methods described in this chapter also have the advantage of being fast and relatively direct indicators of activity. Both characteristics were important in making maps of the input from receptor neurons and in the study of odor-induced oscillations in the *in vivo* vertebrate olfactory bulb.

Two kinds of cameras have been used in our experiments; both have frame rates faster than 1000 fps. One camera is a photodiode array with 464 pixels and the second is a cooled, back-illuminated, 80×80 pixel CCD camera. Even though the spatial resolution of the two cameras differs rather dramatically, the most important difference is in the range of light intensities over which they provide an optimal signal-to-noise ratio. The CCD camera is optimal at low light levels and the photodiode array is optimal at high light levels.

1.2 CALCIUM- AND VOLTAGE-SENSITIVE DYES

The calcium-sensitive dye used here is thought to be located in the axoplasm and changes the neurons' fluorescence in response to changes in the intracellular free calcium. However, the relationship of the dye signals to calcium concentration is generally nonlinear and the dye response often lags behind the change in calcium. In addition, the dyes add to the calcium buffering in the cytoplasm (Baylor et al., 1983; Neher, 2000). Thus, calcium-sensitive dyes provide a measure for calcium concentration in the axoplasm that must be interpreted with care. Furthermore, in our measurements of the map of the input to the olfactory bulb from olfactory receptor neurons, we are using the calcium signal as a measure of action potential activity in the nerve terminals of the receptor neurons; clearly, this measure is somewhat indirect.

The voltage-sensitive dyes described here are membrane-bound chromophores that change their fluorescence in response to changes in membrane potential. In a model preparation, the giant axon from a squid, these fluorescence signals are fast (following membrane potential with a time constant of <10 μsec) and their size is linearly related to the size of the change in potential (Gupta et al., 1981; Loew et al., 1985). Thus, these dyes provide a direct, fast, and linear measure of the change in membrane potential of the stained membranes. There are other optical signals from membrane-bound dyes (e.g., absorption and birefringence), and another class of dyes senses membrane potential by redistribution; these topics are discussed elsewhere (Cohen and Salzberg, 1978). Similarly, the evidence that pharmacological effects and photodynamic damage resulting from the voltage-sensitive dyes are manageable can be found in earlier reviews (Cohen and Salzberg, 1978; Salzberg, 1983; Cohen and Lesher, 1986; Grinvald et al., 1988).

Voltage-sensitive and calcium-sensitive dyes might be expected to have signals with differing localization even if they are distributed equally over the area of a neuron. A voltage-sensitive dye is expected to have a signal everywhere a potential change exists: in the cell body, along the axon, and in the nerve terminals. On the other hand, a calcium-sensitive dye is expected to have a signal relatively restricted to the nerve terminal because the calcium influx is largest there. Results consistent with these expectations were obtained by Wachowiak and Cohen (1999).

1.3 MEASURING TECHNOLOGY FOR OPTICAL RECORDINGS

In the two examples presented below, the fractional fluorescence changes ($\Delta F/F$) were small; they ranged from 10^{-4} to 5×10^{-2}. In order to measure these signals, the noise in the measurements had to be an even smaller fraction of the resting intensity. In the sections that follow, some of the considerations necessary to achieve such a low noise are outlined.

1.3.1 Three Kinds of Noise

Shot Noise. The limit of accuracy with which light can be measured is set by the shot noise arising from the statistical nature of photon emission and detection. The root mean square deviation in the number of photons emitted is the square root of the average number emitted (I). As a result, the signal in a light measurement will be proportional to I while the noise in that measurement will be proportional to the square root of I. Thus, the signal-to-noise ratio (S/N) is proportional to the square root of the number of measured photons; more photons measured means a better signal-to-noise ratio.

The basis for this square-root dependence on intensity is illustrated in Figure 1.1. The result of using a random number table to distribute 20 photons into 20 time windows is shown in Figure 1.1A, while Figure 1.1B shows the same procedure used to distribute 200 photons into the same 20 bins. Relative to the average light level, there is more noise in the top trace (20 photons) than in the bottom trace (200 photons). On the right side of Figure 1.1, the measured signal-to-noise ratios are listed and we show that the improvement from A to B is similar to that expected from the above square-root relationship. This square-root relationship holds for all light intensities, as indicated by the dotted line in Figure 1.2, which plots the light intensity divided by the noise in the measurement (S/N) vs. the light intensity. At high light intensities this ratio is large, and thus small changes in intensity can be detected. For example, at 10^{10} photons/msec a fractional intensity change of 0.1%

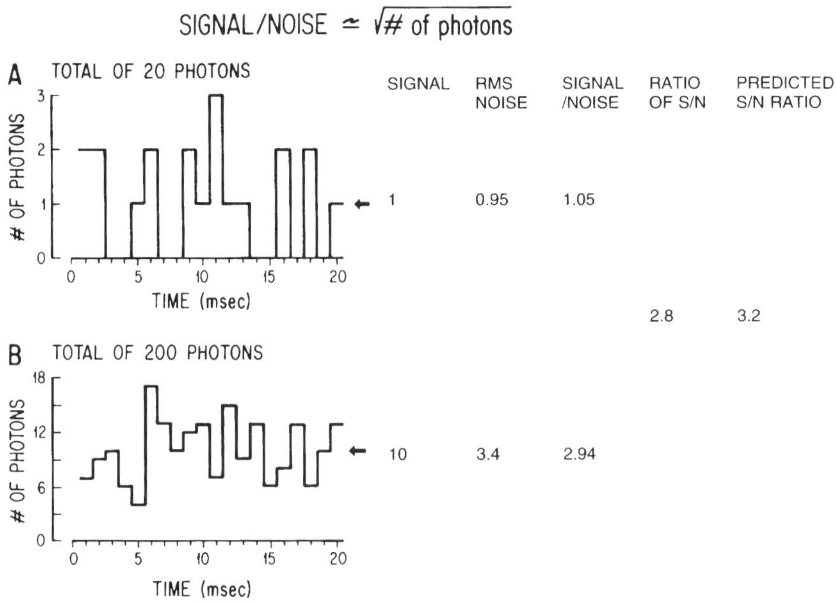

FIGURE 1.1 Plots of the results of using a table of random numbers to distribute 20 photons (top, A) or 200 photons (bottom, B) into 20 time bins. The result illustrates the fact that when more photons are measured, the signal-to-noise ratio is improved. On the right, the signal-to-noise ratio is measured for the two results. The ratio of the two signal-to-noise ratios is 2.8. This is close to the ratio predicted by the relationship that the signal-to-noise ratio is proportional to the square root of the measured intensity. (Redrawn from Wu and Cohen, *Fluorescent and Luminescent Probes for Biological Activity,* Mason, W.T., Ed., Academic Press, London, 1993.)

can be measured with a signal-to-noise ratio of 100. On the other hand, at low intensities the ratio of intensity divided by noise is small and only large signals can be detected. For example, at 10^4 photons/msec the same fractional change of 0.1% can be measured with a signal-to-noise ratio of 1 only after averaging 100 trials. In a shot-noise limited measurement, improvement in the signal-to-noise ratio can be obtained only by 1) increasing the illumination intensity, 2) improving the light-gathering efficiency of the measuring system, or 3) reducing the bandwidth; all of these increase the number of photons in each measurement.

Figure 1.2 compares the performance of two particular camera systems, a photodiode array (solid lines) and a cooled CCD camera (dashed lines) with the shot-noise ideal. The photodiode array approaches the shot-noise limitation over the range of intensities from 3×10^6 to 10^{10} photons/msec. This is the range of intensities obtained in absorption and fluorescence measurements on *in vitro* slices and intact brains in which all of the cells are stained by soaking in a solution of fluorescent dye.

On the other hand, the cooled CCD camera approaches the shot-noise limit over a lower range of intensities, from 3×10^2 to 5×10^6 photons/msec. This is the range of intensities obtained from fluorescence experiments on branches of individual cells and neurons or in experiments where the amount of dye is low. In the discussion

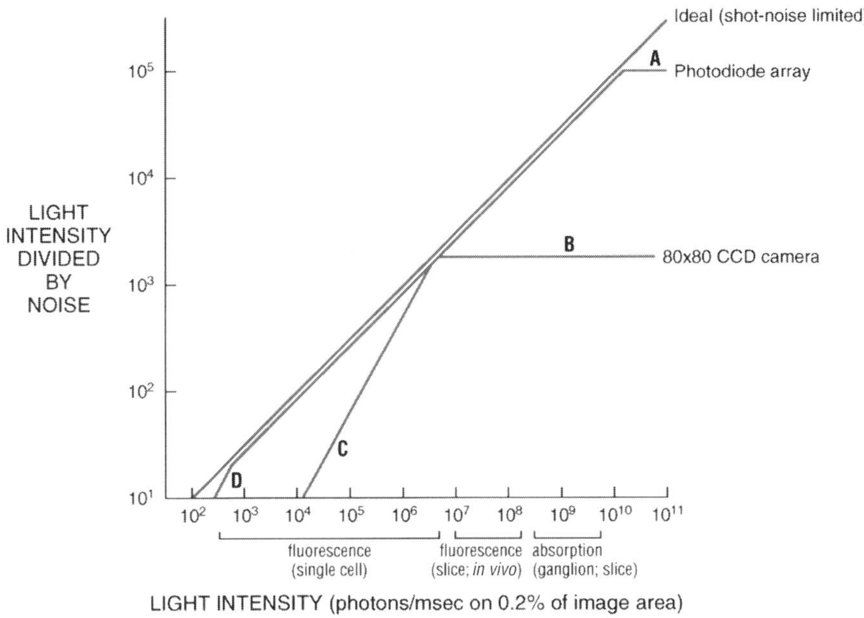

FIGURE 1.2 The ratio of light intensity divided by the noise in the measurement as a function of light intensity in photons/msec/0.2% of the object plane. The theoretical optimum signal-to-noise ratio (dotted line) is the shot-noise limit. Two camera systems are shown, a photodiode array with 464 pixels (solid lines) and a cooled, back-illuminated, 2 kHz frame rate, 80 × 80 pixel CCD camera (dashed lines). The photodiode array provides an optimal signal-to-noise ratio at higher intensities, while the CCD camera is better at lower intensities. The approximate light intensity per detector in fluorescence measurements from a single neuron, in fluorescence measurements from a slice or *in vivo* preparation, and in absorption measurements from a ganglion or a slice is indicated along the *x* axis. The signal-to-noise ratio for the photodiode array falls away from the ideal at high intensities (A) because of extraneous noise and at low intensities and (C) because of dark noise. The lower dark noise of the cooled CCD allows it to function at the shot-noise limit at lower intensities until read noise dominates (D). The CCD camera saturates at intensities above 5×10^6 photons/msec/0.2% of the object plane (B).

that follows, we will indicate aspects of the measurements and characteristics of the two camera systems that cause them to deviate from the shot-noise ideal. The two camera systems we have chosen to illustrate in Figure 1.2 have relatively excellent dark noise and saturation characteristics. Nonetheless, neither camera is ideal. The photodiode array camera has a limited spatial resolution; while the CCD camera has better spatial resolution, it is saturated at light levels obtained in *in vivo* experiments in which all of the membranes are stained directly with a voltage-sensitive dye.

Extraneous Noise. A second type of noise, termed extraneous or technical noise, is more apparent at high light intensities where sensitivity to this kind of noise is high because the fractional shot noise and dark noise are low. One type of extraneous noise is caused by fluctuations in the output of the light source (see below). Two other sources of extraneous noise are vibrations and movement of the preparation. A number of precautions for reducing vibrational noise have been described (Salzberg et al., 1977; London et al., 1987). The pneumatic isolation mounts on many vibration isolation tables are more efficient in reducing vertical vibrations than in reducing horizontal movements. One solution is air-filled soft rubber tubes (Newport Corp., Irvine, CA). Minus K Technology sells Biscuit bench-top vibration isolation tables with very low resonant frequencies. They provide outstanding vibration isolation in both planes. Nevertheless, it has been difficult to reduce vibrational noise to less than 10^{-5} of the total light. For this reason, the performance of the photodiode array system is shown reaching a ceiling in Figure 1.2 (segment A, solid line).

Dark Noise. Dark noise will degrade the signal-to-noise ratio at low light levels. Because the CCD camera is cooled and the photosensitive area (and capacitance) is small, its dark noise is substantially lower than that of the photodiode array system. The excess dark noise in photodiode array accounts for the fact that segment C in Figure 1.2 is substantially to the right of segment D.

1.3.2 Light Sources

Three kinds of sources have been used. Tungsten filament lamps are a stable source, but their intensity is relatively low, particularly at wavelengths less than 480 nm. Arc lamps are somewhat less stable but can provide more intense illumination. Opti-Quip, Inc. provides 150- and 250-watt xenon power supplies, lamp housings, and arc lamps with noise in the range of 1 to 3 parts in 10^4. The 150-watt bulb yielded 2 to 3 times more light at 520 ± 45 nm than a tungsten filament bulb and, in turn, the 250-watt bulb was 2 to 3 times brighter than the 150-watt bulb. The extra intensity is especially useful for fluorescence measurements from single neurons or from weakly stained nerve terminals. Measurements made with laser illumination have been substantially noisier (Dainty, 1984).

1.3.3 Optics

Numerical Aperture. The need to maximize the number of measured photons is a dominant factor in the choice of optical components. In the epifluorescence measurements discussed next, both excitation and emitted light pass through the objective, and the intensity reaching the photodetector is proportional to the fourth power of numerical aperture (Inoue, 1986). Therefore, the numerical aperture of the objective is a crucial factor.

Confocal Microscopes. The confocal microscope (Petran and Hadravsky, 1966) substantially reduces scattered and out-of-focus light that contributes to the image. A recent modification using two-photon excitation of the fluorophore further reduces out-of-focus fluorescence and photobleaching (Denk et al., 1995). With both types of microscope, one can obtain images from *in vivo* vertebrate preparations with much better spatial resolution than that achieved with ordinary microscopy. However, at present many milliseconds are required to record the image from a single x–y plane. Only with line scans can millisecond temporal resolution be obtained. In addition, the very high Z dimension resolution of confocal microscopy can be a drawback if only a very thin section of the preparation is recorded in each frame. The kinds of problems that can be approached using a confocal microscope are limited by these factors.

1.3.4 Cameras

Because the signal-to-noise ratio in a shot noise limited measurement is proportional to the square root of the number of photons converted into photoelectrons (see previous discussion), quantum efficiency is important. Silicon photodiodes have quantum efficiencies approaching the ideal (1.0) at wavelengths where most dyes absorb or emit light (500 to 900 nm). In contrast, only specially chosen vacuum photocathode devices (phototubes, photomultipliers, or image intensifiers) have a quantum efficiency as high as 0.15. Similarly, back-illuminated CCD cameras have a quantum efficiency of >80% at visible wavelengths, while front-illuminated cameras have maximum quantum efficiencies of 20 to 40%.

Many additional factors must be considered in choosing an imaging system. Two important considerations are the requirements for spatial and temporal resolution. Because the signal-to-noise ratio in a shot noise limited measurement is proportional to the square root of the number of measured photons, increases in temporal or spatial resolution will reduce the signal-to-noise ratio. Our discussion considers systems that have frame rates near 1 kHz.

Parallel Readout Arrays. Photodiode arrays with 256 to 1020 elements are now in use in several laboratories (e.g., Iijima et al., 1989; Zecevic et al., 1989; Nakashima et al., 1992; and Hirota et al., 1995). These arrays are designed for parallel readout; each detector is followed by its own amplifier whose output can be digitized at frame rates of 1 kHz. While the need to provide a separate amplifier for each diode element limits the number of pixels in parallel readout systems, it contributes to their very large dynamic range (1 part in 10^5 or 100 db). A discussion of amplifiers has been presented earlier (Wu and Cohen, 1993). Two parallel readout array systems are commercially available: Argus-50 (256 pixels), Hamamatsu Photonics K.K. (www.hpk.co.jp) and NeuroPlex-II (464 pixels), RedShirtImaging LLC (www.redshirtimaging.com).

CCD Cameras. By using a serial readout, the number of amplifiers is greatly reduced. In addition, it is simpler to cool CCD chips to reduce the dark noise. However, because of saturation, CCD cameras presently available do not provide an optimal signal-to-noise ratio for the higher intensities available in some neurobiological experiments (Figure 1.2). This saturation accounts for the bending over of the CCD camera performance at segment B in Figure 1.2. A dynamic range of even 10^3 is not easily achieved with currently available CCD cameras. Thus, these cameras will not be optimal for measurements of absorption or for fluorescence measurements where staining intensity is high. The light intensity would need to be reduced, with a consequent decrease in signal-to-noise ratio. On the other hand, CCD cameras are close to ideal for measurements from branches of individual neurons stained with internally injected dyes. Table 1.1 compares several CCD cameras with frame rates near 1 kHz.

1.4 TWO EXAMPLES

In the experiments described next, we used two basic staining protocols. For the voltage-sensitive dyes, we bathed the turtle olfactory bulb with a solution of the dye for tens of minutes. In contrast, for the calcium-sensitive dye, we infused the

TABLE 1.1
Characteristics of Fast CCD Camera Systems (As Reported by the Manufacturer)

	Frame Rate (Hz) Full Frame	Well Size (×1000 e)	Read Noise (Electrons)	Back Illum.	Bits a to d	Pixels
MiCAM 01[a]	1333	—	—	no	12	92 × 64
Dalsa CA-D1-0128[b]	756	300	360	no	12	128 × 128
SciMeasure[c]	2000	300	20 3 @ 100 Hz	yes	14	80 × 80
TILL Photonics IMAGO[d]	140	35	14	no	12	160 × 120

[a] www.scimedia.co.jp
[b] www.dalsa.com
[c] www.redshirtimaging.com
[d] www.TILL-photonics.com

dye solution (together with a low concentration of detergent) into the nose of the turtle or mouse and then waited several days for the dye to be transported to the nerve terminals in the olfactory bulb. The choice of dye and the staining procedures are described first in each example. We then show how optical recording of activity in the olfactory bulb can be used to address two different aspects of sensory processing. First we describe the use of calcium-sensitive dyes to determine the map of receptor cell input to the olfactory bulb. Second, we describe the use of voltage-sensitive dyes to study the oscillations that occur in the bulb in response to odor presentation. All of the experimental protocols were approved by the Yale University and the Marine Biological Laboratory Institutional Animal Care and Use Committees.

1.4.1 Maps of Input to the Olfactory Bulb from the Olfactory Receptor Neurons Measured with Calcium-Sensitive Dyes

To understand how odorants are distinguished and to address the question of how an odorant is recognized as the same across a broad concentration range, we measured the spatial pattern of the olfactory receptor neuron input to the olfactory bulb in the three-toed box turtle, *T. triunguis,* and in the C57/Bl6 mouse.

The olfactory receptor axon terminals in the turtle and mouse were selectively labeled by anterograde transport of the dextran-conjugated calcium indicator Calcium Green-1 dextran (10 kD molecular weight; Molecular Probes) applied to the olfactory epithelium using a protocol adapted from Friedrich and Korsching (1997). Turtles were first chilled on ice for 1 to 2 h and then placed upside down with the mouth held open, and 25 to 60 µl of a 2% dye solution dissolved in 0.1 M NaCl plus 0.5% Triton-X 100 was injected into each naris. The pharyngeal opening of

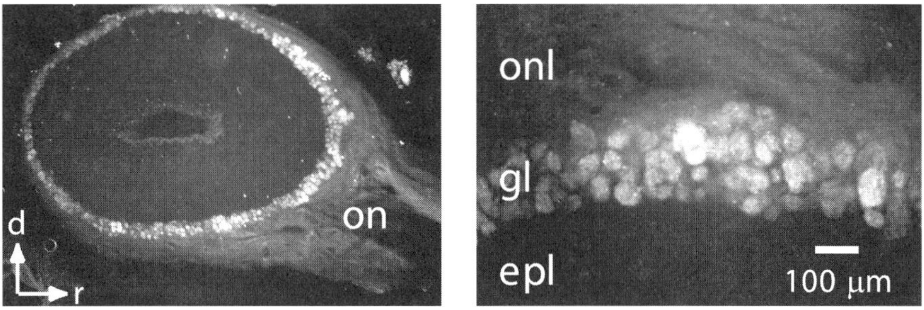

FIGURE 1.3 Calcium Green-1 dextran labeling of turtle olfactory receptor neurons. Sagittal olfactory bulb section showing pattern of dextran labeling. Left: uniform labeling of glomeruli in all olfactory bulb regions. Right: specific labeling of olfactory nerve axons and their terminal branches within glomeruli. Absence of labeling in the external plexiform layer (epl) indicates that the dye is confined to olfactory receptor neurons. Abbreviations: *on*, olfactory nerve; *d*, dorsal; *r*, rostral; *l*, lateral; *onl*, olfactory nerve layer; *gl*, glomerular layer. (M. Wachowiak and L.B. Cohen, unpublished).

each naris was plugged to prevent leakage, and the epithelium was stained for 10 to 20 min, after which the plugs were removed and the mouth closed, allowing the animal to expel the dye from the nose. The turtles were held at room temperature for 4 to 18 days before recording. The nerve terminal labeling was first apparent after about 4 days but appeared most intense after 8 to 10 days. In the nares there was a relatively intense fluorescence in the olfactory cell body layer of the epithelium, while in the olfactory bulb there was a much less intense staining of the olfactory nerve and glomerular layers (Figure 1.3). While glomeruli could be clearly distinguished by the labeled fibers, no fluorescence was visible in subglomerular layers. We observed no differences in the gross appearance of the olfactory epithelium (postmortem examination) or in the size and sensitivity of electroolfactogram (EOG) recordings between stained and unstained animals. Similar methods were used for the mouse and similarly selective staining was obtained.

Because the amount of dye that reaches the nerve terminals is small, the light intensity in such a measurement is low and a better signal-to-noise ratio is obtained with a cooled CCD camera. We formed a magnified (4× or 15×) image of the bulb on an 80 × 80 CCD camera. For additional details of the methods, see Wachowiak et al. (2001).

We recorded the changes in Calcium Green fluorescence that resulted from a 1- to 2-sec odorant pulse delivered to the nose. In both the turtle and the mouse the signals had approximately the same time course everywhere in the bulb; we therefore characterized the response by the amplitude of the signal as a function of its position on the bulb.

Turtle. We tested several odorants on the same animal. In every instance, the maps were different for different odorants. In some cases a substantial overlap occurred in the activated area, but at higher magnification, the patterns were easily distinguished (data not shown). Thus, the maps of input from the olfactory receptor neurons could be used for odor identification.

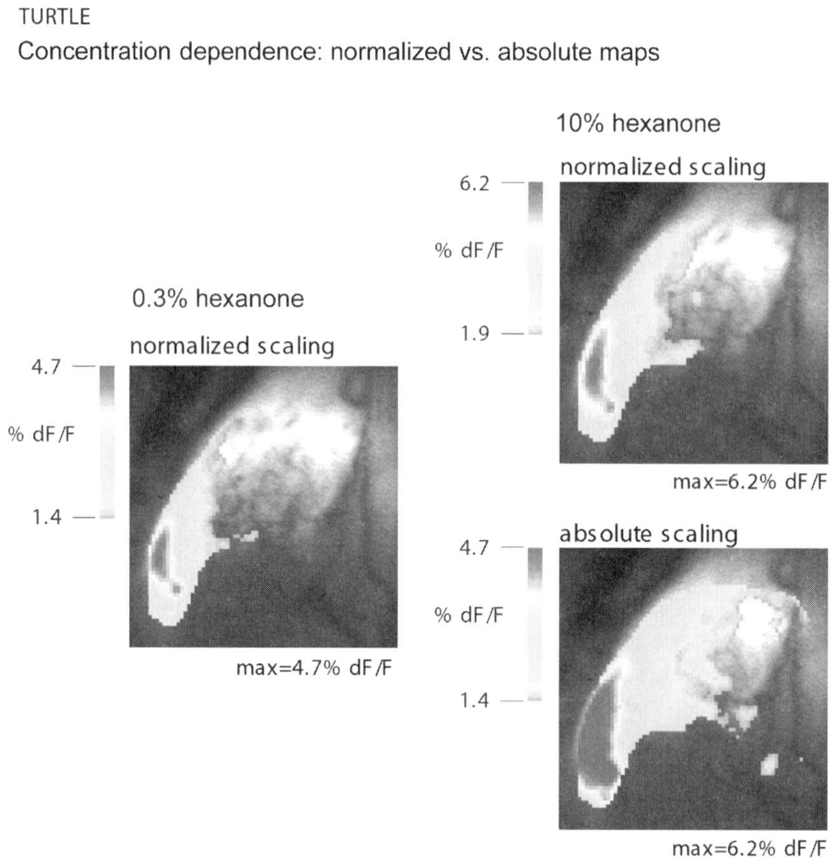

FIGURE 1.4 (See Color Figure 1.4.) Normalized maps of receptor neuron input to the turtle olfactory bulb are concentration-invariant. The left panel shows a pseudocolor map of the response to a 0.3% dilution of saturated vapor of 2-hexanone. The map is normalized to the maximum signal amplitude for this trial. The right panels show pseudocolor maps of a response to a 10% dilution of 2-hexanone. The map on the top is normalized to its maximum signal amplitude. The map on the bottom (absolute scaling) shows the same data using the same scaling as for the 0.3% trial. The figure shows a concentration-dependent increase in the number of glomeruli activated above a given absolute level but shows a concentration invariance in the relative levels of input to all glomeruli activated by an odorant. 4× image magnification. The field of view is approximately 4 × 4 mm.

Color Figure 1.4* shows three pseudocolor representations of activity in response to hexanone at two concentrations in the turtle. Red represents a large signal in each measurement and blue represents a signal 30% as large. The left-hand image shows the response to hexanone at a concentration that was 0.3% of saturation. The largest signal in the response ($\Delta F/F$ = 4.7%) was colored red (normalized scaling). Both right-hand images show the response to 10% hexanone using different

* Color figures follow page 112.

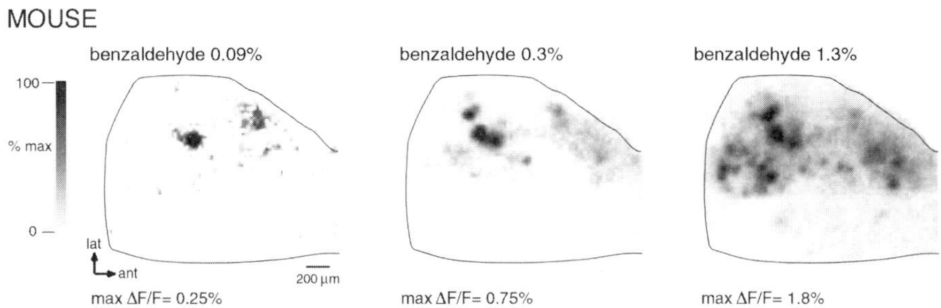

FIGURE 1.5 Normalized maps of receptor neuron input to the mouse olfactory bulb are *not* concentration invariant. The three panels show the normalized response to three concentrations of benzaldehyde. The maps are normalized to the maximum signal amplitude for each measurement.15× image magnification. The field of view is approximately 1 × 1 mm. (M. Wachowiak and L.B. Cohen, unpublished)

scaling procedures. The bottom image shows the response using the same scale as that used for the response to 0.3% hexanone (absolute scaling) where signals of 4.7% and higher are colored red. This image is qualitatively different from the 0.3% image. In contrast, the top image shows the response using normalized scaling; again, the largest signal was colored red. This image is very similar to the image on the left, even though the concentration of odorant differed by a factor of 30. We have observed similar results over a concentration range of up to 600 (data not shown). Thus, normalized maps of input to the turtle olfactory bulb appear to be approximately concentration invariant. We hypothesize that concentration-invariant odorant identification could be achieved if higher olfactory centers "read" the normalized maps of the input to the turtle olfactory bulb.

Mouse. In contrast to the results from the turtle, when similar experiments were carried out in C57/Bl6 mice, the response maps were qualitatively and statistically different at low and high concentrations (Figure 1.5). At low concentrations (<0.3% of saturated vapor) fewer glomeruli were activated by benzaldehyde, while at high concentration more than half of the imaged glomeruli were activated. The reason for the difference in results between turtle and mouse is unclear. The significant concentration-dependent recruitment of input to new glomeruli in the mouse, as compared to the turtle, might reflect differences in the specificity of receptor neurons to a range of odorants, with more broadly tuned receptor neurons showing less recruitment and a higher degree of concentration invariance. It is also unclear whether this apparent difference in the concentration dependence of maps of olfactory bulb input implies that these animals use different strategies for processing olfactory information. One possibility is that the mouse also forms a concentration-invariant map, but at a later stage of olfactory processing.

1.4.2 Oscillations in the Olfactory Bulb in Response to Odors Measured with Voltage-Sensitive Dyes

Choosing Voltage-Sensitive Dyes. Several voltage-sensitive dyes (see, for example, Figure 1.6) have been used to monitor changes in membrane potential in a variety

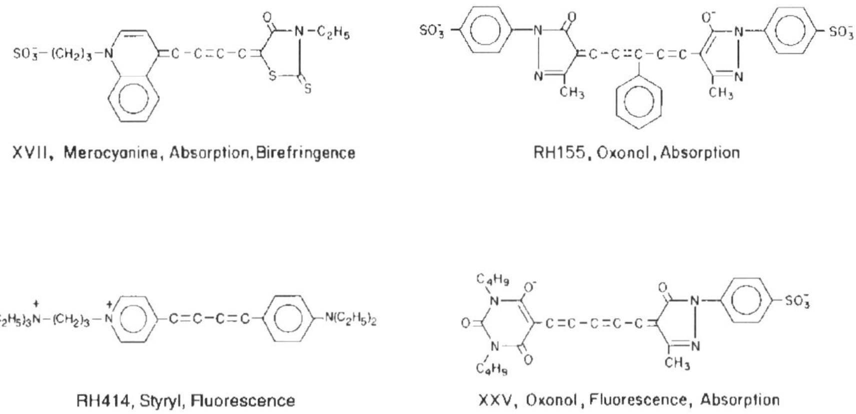

FIGURE 1.6 Examples of four different chromophores used to monitor membrane potential. The merocyanine dye, XVII (WW375), and the oxonol dye, RH155, are commercially available as NK2495 and NK3041 from Nippon Kankoh–Shikiso Kenkyusho Co. Ltd., Okayama, Japan. The oxonol, XXV (WW781), and styryl, di-4-ANEPPS, are available commercially as dye R-1114 and D-1199 from Molecular Probes, Junction City, OR.

of preparations. This figure illustrates four different chromophores. For each chromophore approximately 100 analogues have been synthesized in an attempt to optimize the signal-to-noise ratio that can be obtained in a variety of preparations. This screening was made possible by synthetic efforts of three laboratories: Jeff Wang, Ravender Gupta, and Alan Waggoner, then at Amherst College; Rina Hildesheim and Amiram Grinvald at the Weizmann Institute; and Joe Wuskell and Leslie Loew at the University of Connecticut Health Center. For each of the four chromophores illustrated in Figure 1.6, there were 10 or 20 dyes that gave approximately the same signal size on squid axons (Gupta et al., 1981). However, dyes that gave nearly identical signals on squid axons gave very different responses on other preparations; thus, tens of dyes usually must be tested to obtain the largest possible signal. Often, dyes that worked well in squid did poorly in other preparations because they did not penetrate through connective tissue or along intercellular spaces to the membrane of interest.

For the *in vivo* measurements of oscillations in the turtle olfactory bulb, the animals were first anesthetized by placing them in ice for 2 h. A craniotomy was performed over the olfactory bulb. The dura and arachnoid matter were then carefully removed to facilitate staining. The exposed olfactory bulb was stained by covering it with dye solution (RH414, 0.02 to 0.2 mg/ml; Grinvald et al., 1994; Molecular Probes, Eugene, OR) for 30 to 60 min. Excess dye was then washed away with turtle saline. In some experiments the dye signals remained stable for several hours. The direct staining with voltage-sensitive dye results in a high light intensity; therefore we used the photodiode array camera (see discussion in Section 1.6). For additional details of the methods, see Lam et al. (1999).

Multiple Components of the Odor-Induced Response. Odor stimuli have long been known to induce stereotyped local field-potential responses consisting of “sinusoidal" oscillations of 10 to 80 Hz riding on top of a slow "DC" signal. Since its discovery in the hedgehog (Adrian, 1942), odor-induced oscillations have been seen in phylogenetically distant species including locust (Laurent and Naraghi, 1994), turtle (Beuerman, 1975), and monkey (Hughes and Mazurowski, 1962). Because the optical measurements have a spatial resolution about 25 times better than local field-potential measurements (Zochowski et al., 2000), we obtained a more detailed visualization of the spatio-temporal characteristics of the oscillations.

We found three different oscillations in the bulb in response to odorants. These oscillations differed in amplitude, location, frequency, latency, shape, and duration, but they were similar for the two odorants tested, isoamyl acetate and cineole. In Figure 1.7, the recordings from seven selected diodes in a single trial are shown. The location of these diodes is indicated on the image in the left panel by the

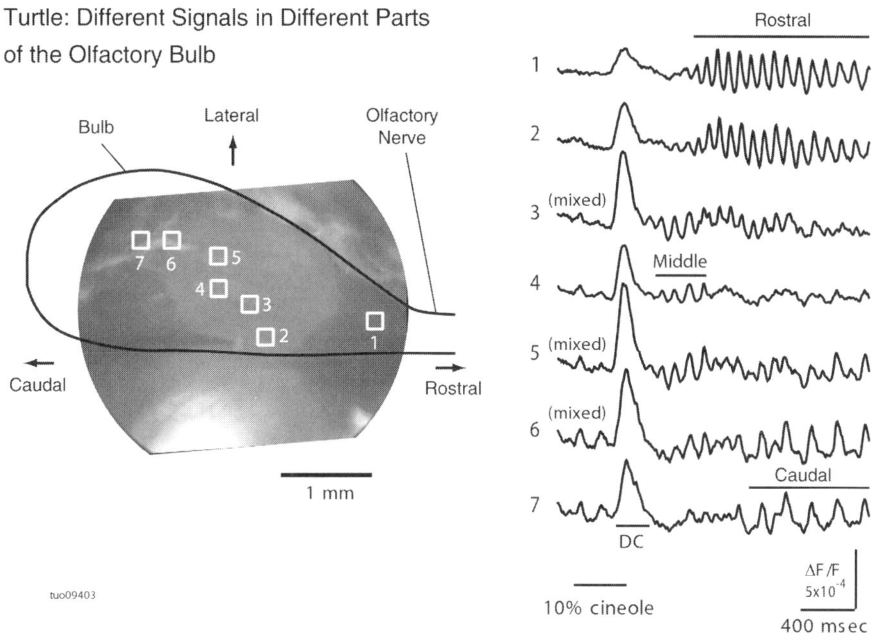

FIGURE 1.7 Simultaneous optical recordings from seven different areas of an olfactory bulb. An image of the olfactory bulb is shown on the left. Signals from seven selected pixels are shown on the right. The positions of these pixels are labeled with squares and numbers on the image of the bulb. All seven signals have a filtered version of the DC signal at the time indicated by the bar labeled DC. The oscillation in the rostral region has a high frequency and relatively long latency and duration (detectors 1 and 2). The oscillation from the middle region has a high frequency and short latency and duration (detector 4). The oscillation from the caudal region has a lower frequency and the longest latency (detector 7). The signal from detectors between these regions (3, 5, and 6) appears to contain a mixture of two components. The horizontal line labeled "10% cineole" indicates the time of the command pulse to the odor solenoid. The data are filtered by a high-pass digital RC (5 Hz) and low-pass Gaussian (30 Hz) filters. (Modified from Lam et al., *J. Neurosci.*, 20: 749, 1999.)

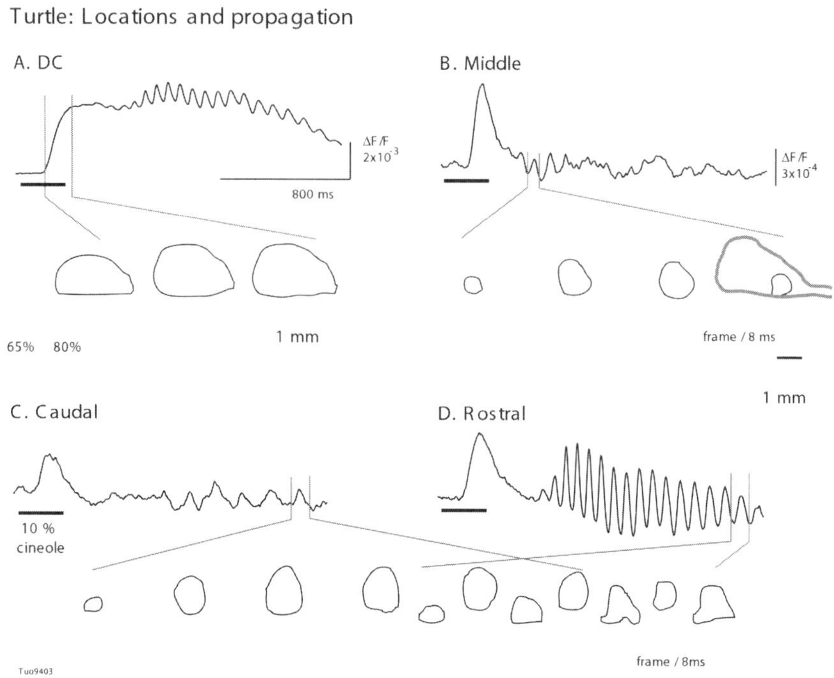

FIGURE 1.8 (See Color Figure 1.8.) The locations and propagation of the four components from the trial shown in Figure 1.7. Multiframe pseudocolor displays of the signals are overlaid on the image of the olfactory bulb. The red color and the black contour lines label the areas where the signals are larger than 80 and 20% of the peak. The DC component in this animal covers almost the entire bulb. The other three panels show the position and propagation of one cycle (indicated by the red and green lines) of the three different oscillations. The black horizontal bars indicate the time of the odor command pulse. The data are filtered by a high-pass digital RC (5 Hz) and low-pass Gaussian (30 Hz) filters. (Modified from Lam et al., *J. Neurosci.*, 20: 749, 1999.)

numbered squares on the left. In rostral locations (detectors 1 and 2), we found a relatively large, long-lasting oscillation with a relatively high frequency (~15 Hz). On a diode from a middle location (detector 4), there was a relatively small, brief, short-latency oscillation (~14 Hz). On detector 7 from the caudal bulb, the oscillation had a lower frequency (~7 Hz) and longer latency. We named the oscillations for the regions of the dorsal bulb in which they occurred: rostral, middle, and caudal. In areas between two regions, the recorded oscillations were combinations of two signals, i.e., rostral/middle in detector 3 and middle/caudal in detectors 5 and 6. The three oscillations also had different shapes — the rostral and caudal oscillations had relatively sharp peaks while the middle oscillation was more sinusoidal.

In addition to the three oscillations, a DC signal, which appears as a single peak after high-pass filtering in Figure 1.7, was observed over most of the ipsilateral olfactory bulb. Color Figure 1.8A shows the time course of an unfiltered recording from the rostral region. Following the start of the odor pulse, the optical signal rose to a plateau and then continued for a period of seconds. After a delay, the rostral oscillations appeared on the DC response.

Color Figure 1.8 shows the time courses of the signals from four detectors from this trial together with multiple-frame images indicating the position and propagation during one cycle. The rostral, middle, and caudal oscillations are seen more clearly after the DC signal was reduced with a high-pass filter (Color Figure 1.8B, C, D). In these multiframe images, the red color and area enclosed by the black line indicate the areas where the signals are larger than 80 and 20% of the maximum signal. The DC signal covered most of the olfactory bulb. The rostral signal (D) propagated in the caudal direction, the middle signal (B) did not appear to propagate, and the caudal signal (C) propagated in a lateral-caudal direction.

Oscillations are not restricted to the olfactory system. Despite their ubiquity, the roles and functions of oscillation are not well understood. However, in view of their ubiquity and the very large number of neurons that they involve, it is reasonable to speculate that oscillations may have an important role in odor processing. Our data shows that the odor-induced oscillations in the olfactory bulb are substantially more complicated than had been anticipated and that multiple functional population domains are processing olfactory input in parallel.

1.5 NOISE FROM *IN VIVO* PREPARATIONS

Measuring signals from *in vivo* preparations is often more difficult because of noise from the heartbeat and respiration. In the instances described earlier, this kind of noise was not a particular problem because the noise and the signals were at different frequencies and could be separated by frequency filtering. In the turtle the heartbeat and respiration frequencies are about 0.2 Hz, slower than the optical signals, and in the mouse they are about 5 Hz, faster than the signals that we measured. Obviously, this simple solution is not general. Two methods for reducing the movement artifacts from the heartbeat are, together, quite effective. First, a subtraction procedure is used in which two recordings are made but only one of the trials has a stimulus (Orbach et al., 1985). Both recordings are triggered from the upstroke of the electrocardiogram, so both should have similar heartbeat noise. When the trial without the stimulus is subtracted from the trial with the stimulus, the heartbeat artifact is reduced. Second, an airtight chamber is fixed onto the skull surrounding the craniotomy (Blasdel and Salama, 1986). When this chamber is filled with silicon oil and closed, the movements due to heartbeat and respiration are substantially reduced. Using both methods reduces the noise from these movement artifacts enough so that they are no longer the main source of noise in fluorescence measurements.

1.6 FUTURE DIRECTIONS

Because the light-measuring apparatus is already reasonably optimized (see previous discussion), any improvement in sensitivity will need to come from the development of better dyes or investigating signals from additional optical properties of the dyes. The dyes in Figure 1.6 and the vast majority of those synthesized are of the general class named polyenes (Hamer, 1964). It is possible that improvements in signal size can be obtained with new polyene dyes (see Waggoner and Grinvald (1977) and Fromherz (1991) for a discussion of maximum possible fractional changes in absorption and fluorescence). On the other hand, the fractional change on squid axons has not increased in recent years (Gupta et al., 1981; Cohen et al., unpublished results), and most improvements (e.g., Grinvald et al., 1982; Momose–Sato et al., 1995; Tsau et al., 1996; Antic and Zecevic, 1995) have involved synthesizing analogues that work well on new preparations.

The best of the styryl and oxonol polyene dyes have fluorescence changes of 10 to 20%/100 mv in situations where the staining is specific to the membrane whose potential is changing (Grinvald et al., 1982; Loew et al., 1992; Rohr and Salzberg, 1994). Recently, Gonzalez and Tsien (1995) introduced a new scheme for generating voltage-sensitive signals using two chromophores and energy transfer. While these fractional changes were also in the range of 10%/100 mv, more recent results are about 30% (Gonzalez and Tsien, personal communication). However, because one of the chromophores must be very hydrophobic and does not penetrate into brain tissue, it has not been possible to measure signals with this pair of dyes in intact tissues (Gonzalez and Tsien; Obaid and Salzberg; personal communications). Cacciatore et al. (1999) have used a less hydrophobic oxonol to improve the dye penetration and measured changes in membrane potential in leech neurons during swimming. However, reducing the hydrophobicity also resulted in a substantially slower (~100 msec) response time for the optical signal.

Neuron-Type Specific Staining. An important new direction is the development of methods for neuron-type specific staining. Three quite different approaches have been tried. First, the use of retrograde staining procedures has recently been investigated in the embryonic chick and lamprey spinal cords (Tsau et al., 1996). An identified cell class (motoneurons) was selectively stained. In lamprey experiments, spike signals from individual neurons were sometimes measured (Hickie et al., 1996). Further efforts at optimizing this staining procedure are needed. The second approach is based on the use of cell-type specific staining developed for fluorescein by Nirenberg and Cepko (1993). It might be possible to use similar techniques to selectively stain cells with voltage-sensitive or ion-sensitive dyes. Third, Siegel and Isacoff (1997) constructed a genetically encoded combination of a potassium channel and green fluorescent protein. When introduced into a frog-oocyte, this molecule had a (relatively slow) voltage-dependent signal with a fractional fluorescence change of 5%. More recently, Sakai et al. (2001) and Ataka and Pieribone (2002) have devised GFP-channel constructs with much faster kinetics. Neuron-type specific staining would make it possible to determine the role of specific neuron types in generating the input–output function of a brain region.

Roger Tsien and collaborators have developed a genetically encoded calcium indicator based on fluorescence energy transfer between CFP and YFP bound to calmodulin (Miyawaki et al., 1997). While we are not aware of the use of these molecules in *in vivo* preparations, improvements in the probes (Griesbeck et al., 2001) may improve their utility.

Optical recordings already provide unique insights into brain activity and organization. Clearly, improvements in signal size or selectivity would make these methods more powerful.

ACKNOWLEDGMENTS

The authors are indebted to their collaborators Vicencio Davila, Amiram Grinvald, Kohtaro Kamino, David Kleinfeld, Les Loew, Bill Ross, Guy Salama, Brian Salzberg, Alan Waggoner, Jian-young Wu, Joe Wuskell, and Dejan Zecevic for numerous discussions about optical methods. Work was supported by grants NS08437-DC052591717, a Brown-Coxe Fellowship from the Yale University School of Medicine, and an NRSA fellowship, DC 00378.

ABBREVIATIONS

NA numerical aperture
LFP local field potential
EOG electroolfactogram

REFERENCES

Adrian, E.D., Olfactory reactions in the brain of the hedgehog, *J. Physiol.*, 100, 459, 1942.

Antic, S. and Zecevic, D., Optical signals from neurons with internally applied voltage-sensitive dyes, *J. Neurosci.*, 15, 1392, 1995.

Ataka, K. and Pieribone, V.A., A genetically-targetable fluorescent probe of channel gating with rapid kinetics, *Biophys. J.*, 82, 509, 2002.

Baylor, S.M., Chandler, W.K., and Marshall, M.W., Sarcoplasmic reticulum calcium release in frog skeletal muscle fibres estimated from Arsenazo III calcium transients, *J. Physiol.*, 344, 625, 1983.

Beuerman, R.W., Slow potentials of the turtle olfactory bulb in response to odor stimulation of the nose, *Brain Res.*, 97, 61, 1975.

Blasdel, G.G. and Salama, G., Voltage-sensitive dyes reveal a modular organization in monkey striate cortex, *Nature*, 321, 579, 1986.

Cacciatore, T.W. et al., Identification of neural circuits by imaging coherent electrical activity with FRET-based dyes, *Neuron*, 23, 449, 1999.

Cohen, L.B. and Lesher, S., Optical monitoring of membrane potential: methods of multisite optical measurement, *Soc. Gen. Physiol. Ser.*, 40, 71, 1986.

Cohen, L.B. and Salzberg, B.M., Optical measurement of membrane potential, *Rev. Physiol. Biochem. Pharmacol.*, 83, 35, 1978.

Dainty, J.C., *Laser Speckle and Related Phenomena*, Springer-Verlag, New York, 1984.

Denk, W., Piston, D.W., and Webb, W., Two-photon molecular excitation in laser-scanning microscopy, in *Handbook of Biological Confocal Microscopy*, Pawley, J.W., Ed., Plenum Press, New York, 1995, 445.

Friedrich, R. and Korsching, S., Combinatorial and chemotopic odorant coding in the zebrafish olfactory bulb visualized by optical imaging, *Neuron*, 18, 737, 1997.

Fromherz, P. et al., Fluorescent dyes as probes of voltage transients in neuron membranes: progress report, *Ber. Bunsenges. Phys. Chem.*, 95, 1333, 1991.

Gonzalez, J.E. and Tsien, R.Y., Voltage sensing by fluorescence energy transfer in single cells, *Biophysical J.*, 69, 1272, 1995.

Griesbeck, O. et al., Reducing the environmental sensitivity of yellow fluorescent protein: mechanism and applications, *J. Biol. Chem.*, 2001, in press.

Grinvald, A. et al., Improved fluorescent probes for the measurement of rapid changes in membrane potential, *Biophys. J.*, 39, 301, 1982.

Grinvald, A. et al., Cortical point-spread function and long-range lateral interactions revealed by real-time optical imaging of Macaque monkey primary visual cortex, *J. Neurosci.*, 18, 9977, 1994.

Grinvald, A. et al., Optical imaging of neuronal activity, *Physiol. Rev.*, 68, 1285, 1988.

Gupta, R.K. et al., Improvements in optical methods for measuring rapid changes in membrane potential, *J. Memb. Biol.*, 58, 123, 1981.

Hamer, F.M., *The Cyanine Dyes and Related Compounds*, John Wiley & Sons, New York, 1964.

Hickie, C. et al., Optical monitoring of activity from individual and identified populations of neurons retrogradely labeled with voltage-sensitive dyes, *Abs. Soc. Neurosci.*, 22, 321, 1996.

Hirota, A. et al., A new simultaneous 1020-site optical recording system for monitoring neural activity using voltage-sensitive dyes, *J. Neurosci. Methods*, 56, 187, 1995.

Hughes, J.R. and Mazurowski, J.A., Studies on the supracallosal mesial cortex of unanesthetized, conscious mammals. II Monkey. B. Responses from the olfactory bulb, *Electroencephalogr. Clin. Neurophysiol.*, 14, 635, 1962.

Iijima, T., Ichikawa, M., and Matsumoto, G., Optical monitoring of LTP and related phenomena, *Abst. Soc. Neurosci.*, 15, 398, 1989.

Inoue, S., *Video Microscopy*, Plenum Press, New York, 1986, 128.

Lam, Y.-W. et al., Odors elicit three different oscillations in the turtle olfactory bulb, *J. Neurosci.*, 20, 749, 1999.

Laurent, G. and Naraghi, M., Odorant-induced oscillations in the mushroom bodies of the locust, *J. Neurosci.*, 14(5 Pt 2), 2993, 1994.

Loew, L.M. et al., A napthyl analog of the aminostyryl pyridinium class of potentiometric membrane dyes shows consistent sensitivity in a variety of tissue, cell, and model membrane preparations, *J. Memb. Biol.*, 130, 1, 1992.

Loew, L.M. et al., Charge-shift probes of membrane potential. Characterization of aminostyrylpyridinium dyes on the squid giant axon, *Biophys. J.*, 47, 71, 1985.

London, J.A., Zecevic, D., and Cohen, L.B., Simultaneous optical recording of activity from many neurons during feeding in *Navanax*, *J. Neurosci.*, 7, 649, 1987.

Miyawaki, A. et al., Fluorescent indicators for Ca2+ based on green fluorescent proteins and calmodulin, *Nature*, 388, 882, 1997.

Momose–Sato, Y. et al., Evaluation of optimal voltage-sensitive dyes for optical measurement of embryonic neural activity, *J. Memb. Biol.*, 144, 167, 1995.

Nakashima, M. et al., 448-detector optical recording system: development and application to *Aplysia* gill-withdrawal reflex, *IEEE Trans. Biomed. Eng.*, 39, 26, 1992.

Neher, E., Some quantitative aspects of calcium fluorimetry, in *Imaging Neurons: A Laboratory Manual*, Yuste, R., Lanni, F., and Konnerth, A., Eds., Cold Spring Harbor Laboratory Press, Cold Spring Harbor, New York, 2000, p. 31.1.

Nirenberg, S. and Cepko, C., Targeted ablation of diverse cell classes in the nervous system *in vivo*, *J. Neurosci.*, 13, 3238, 1993.

Orbach, H.S., Cohen, L.B., and Grinvald, A., Optical mapping of electrical activity in rat somatosensory and visual cortex, *J. Neurosci.,* 5, 1886, 1985.

Petran, M. and Hadravsky, M., Czechoslovakian patent 7720, 1966.

Rohr, S. and Salzberg, B.M., Multiple site optical recording of transmembrane voltage in patterned growth heart cell cultures: assessing electrical behavior, with microsecond resolution, on a cellular and subcellular scale, *Biophys. J.,* 67, 1301, 1994.

Sakai, R., Repunte-Canonigo, V., Raj, C.D., and Knopfel, T., Design and characterization of a DNA-encoded, voltage-sensitive fluorescent protein, *Eur. J. Neurosci.,* 13, 2314, 2001.

Salzberg, B.M., Optical recording of electrical activity in neurons using molecular probes, in *Current Methods in Cellular Neurobiology*, Barker, J.L. and McKelvy, J.F., Eds., John Wiley & Sons, New York, 139, 1983.

Salzberg, B.M. et al., Optical recording of neuronal activity in an invertebrate central nervous system: simultaneous monitoring of several neurons, *J. Neurophysiol.,* 40, 1281, 1977.

Siegel, M.S. and Isacoff, E.Y., A genetically encoded optical probe of membrane voltage, *Neuron,* 19, 735, 1997.

Tsau, Y. et al., Dye screening and signal-to-noise ratio for retrogradely transported voltage-sensitive dyes, *J. Neurosci. Methods*, 70, 121, 1996.

Wachowiak, M. and Cohen, L.B., Presynaptic inhibition of primary olfactory afferents mediated by different mechanisms in the lobster and turtle, *J. Neurosci.*, 19, 8808, 1999.

Wachowiak, M., Cohen, L.B., and Zochowski, M., Spatial representations of odorant quality and intensity by receptor neuron input to turtle olfactory bulb glomeruli, 2001, submitted for publication.

Waggoner, A.S. and Grinvald, A., Mechanisms of rapid optical changes of potential sensitive dyes, *Ann. NY Acad. Sci.*, 303, 217, 1977.

Wu, J.Y. and Cohen, L.B., Fast multisite optical measurements of membrane potential, in *Fluorescent and Luminescent Probes for Biological Activity,* Mason, W.T., Ed., Academic Press, London, 389, 1993.

Zecevic, D. et al., Hundreds of neurons in the *Aplysia* abdominal ganglion are active during the gill-withdrawal reflex, *J. Neurosci.*, 9, 3681, 1989.

Zochowski, M. et al., Imaging membrane potential with voltage-sensitive dyes, *Biolog. Bull.,* 198, 1, 2000.

2 Visualizing Adult Cortical Plasticity Using Intrinsic Signal Optical Imaging

Ron D. Frostig, Cynthia Chen–Bee, and Daniel B. Polley

CONTENTS

0-8493-2389-4/02/$0.00+$1.50
© 2002 by CRC Press LLC

2.1 INTRODUCTION

Our understanding of the functional organization and plasticity of the adult sensory cortex has been transformed in the last 20 years. The view that cortical functional representations of the sensory surface in adult animals are fixed, especially for the primary sensory cortices, has been replaced by the view that they are dynamic and continuously modified by the animal's experience. Such "experience-dependent" plasticity in adult cortical functional representations has been demonstrated in a range of mammalian species across many primary cortices and is believed to follow one general principle: a greater amount of cortical territory is allocated to a peripheral input source that is in greater use. This use-dependent or experience-driven plasticity of adult functional representations has been implicated in a range of fundamental processes, including rehabilitation following peripheral sensory loss or damage, recoveries from central nervous system damage, improvements in sensory motor skills with practice, and learning and memory, thus underscoring its importance to an organism's survival. For recent reviews on adult cortical plasticity, see References 1 through 3.

While much progress has been made in understanding phenomenology and underlying mechanisms, research on the plasticity of adult cortical functional representations is still in its infancy. So far, several techniques have been available for the study of adult cortical plasticity. Ideally, a technique should have the ability to sample 1) neuronal spiking activity; 2) noninvasively; 3) simultaneously from large cortical regions; 4) with high, three-dimensional spatial resolution; 5) with high temporal resolution; and 6) in the awake animal. This ideal technique would enable the direct, noninvasive (hence, long-term and nondamaging) assessment of a cortical functional representation (comprising many neurons collectively occupying a volume of cortex) with sufficient spatial and temporal resolution to track real-time changes in the functional representation in the awake, behaving animal. Unfortunately, such an ideal technique does not exist, but each of the most widely used current techniques for studying cortical plasticity offers a particular subset of these described advantages.

The most popular technique is the use of a single microelectrode to record from the cortex. It offers the advantage of recording action potentials directly from cortical neurons with high spatial (point) and temporal (msec) resolution sufficient to follow real-time changes in neuronal activity at any location along a volume of cortex; the disadvantage is that recordings are invasive to the cortex. In order to assess the functional representation of a sensory organ (e.g., a finger or a whisker), neurons are recorded from different cortical locations and the functional representation of the organ is then defined as the cortical region containing neurons responsive to that organ (i.e., neurons that have receptive fields localized at the sensory organ). A change in the spatial distribution of neurons responsive to a given sensory organ is then taken as evidence for plasticity in the functional representation of that sensory organ.[4]

Because a cortical functional representation comprises many neurons distributed over a volume of cortex, the use of a single microelectrode to characterize the plasticity of a functional representation requires many recordings across a large cortical region. These recordings can be obtained only in a serial fashion and require many hours to complete; thus, the animal is typically anesthetized. Because the number of recordings will be limited due to cortical tissue damage incurred during the experiment and to time constraints, the characterization of a functional representation is dependent on the extrapolation between recording locations. Also, because of its invasiveness, this technique is not ideal for assessment of the same functional representation before and after a manipulation within the same animal. The recent development of simultaneous recordings from a chronically implanted array of microelectrodes[5,6] provides relief to some of the challenges described here and has great potential for long-term recordings from the same animal — although recordings still damage the cortical tissue, the sample size is still small, and extrapolation is still needed between recording locations.

Another popular technique is 2-deoxyglucose autoradiography (2-DG),[7] capable of sampling simultaneously from a large cortical region with high spatial resolution (microns). The main appeal is the ability of 2-DG to provide visualization of a cortical functional representation, in contrast to an extrapolation of a functional representation based on multiple, discrete sampling with microelectrodes. This method also offers the ability to characterize a functional representation across different layers of the cortex. However, 2-DG is unable to track real-time changes in a functional sensory representation and is limited to only one assessment per animal. Also, 2-DG measures the level of neuronal metabolic activity as an indirect measure of neuronal spiking activity. Although strong metabolic activity is correlated with activated brain areas that exhibit strong spiking activity, it remains unclear whether it is also a faithful indicator of activation of brain areas that exhibit lower spiking activity.[8]

In this chapter we discuss in detail our principal technique for studying plasticity of adult cortical functional representations called optical imaging of intrinsic signals[9-11] (recently reviewed in Grinvald et al.[12]), specifically the imaging of rat primary somatosensory cortex. Irrespective of which animal model is used, optical imaging of intrinsic signals (also known as intrinsic signal imaging or ISI) is similar to 2-DG in that both lack the ability to measure neuronal spiking activity directly and track data in millisecond temporal resolution, but both are capable of visualizing functional representations via the simultaneous sampling of a large cortical region with high spatial resolution. We find the combined use of ISI with rat primary somatosensory cortex particularly appealing because:

1. This animal model offers many advantages for the study of adult cortical plasticity.
2. It can be noninvasive to the cortical tissue and thus offers the opportunity to image chronically from the same animal.
3. In contrast to 2-DG, it is capable of visualizing functional representations in real time (albeit providing only an aerial, i.e., two-dimensional, view of functional representations) and exploits a signal source that appears to be more closely related to underlying spiking activity.

The remainder of the chapter is devoted to describing the background, theory, and methodological issues relevant to imaging cortical functional representations, the successful application of ISI for studying the plasticity of a single whisker functional representation in the adult rat, and potential future directions for ISI.

Other nonoptical imaging methods suited for plasticity studies and applied mostly to humans (e.g., fMRI, PET, and MEG) are beyond the scope of this chapter.

2.2 BACKGROUND AND THEORY

2.2.1 Advantages of Applying ISI to the Rat Primary Somatosensory Cortex

One main advantage is the relative ease in imaging cortical activity in the rat. Unlike imaging experiments in cats and monkeys, the surgical preparation for imaging in rats is minimally invasive to the animal[13] because there is no longer a need to 1) control for respiration and heart rate artifacts via paralysis and artificial ventilation in addition to the synchronization of data collection to heart and respiration; 2) remove the skull and dura in order to image cortical activity, leaving the brain intact and in optimal condition throughout the experiment; and 3) build and affix an elaborate "recording chamber" to the skull with screws and cement as is typically done for imaging in cats and monkeys. In addition, because the skull and dura are left intact, the opportunity is available to image the same rat repeatedly over an extended time period (chronic imaging), which can be advantageous for the study of cortical plasticity within the same animal. It is worthwhile to note that, in contrast to rats, the imaging of mouse somatosensory cortex can be achieved through the unthinned skull, which is sufficiently transparent.[14]

The other main advantage is related to the use of the posterior medial barrel subregion (PMBSF) — a region within the primary somatosensory cortex that processes sensory information from the large facial whiskers (or vibrissae) — as an animal model of cortical function and plasticity. The body of literature accumulated thus far on the PMBSF suggests that this system shares many features with sensory cortex in higher mammals, such as topographical organization of whisker inputs to the cortex and functional "columns" dedicated to processing tactile information from the whiskers, and provides a rich source of information from which hypotheses can be formulated about cortical plasticity.[15,16] The popularity of the PMBSF for the study of cortical plasticity stems in part from the relative ease in manipulating the whiskers. Precise stimulation can be delivered to a specific whisker or a specific combination of whiskers, and sensory deprivation can be easily achieved by trimming or plucking whiskers and can be easily reversed by whisker regrowth following termination of whisker trimming or plucking. Because of these advantages, the combination of ISI with rat PMBSF can be a powerful means for investigating adult cortical plasticity.

2.2.2 Intrinsic Signals and Their Sources

ISI is based on the finding that when the brain is illuminated, neuronal activity causes changes in the intensity of the light reflected from the brain. Accordingly, patterns of evoked activity can be detected and quantified by measuring reflectance patterns from the living brain.[9] These stimulus-evoked optical changes are referred to as *intrinsic signals* (IS) to differentiate them from other optical signals obtained using external probes such as voltage-sensitive dyes or calcium indicators. (See other chapters in this book for imaging with external probes.) The IS originates from activity-dependent changes in several intrinsic sources in the brain, including oxygen consumption affecting hemoglobin saturation level, changes in blood volume affecting tissue light absorption and hemoglobin saturation level, and changes in tissue light scattering related to physiologic events accompanying neuronal activity such as ion and water movement in and out of neurons, changes in the volume of neuronal cell bodies, capillary expansion, and neurotransmitter release (reviewed in References 12 and 17).

The relative contribution of these various sources to the IS is dependent on the wavelength of light used to illuminate the cortex. Roughly, blood-related sources dominate the IS in the blue to red parts of the spectrum, while tissue light scattering dominates the IS in the infrared part.[10] Irrespective of which source dominates the IS, it has been demonstrated that the visualization of brain activity by imaging at different wavelengths is similar; thus, any wavelength can be used for functional imaging. However, in terms of signal-to-noise and contrast between active and nonactive regions, the best images of cortical activity are routinely obtained using red light illumination[10] and, thus, the remainder of the present chapter is restricted to imaging of IS using red light illumination. Referring to Figure 2.1A, the temporal profile of the IS under red light illumination is such that, following stimulus delivery, the initial increase in IS is due to an increase in the concentration of deoxyhemoglobin resulting from an increased demand for oxygen from capillaries near the activated neurons. The continued increase and change to later components of the IS are due to the remaining sources: changes in blood volume, blood flow, and tissue scattering.

2.2.3 Stimulus-Evoked IS

A representative IS obtained before, during, and after a stimulus from a cortex illuminated with red light (630 nm) is shown in Figure 2.1A. In the absence of any stimulation, the IS exhibits some baseline level of activity fluctuations. A reliable increase in intrinsic signal is observed at ~500 msec after the onset of a stimulus delivery and typically peaks at ~1.5 to 2 sec post-stimulus onset before returning to baseline. This part of the signal is also referred to as the "initial dip" by the fMRI community, which does not plot signals with a *y*-axis inversion. After returning to baseline, the IS continues to decrease to a minimum level below baseline before repeating another peak in activity that can be larger in magnitude than the initial peak. This oscillatory mode is typical to ISs recorded under red light illumination, with an increase in the concentration of deoxyhemoglobin contributing to the initial increase in IS and the remaining blood-related sources (e.g., changes in blood volume, blood flow, and tissue scattering) contributing to the continued increase and later components of the IS.[18,19]

The initial increase in IS is particularly relevant for imaging brain activity because changes in deoxyhemoglobin concentration occurring at this time point are tightly coupled with the demand for oxygen from capillaries near activated neurons. Note that the amplitude of changes in IS following stimulus onset is quite small, typically between 1 to 10×10^{-4} of the total amount of light reflected from the cortex. Thus, on first inspection, ISs do not seem very promising for imaging brain activity because of their weak amplitude and slower time course (compared to msec domain

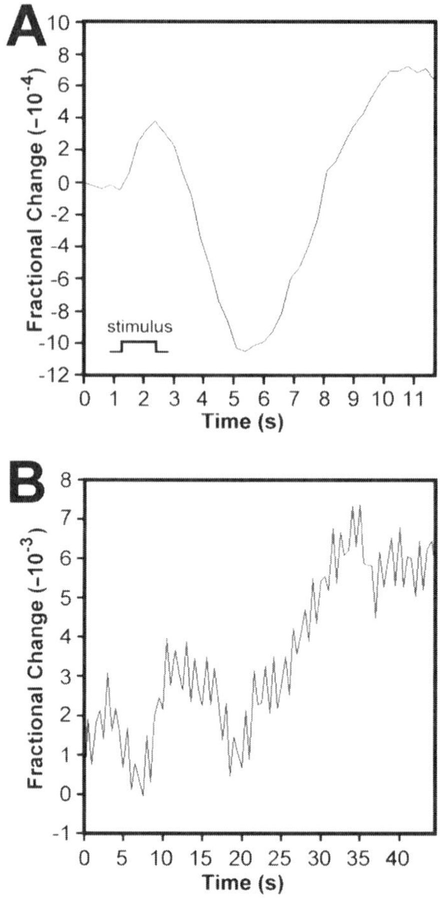

FIGURE 2.1 Temporal profile of evoked and spontaneous intrinsic signal (IS) obtained from the rat barrel cortex. IS was collected in either 300-msec (A) or 500-msec (B) frames and plotted on the *y*-axis in fractional change units relative to the first collected frame. By convention, decreasing light reflectance is plotted as upgoing. (A) Stimulus-evoked cortical IS from the left PMBSF of a rat. IS was sampled over a 0.16-mm^2 cortical area in 300-msec frames after averaging 32 trials. On the *x*-axis, the timepoint of 0 contains IS collected during 0 to 299 msec, with timepoints 1.2 to 2.4 sec containing IS collected during the 1.2-sec stimulation of whisker C2 at 5 Hz. An initial increase was observed within ~0.5 sec after stimulus onset, while a return to baseline was observed within ~2 sec after stimulus onset. As compared to the slow changes in the spontaneous IS shown in B, the amplitude of the fast changes in the stimulus-evoked cortical IS (i.e., prior to the large undershoot) was smaller by an order of magnitude. (B) Spontaneous cortical IS from the right PMBSF of a rat. IS was sampled over a 0.16-mm^2 cortical area in 500 msec frames during a single 45.5-sec trial. On the *x*-axis, the timepoint of 0 contains IS collected during 0 to 499 msec, while the timepoint of 5 sec contains IS collected during 5000 to 5499 msec, and so forth. The magnitude of changes in IS remained similar when the frame duration was increased to 5 sec and the trial interval was increased to 455 sec (data not shown). (Modified with permission from Chen–Bee, C.H. et al., *J. Neurosci. Methods,* 68, 27, 1996.)

of action potentials). Nevertheless, as described later, by simple manipulations of the IS data, clear images of evoked cortical activity are obtained with very high spatial resolution (~50 μm) from large areas of the cortex.

2.2.4 The Imaging Camera

Successful imaging of IS requires a camera capable of detecting small changes in light reflectance, i.e., a camera system with a good signal-to-noise ratio. As no fluorescent probes are involved, bright illumination can be utilized for IS imaging. The ability to image in bright illumination is advantageous as it helps to eliminate nonbiological sources of noise such as shot noise (known also as photon statistical noise) — a noise source that stems from the quantal properties of light and is the major source of nonbiological noise under bright illumination. Because the shot noise depends on the square root of light intensity, it is desirable to work as close as possible to the saturation level of the camera. Thus, cameras that can register high light intensities without compromising their linearity are advantageous for ISI.

We use a slow-scan, cooled, digital charge-coupled device (CCD) camera that contains pixels with large well capacities (~500,000 photons per pixel); the well capacity indicates the number of electrons that can be accumulated by one pixel of the CCD chip before an overflow of charge. Such large well capacities enable the use of high-intensity illumination, effectively reducing shot noise and consequently improving the signal-to-noise ratio. Further improvement in signal-to-noise ratio is accomplished after performing a 2×2 pixel binning, resulting in the final array of 192×144 pixels. To enhance the signal-to-noise ratio further, averaging of trials is used, as described next.

The CCD camera is positioned above the cortex to be imaged and thus offers an aerial or two-dimensional collection of light reflection from the illuminated cortex. Therefore, data collection comprises the cumulative activity integrated orthogonally to the cortical surface. The camera operates in a "frame-transfer" mode that enables the continuous capture of sequence of frames for imaging before, during, and after sensory stimulation (Figure 2.1A). Penetration of light into the cortex depends on the wavelength of the illumination; longer wavelengths penetrate deeper into the cortical tissue and thus the use of red illumination is also advantageous in terms of penetration. However, at present it is still difficult to estimate to what extent the red light penetrates *in vivo* and, consequently, to what extent different cortical layers, especially the deeper layers, contribute to the collected data.

2.2.5 Biological Noise

Two types of biological noise are typically present when imaging IS with red illumination, both of which exhibit magnitude changes much larger than those of stimulus-evoked IS: 1) global, spontaneous fluctuations in oxygenation and 2) local fluctuations overlying surface blood vessels. Spontaneous fluctuations in IS are up to ten times greater and occur on a slower time scale (oscillations of ~0.05 to 0.1 Hz or one complete cycle every 10 to 20 sec) as compared to stimulus-evoked IS (Figure 2.1B). Because stimulus delivery evokes only a small change on top of the large spontaneous IS fluctuations, the successful imaging of stimulus-evoked IS requires that these spontaneous fluctuations be somehow averaged out. As they are not time-locked to stimulus delivery, spontaneous IS fluctuations can be minimized by averaging a set of stimulation trials. The number of imaging trials needed (32 to 128 trials) is comparable to that of single unit recording experiments. Any residual presence of spontaneous fluctuations can then be addressed at the level of data analysis.

On the other hand, the averaging of stimulation trials is not effective in minimizing local fluctuations that overlie surface blood vessels because they are also time-locked to stimulus delivery. However, as they typically follow a slower time course, contributions from IS overlying surface vessels can be separated from stimulus-evoked IS by limiting analysis to data collected <1.5 sec after stimulus onset (Figure 2.2).

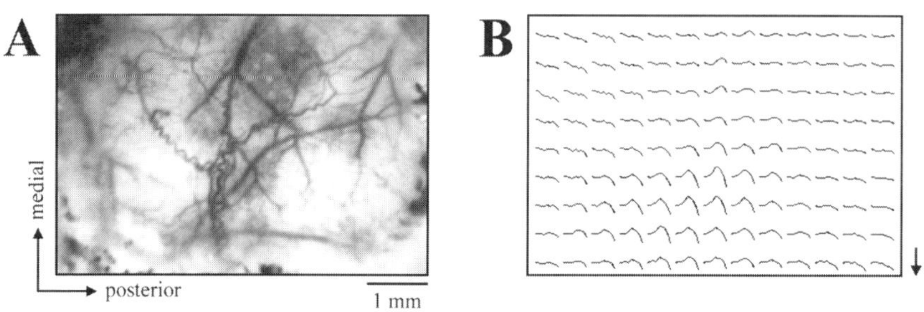

FIGURE 2.2 Intrinsic signal imaging of the rat PMBSF. (A) The imaged area within the left PMBSF of an adult rat as viewed through a thinned skull. Dark streaks correspond to large blood vessels found on the cortical surface or dura mater. Orientation and scale bar apply to all panels. (B) Array of intrinsic signals (IS) evoked by 5-Hz stimulation of whisker D1 for 1 sec from a collection of 128 trials. Increasing IS (i.e., decreasing light reflectance) is plotted as upgoing (downward pointing scale bar = 1×10^{-3}), with each plot corresponding to the 4.5-sec time course as averaged for the underlying 0.46×0.46 mm area. The region exhibiting an increase in IS overlying the cortical tissue is located in the lower center of the total imaged area, with a concurrent but slower increase in IS overlying nearby large surface blood vessels (upper center). (C) Images of evoked IS in 500-msec frames from same data presented in (B). Each image was generated after dividing a given post-stimulus frame (indicated at top left of each image; 0 sec = stimulus onset) by a frame collected immediately prior to stimulus onset and applying an 8-bit, linear grayscale map to the processed data so that increased IS greater than -1.5×10^{-4} is mapped to a grayscale value of black. Evoked IS overlying the cortical tissue (black patch in lower center) is present starting 0.5 sec post-stimulus onset and remains elevated through 2.0 sec post-stimulus onset before diminishing in strength. In contrast, evoked IS overlying large surface blood vessels (black streaks in upper center) follows a slower time course, with minimal activity present 1.0 sec post-stimulus onset that increases and remains elevated past 2.5 sec post-stimulus onset. (D–E) Visualization of evoked IS collected from either 0.5 up to 2.5 sec (D) or 0.5 up to 1.5 sec (E) post-stimulus onset. The resultant image for 0.5 up to 2.5 sec (D) or 0.5 up to 1.5 sec (E) post-stimulus onset is equivalent to averaging all four frames or the first two frames shown in (C), respectively. (Each frame has been divided by the frame collected immediately prior to stimulus onset.) When data processing includes the time epoch of 0.5 up to 2.5 sec post-stimulus onset (D), evoked IS overlying both the cortical tissue (black patch in lower center) and large surface blood vessels (black streaks in upper center) are visualized. However, data processing of the shorter post-stimulus time epoch (E) generates an image of evoked IS overlying the cortical tissue with limited presence of evoked IS overlying the large surface blood vessels. Grayscale bar applies to all figures (C–E). (Modified with permission from Chen–Bee, C.H. et al., *J. Neurosci. Methods,* 97, 157, 2000.)

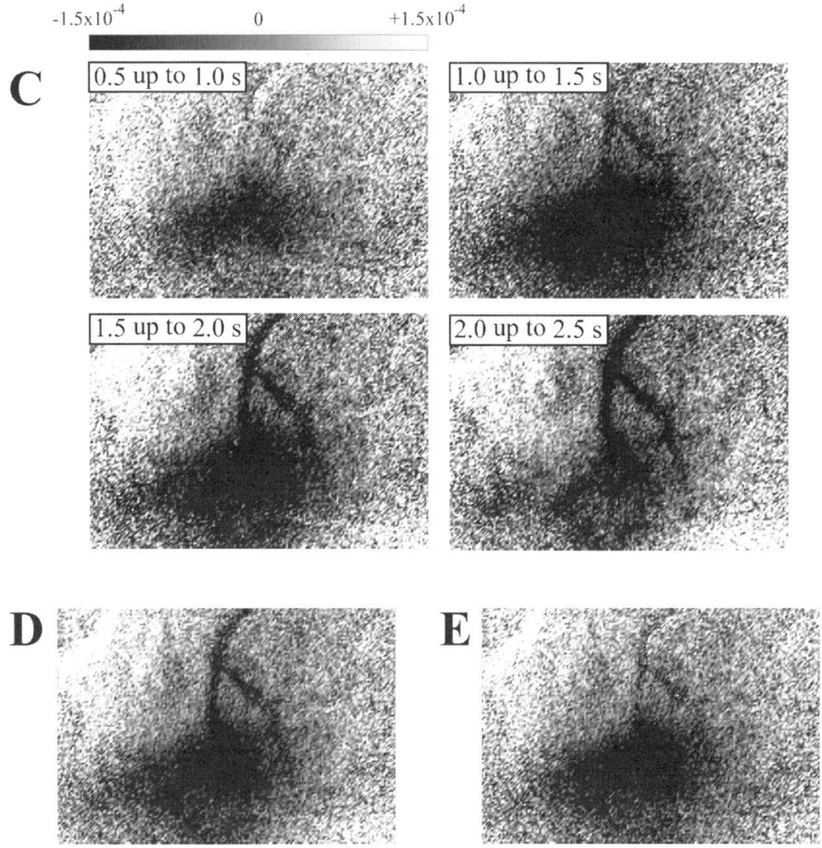

FIGURE 2.2 *Continued.*

2.2.6 Imaging Plasticity in Sensory Cortex

In recent years most of ISI research has been focused on imaging functional columns — groups of vertically organized neurons that exhibit preference in responding to a specific stimulus (e.g., orientation columns, ocular dominance columns, and other columnar organizations in the visual cortex) and believed to be an important module of the functional organization of cortex. Thus, the study of cortical plasticity can be conducted at the level of functional columns. Imaging plasticity of the columnar organization requires differential imaging of a cortical area after the application of different stimuli that activate different columns of cortical neurons. Division or subtraction of differences in activation patterns by different stimuli is used to highlight the cortical location of columns with a preferred response to a specific stimulus. Plasticity is then typically assessed by quantifying changes in the maps of cortical columns following sensory manipulation.[20,21]

Imaging the functional representation of a peripheral organ provides an alternative means to study plasticity in the adult cortex. An organ functional representation is defined as the total cortical area activated by stimulation of the sensory organ and visualized by subtracting or dividing background (spontaneous) activity that precedes the stimulation from the post-stimulus activity rather than employing a differential imaging strategy to create the columnar preference maps. This approach is preferable in cortical regions such as the somatosensory cortex, in which the sensory representations of the organs overlap and do not exhibit clear columnar organization for stimulus submodality (e.g., orientation) to the extent seen in visual cortex. Plasticity of a sensory organ's functional representation is typically assessed by quantifying changes in the cortical area activated by the sensory organ following manipulation. The focus of the present chapter is on imaging functional representation for the study of cortical plasticity.

2.2.7 Relationship between IS and Neuronal Activity

ISI provides an indirect measure of neuronal activity; thus, it is important to determine the relationship between the IS and underlying neuronal activity. In the rat PMBSF, we have found that the pattern of IS activity is very similar to that of the underlying spiking neurons. Specifically, the peak of IS activity corresponds spatially to the location where neurons respond in the strongest fashion to the same sensory stimulus. When sampling at progressively larger distances away from the peak, both optical and spiking activity become progressively weaker with increasing distance from peak activity, and areas with no optical activity do not exhibit any spiking activity, either.[22-24] As will be described next, this correspondence has been upheld for at least one instance of deprivation-induced plasticity in rat PMBSF. Further research is needed to establish the exact correspondence between these two measures of cortical activity so that future experiments can be performed without repeated single-unit verification.

2.3 METHODOLOGY

2.3.1 Anesthesia

At the start of an experiment, rats are inducted with an inhalant (Isoflurane) followed by a bolus injection of sodium pentobarbital (Nembutal). They are then maintained for the remainder of the experiment with supplemental injections of Nembutal as needed. Changes in the anesthesia level can affect the quality of imaging data; too light a level can lead to imaging data with increased contributions from surface vasculature, while too deep a level can lead to a decrease in the ratio of stimulus-evoked signal to spontaneous background activity. This relationship is presumably the outcome of imaging with red light illumination, where intrinsic sources dominating the IS are metabolic mechanisms related to oxygen metabolism. Thus, maintaining a constant level of anesthesia is essential for a successful imaging experiment. To assess the level of anesthesia, we examine many physiological indicators, including withdrawal and corneal reflexes, body temperature, heart rate, and the level of oxygenated blood as indicated by coloring of the eyes and extremities.

2.3.2 Surgical Preparation

The scalp is retracted above the skull overlying the cortical region of interest, and the exposed skull is carefully thinned to a thickness of approximately 150 μm using a high-speed drill. To maintain its transparency, the thinned skull is then bathed in saline contained within a simple petroleum jelly chamber erected around the thinned skull and sealed with a glass coverslip. If the rat is revived for chronic imaging, all petroleum jelly and other exogenous materials are removed from the skull, the scalp is sutured, and topical and systemic antibiotics are administered immediately after the first imaging session to decrease the risk of infection. Rats are able to resume normal behaviors (e.g., grooming, feeding, and drinking) within a day after surgery despite the removal of some muscle tissue during the initial surgery. We have found that the thinned skull remains in excellent condition throughout the interim between imaging sessions, provided that the scalp is kept moist with saline throughout the duration of the initial imaging experiment. In nearly every case we have found that the thinned region of the skull and underlying vessels in the dura and pia mater appear largely unchanged from the first imaging session.

2.3.3 Imaging Setup

A schematic of our basic setup is shown in Figure 2.3. The anesthetized rat is held in a stereotactic frame (not shown). Illumination is provided by a standard 100-W tungsten-halogen light source stabilized with a regulated power supply in order to minimize voltage ripples to less than 10^{-5}. Stabilization of the light source is essential because magnitude changes in stimulus-evoked IS are on the order of $\sim 10^{-4}$. The light then passes a protective heat filter and a red bandpass filter (630 ± 15 nm)

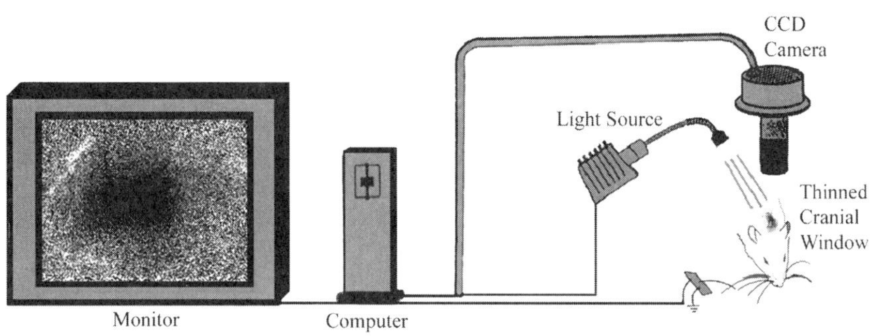

FIGURE 2.3 Setup for *in vivo* intrinsic signal optical imaging of cortical functional representations. Images are taken through the animal's thinned skull. The cortex is illuminated with a stabilized 630-nm light. The CCD camera captures a sequence of images before, during, and after a computer-controlled mechanical stimulation to a whisker. The sequence of digital images is sent to a computer that also controls the experiment. After averaging of several image sequences, analyzed data are displayed as an image using an 8-bit linear scale on a computer monitor. The black patch represents activity evoked by 5-Hz stimulation to a single whisker. White streaks are blood vessel artifacts. Typically, one computer acquires the data and controls the experiment, while the data are sent to a second computer for online analysis.

before being directed by flexible light guides to illuminate the cortex through the thinned skull. A sensitive (12 bit) slow-scan cooled CCD camera is used to capture the light reflected from the cortex. To achieve the optimal viewing area and working distance, this CCD camera is fitted with a 50 mm lens that is inverted and attached to an extender. The camera is focused 300 to 500 μm below the cortical surface to minimize contributions from surface blood vessels. A computer controls data collection and stimulus delivery for a set of trials. The collected data are then transferred to a second computer so that incoming data can be analyzed concurrently with the collection of the next set of trials.

2.3.4 Data Collection

For each trial, the camera collects an uninterrupted sequence of frames. Each frame represents light collected from a 6.8 × 5.1 mm cortical area with a 192 × 144 pixel array; thus, one pixel corresponds to a 12.5-μm^2 cortical area. The camera has the capacity to capture a specified number of consecutive frames ranging from 2 to 100+ frames and a specified frame duration as brief as 200 msec. At the single trial level, we routinely collect a sequence of either 9 or 11 frames, both with a frame duration of 500 msec (i.e., two frames per second). Hence, the total duration of a trial is either 4.5 or 5.5 sec. In a stimulation trial, 1 sec of spontaneous activity is collected prior to stimulus delivery. Stimulation consists of five 1.9° deflections of a single whisker in the rostral-caudal direction at a frequency of 5 pulses per second (i.e., 5 Hz). Whisker deflections are achieved by a computer-controlled copper wire attached to a stepping motor. Data are collected in blocks of 32 stimulation trials randomly interlaced with no stimulation, control trials, with an intertrial interval of 15 sec. For each type of stimulation, four blocks of 32 stimulation trials each are then summed together to form a complete data file.

2.3.5 Data Analysis and Quantification

A summary is provided in Figure 2.4. To determine the overall baseline level of activity from the imaged cortex prior to stimulus onset, a prestimulus ratio value is calculated for each pixel within the 192 × 144 pixel array via dividing the latter of the two prestimulus frames by the earlier prestimulus frame (see Figure 2.4B). The median, rather than the mean, ratio value (referred to hereafter as prestimulus baseline) is then used to indicate the baseline level of activity because it is less sensitive to outlier values due to random or biological noise. Analysis of stimulus-evoked IS is restricted to 1 sec of data collected from 0.5 up to 1.5 sec after stimulus onset to ensure that only the rising phase of the IS is captured with minimal contribution from surface blood vessels. A poststimulus ratio value is assigned to each pixel within the 192 × 144 pixel array by first averaging this 1 sec of post-stimulus data and then dividing the averaged data by 0.5 sec of data collected immediately prior to stimulus onset (see Figure 2.4C). As more active cortex reflects less light than less active cortex, a pixel whose post-stimulus ratio value is smaller than the prestimulus baseline would correspond to a cortical region exhibiting increased activity. Prior to quantifying the evoked activity area, a Gaussian filter

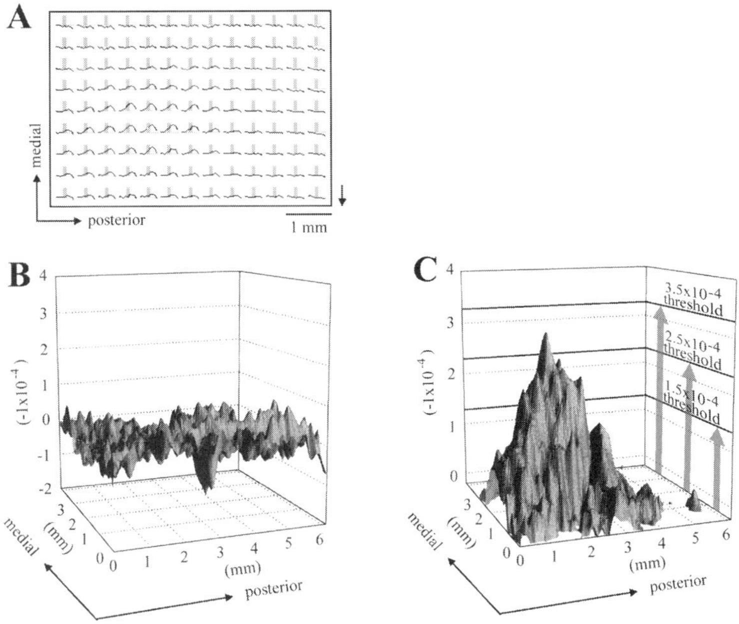

FIGURE 2.4 Areal extent quantification with the use of absolute thresholds combined with the prestimulus baseline. (A) Array of IS evoked by stimulation of a single whisker E2 after collection of 128 trials (downward pointing scale bar = fractional change of 1×10^{-3}). Each plot corresponds to the 4.5-sec time course as averaged for the underlying 0.46×0.46 mm area, with the time epoch of 0.5 up to 1.5 sec post-stimulus onset highlighted in gray. Note the region of evoked IS is located in the left center of the total imaged area. (B) Determination of prestimulus baseline. Data collected from 1.0 up to 0.5 sec prior to stimulus onset are converted to ratio values relative to data collected from 0.5 up to 0 sec prior to stimulus onset such that the processed data may be thought of as a "mountain of activity," with the x- and y-axes indicating cortical location and the z-axis indicating strength of prestimulus IS. The median ratio value (0.21×10^{-4} for this example) is used as a measure of the average prestimulus activity over the entire imaged area. Ratio values are filtered (Gaussian half width 5) prior to plotting. (C) Setting absolute thresholds when used in combination with the prestimulus baseline. Data collected from 0.5 up to 1.5 sec post-stimulus onset are converted to ratio values relative to prestimulus data such that the processed data may be thought of as a "mountain of evoked activity," with the x- and y-axes indicating cortical location and the z-axis indicating strength of post-stimulus IS. Ratio values are filtered (Gaussian half-width 5) prior to plotting. Thresholds are set at absolute increments away from the prestimulus baseline (0.21×10^{-4} as indicated by z-axis minimum). Three arbitrary increments are illustrated here: 1.5, 2.5, and 3.5×10^{-4} away from prestimulus baseline. (D–F) Visualizing the quantified area of evoked IS using absolute thresholds combined with the post-stimulus baseline. An 8-bit, linear grayscale mapping function is applied to the nonfiltered ratio values so that the quantified area (enclosed by a white border) is visualized as a black patch within the total imaged area for each of the three absolute thresholds. Note that an area was not quantified with the highest (3.5×10^{-4}) threshold (D), as the peak ratio value for this example is 3×10^{-4}. Orientation and horizontal scale bar in (A) also apply to (D–F). (Modified with permission from Chen–Bee, C.H. et al., *J. Neurosci. Methods,* 97, 157, 2000.)

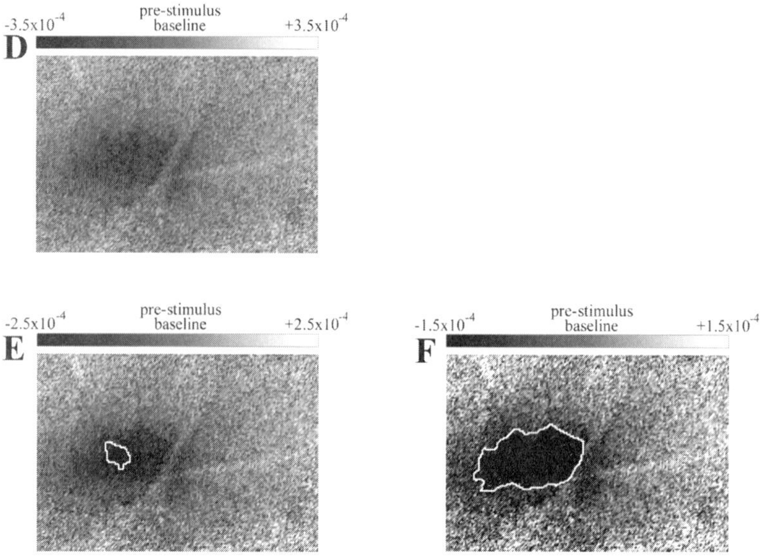

FIGURE 2.4 *Continued.*

(half-width = 5) is applied to the post-stimulus ratio values to remove high frequency noise likely attributable to shot noise. The filtered post-stimulus ratio values are then thresholded at absolute increments away from the prestimulus baseline, the number of qualifying pixels are counted, and the total pixel count is translated to a unit of area (mm^2; see Figure 2.4D to F).

For two-dimensional visualization of the evoked activity area as a dark patch, an 8-bit linear gray scale can be applied to the post-stimulus ratio values such that those pixels with post-stimulus ratio values similar to the prestimulus baseline are assigned a middle shade of gray, those with smaller post-stimulus ratio values (i.e., corresponding to increased activity) are assigned darker shades of gray (darkest shade = black), and those with larger post-stimulus ratio values are assigned lighter shades of gray (see Figure 2.4D to F). For more details on data analysis and quantification, see References 25 and 26.

2.4 SUCCESSFUL APPLICATION OF ISI FOR STUDYING ADULT PLASTICITY OF FUNCTIONAL REPRESENTATIONS

In this section, we discuss a study conducted in our laboratory to illustrate the successful application of ISI for studying adult plasticity of functional representations.[23] This study further addressed the competition hypothesis of cortical plasticity, which states that sensory inputs into the cortex are constantly competing for cortical territory and any advantage to one of the sensory inputs increases its ability to gain cortical territory at the expense of neighboring inputs. Previous selective deprivation

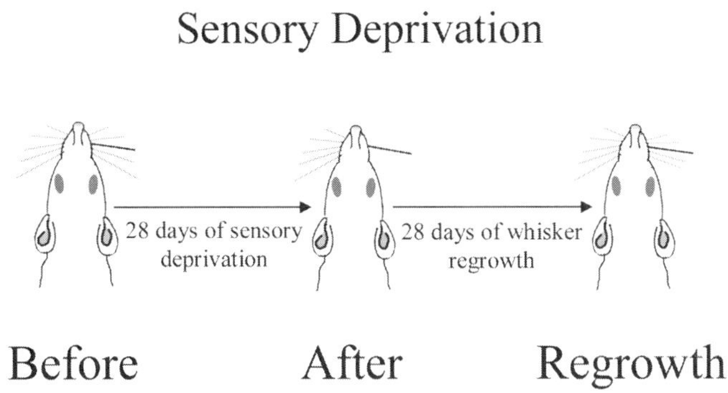

FIGURE 2.5 Sensory deprivation paradigm. Imaging of whisker C2 (black) functional representation was taken prior to the removal of all other whiskers on the snout (Before). A second imaging session of the C2 functional representation was taken after a month of deprivation (After). Immediately following the second imaging session the deprivation was terminated and all whiskers were allowed to regrow. In a few cases, a third imaging session of the C2 functional representation, from the same animal, was taken a month after the termination of deprivation (Regrowth). (Modified with permission from Polley, D.B. et al., *Neuron,* 24, 623, 1999.)

experiments in the somatosensory system of the adult rat have verified the predictions of the competition hypothesis. For example, a popular experimental design is to remove (either by plucking or trimming) all but a single whisker (the spared whisker) on one side of the snout in order to assess potential changes in the spared whisker's territory within the PMBSF. Such studies report an expansion of the spared whisker functional representation, a finding consistent with the competition hypothesis.[27-29]

Because the removal of almost the entire whisker array on one side of the snout can cause significant behavioral asymmetries in whisker-use behavior,[30] combined with reports that behavioral usage of a somatosensory organ can affect its functional representation,[31-34] we wondered whether selective removal of all but the spared whisker can still induce changes in the rat's whisker usage and, if so, whether this can affect the spared whisker's functional representation. To address this issue, we exploited the advantages offered by ISI and imaged the functional representation of the spared whisker at two time points per animal, once before and once after a 1-month sensory deprivation of all but the spared whisker and, in a few cases, after an additional month of the deprived whiskers' regrowth (Figure 2.5).

We used ISI successfully to capture the expected expansion in the size of the spared whisker functional representation after the 1-month sensory deprivation (Figure 2.6A). In the group of rats that underwent assessment for potential behavioral asymmetries in whisker usage due to whisker deprivation, we detected an unexpected contraction in the size of the spared whisker functional representation (Figure 2.6B). Subsequent analysis revealed that this contraction did not correlate with any behavioral asymmetry in whisker usage (no asymmetry was detected) but instead correlated with the total time spent actively using the whiskers during the behavioral assessment. In rats that received only the behavioral assessment, we observed no systematic change in the size of the representation of the whisker that was spared

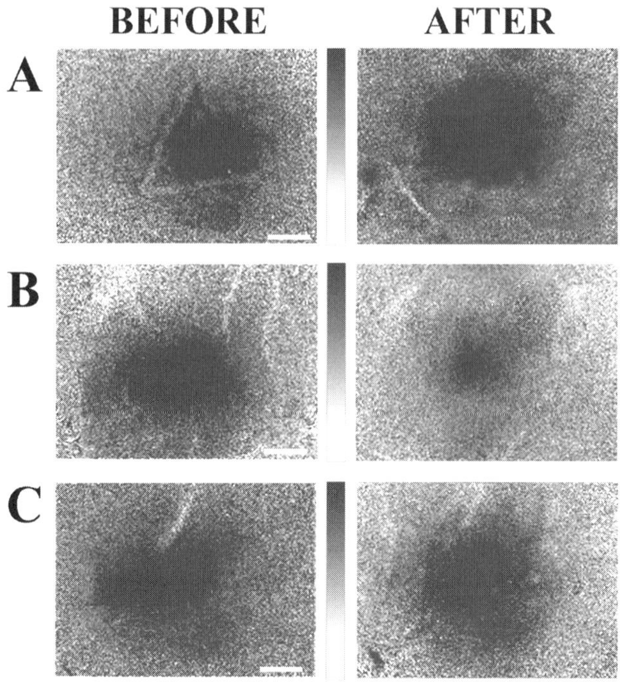

FIGURE 2.6 Interaction between sensory deprivation and behavioral assessment as reflected by changes in the size of the C2 functional representation before and after 1 month of treatment. (A–C) Representative cases for three groups: (A) Deprivation, (B) Deprivation with behavioral assessment, and (C) Behavioral assessment only. Unfiltered ratio values are converted to grayscale values so that the functional representation of the C2 whisker is visualized as a dark coherent area. The middle of a linear grayscale bar represents prestimulus baseline activity, while the black and white values are set to a decrease or increase of $\geq 2.5 \times 10^{-4}$ from baseline, respectively. Black and white streaks correspond to large blood vessels. Scale bars = 1mm. (D) The average area (mean + SE) of the C2 whisker's functional representation as compared at a single (2.5×10^{-4}) activity threshold between the first and second imaging sessions. The area is similar among the three treatment groups in the first imaging session. Between the first and second imaging session, the area is observed to expand for the deprivation group (circle), contract for deprivation with the behavioral assessment group (diamond), and not change for the behavior-only group (triangle). (E) Average percent change in the area of the C2 whisker's functional representation as quantified using multiple activity thresholds between the first and second imaging sessions (mean ±SE). (Modified with permission from Polley, D.B. et al., *Neuron,* 24, 623, 1999.)

in the other groups (Figure 2.6C) Verification with single unit recordings within a subset of imaged animals revealed that the observed plasticity in neuronal activity was complementary with the observed plasticity in IS activity. In another subset of the rats, ISI was used successfully to image the spared whisker functional representation for a third time, 1 month after termination of whisker deprivation, thereby determining that an expanded or contracted functional representation can return to its original size (Figure 2.7).

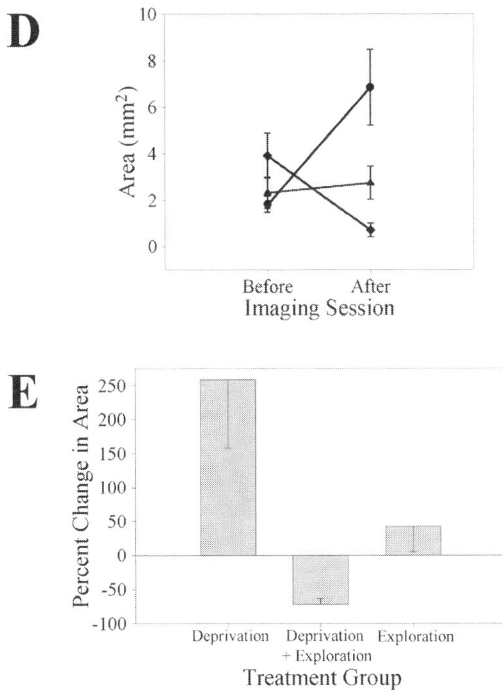

FIGURE 2.6 *Continued.*

The preceding example of plasticity in the adult somatosensory cortex of the rat serves to emphasize the apparent advantage of visualizing plasticity by chronic imaging from the same rat: the ability to image in a noninvasive way to the brain the cortical functional representation of an organ before, during, and after a manipulation to the sensory system. Based on the multiple sampling from the same rat, we were able to pursue plasticity-related *in vivo* investigations regarding the consequences of deprivation before and after 1 month of deprivation as well as the consequences of reversing the deprivation.

2.4.1 Successful Application of ISI in Other Types of Plasticity Experiments

In the previous section, a long-term plasticity study was provided to detail how ISI can be used successfully to study adult cortical plasticity. Here, we mention briefly other types of plasticity experiments that can also benefit from the use of ISI. For example, changes in the functional representations of a whisker in the adult brain have been followed in real time on the scale of minutes in response to topical application of neuromodulators, such as neurotrophins,[35] and cholinergic agonists and antagonists.[36] These studies exploit the ability of ISI to assess simultaneously the response of the entire functional representation of a whisker to a neuromodulator rather than the extrapolation from single unit recordings of a selected sample of cells that respond to neuromodulators following direct iontophoresis.

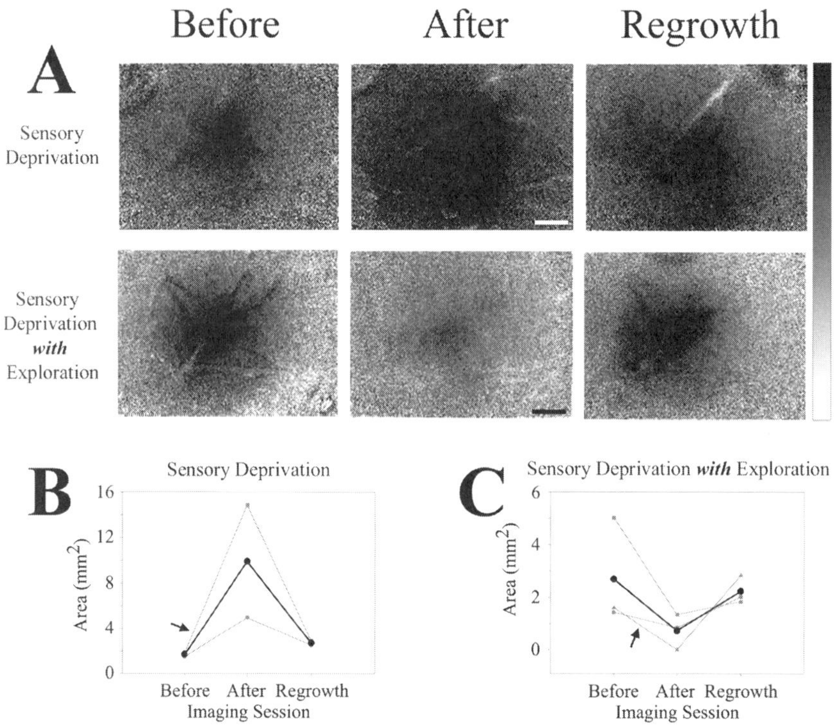

FIGURE 2.7 Plasticity of the spared whisker's functional representation is reversible upon restoration of normal sensory input. (A) Unfiltered ratio images from a sensory-deprived animal that remained in its home cage (top row) or was given an opportunity for behavioral assessment (bottom row) before deprivation (Before), after 28 days of deprivation (After), and after 28 days of whisker regrowth (Regrowth). Ratio values are converted to grayscale values in which the prestimulus baseline is shown as gray and the black and white values on the grayscale bar are set to a decrease or increase of $\geq 2.5 \times 10^{-4}$ from baseline values, respectively. Scale bar =1 mm and applies to all images. (B and C) Areal extent of the C2 whisker's functional representation before deprivation, after deprivation, and following deprived whisker regrowth for two animals that remained in the home cage (B) and for three animals that received behavioral assessment (C) (arrow indicates representative case shown in A; group mean shown in black). For both groups, the area of the functional representation is nearly restored to predeprivation levels following the regrowth of the whiskers. Note differences in y-axes. (Modified with permission from Polley, D.B. et al., *Neuron,* 24, 623, 1999.)

ISI has also been used to study the consequences of knocking out a gene on the functional representations of single whiskers as well as the arrangement of multiple whisker functional representations within the somatosensory cortex (the whiskers' functional map) of the adult mouse.[14] This study focused on the functional consequences of knocking out gene coding for a putative mapping label (Ephrin A5) hypothesized to participate in guidance of axons to the somatosensory cortex from the thalamus. Imaging allowed us to observe and quantify distortions in the whiskers' functional map that corresponded to the gradient of Ephrin A5 and also revealed potential changes in functional connectivity within the cortex. This helped to establish Ephrin A5 as a cortical mapping label contributing to the functional organization of the somatosensory cortex of the mouse. This study, combining for the first time gene manipulations with *in vivo* functional imaging, demonstrates the advantage of combining both approaches to the study of gene function: the ability to study the functional consequences of gene deletion at the level of single whisker representations as well as the level of the whiskers' functional map.

While a main goal of the present chapter is to emphasize the advantages offered by ISI, we are appreciative of the advantages offered by the more traditional techniques used to study plasticity. In fact, several types of questions regarding cortical plasticity can be answered only by using traditional techniques such as single-unit recording techniques. Whenever possible, we attempt to apply many techniques to the same animal when pursuing the characterization of plasticity and its underlying mechanisms.

We feel that the use of an imaging technique such as ISI is an optimal first step, as it provides a clear image of the "macro" or population vantage point of cortical activity and can follow cortical plasticity within the same animal. The functional images obtained with ISI can then be used to guide the precise placement of microelectrodes for neuronal recordings, micropipettes for injection of neuronal tracers, and iontophoresis of drugs and microdialysis probes at "hot spots" of activity, areas of reduced activity, and areas of no activity. Consequently, the investigator can benefit from using the optical activity imaging map to ask more targeted questions about the physiology, anatomy, and pharmacology underlying cortical plasticity. Thus, the optical map of activity can be seen as a central component of a multidisciplinary approach to the study of cortical structure and function and their relationships in the normal and plastic cortex.

2.5 FUTURE DIRECTIONS

The most desirable direction for the study of plasticity is to be able to use ISI in a noninvasive fashion in freely behaving animals. This will enable the imaging of changes in the cortex on a continuous basis without the need for anesthesia. Ideally, the camera chip is placed directly on the animal's thinned skull and transmits the optical data to the host computer by telemetry, but flexible connections that do not restrain the animal's movement could also serve this purpose. As prices of the CCD chips drop and the trend of miniaturization of electronic circuits that can support such imaging continues, the ability for long-term, real-time imaging of cortical function within the same animal may be realized in the near future. In fact, other chapters in this book already address the first steps toward realizing this technical challenge. ISI would also benefit from an improved ability to resolve signals from different depths in the cortex so as to provide functional imaging of single cortical layers and, consequently, the ability to produce detailed 3-D maps of cortical activation.

Another promising direction would be the development of a multipurpose imaging system combining the advantages of ISI with the advantages offered by other imaging techniques. Combining the high-resolution spatial maps obtained with ISI, for example, with the high temporal resolution offered by dye-based optical imaging (e.g., voltage-sensitive dyes) and the detailed images of neuronal structure and function provided with multiphoton imaging will allow researchers to address questions about cortex on multiple spatial and temporal scales. By visualizing cortical activity on many levels, from single dendritic spine to assemblies of thousands and millions of neurons, one may gain optimal insight into the mechanisms of cortical function and its plasticity.

ACKNOWLEDGMENTS

We thank our previous collaborators Susan Masino, Michael Kwon, Neal Prakash, Jonathan Bakin, Yehuda Dory, Barbara Brett–Green, and Silke Penschuck for their original contributions. Work was supported by NIH grants NINDS NS-34519, NINDS NS-39760, and NSF IBN-9507636.

REFERENCES

1. Buonomano, D.V. and Merzenich, M.M., Cortical plasticity: from synapses to maps, *Annu. Rev. Neurosci.,* 21, 149, 1998.
2. Kaas, J.H., The reorganization of sensory and motor maps after injury in adult mammals, in *The New Cognitive Neurosciences*, Gazzaniga, M.S., Ed., MIT Press, Cambridge, 2000, chap. 15.
3. Recanzone, G.H., Cerebral cortical plasticity: perception and skill acquisition, in *The New Cognitive Neurosciences*, Gazzaniga, M. S., Ed., MIT Press, Cambridge, 2000, chap. 16.
4. Merzenich, M.M. et al., Topographic reorganization of somatosensory cortical areas 3b and 1 in adult monkeys following restricted deafferentation, *Neuroscience,* 8, 33, 1983.
5. Nicolelis, M.A. et al., Reconstructing the engram: simultaneous, multisite, many single neuron recordings, *Neuron,* 18, 529, 1997.
6. deCharms, R.C., Blake, D.T., and Merzenich, M.M., A multielectrode implant device for the cerebral cortex, *J. Neurosci. Methods,* 93, 27, 1999.
7. Kennedy, C. et al., Mapping of functional neural pathways by autoradiographic survey of local metabolic rate with (14C)deoxyglucose, *Science,* 187, 850, 1975.
8. Hess, A. and Scheich, H., Optical and FDG mapping of frequency-specific activity in auditory cortex, *Neuroreport,* 7, 2643, 1996.
9. Grinvald, A. et al., Functional architecture of cortex revealed by optical imaging of intrinsic signals, *Nature,* 324, 361, 1986.
10. Frostig, R.D. et al., Cortical functional architecture and local coupling between neuronal activity and the microcirculation revealed by *in vivo* high-resolution optical imaging of intrinsic signals, *Proc. Natl. Acad. Sci. U.S.A.,* 87, 6082, 1990.
11. Ts'o, D.Y. et al., Functional organization of primate visual cortex revealed by high resolution optical imaging, *Science,* 249, 417, 1990.
12. Grinvald, A. et al., *In-vivo* optical imaging of cortical architecture and dynamics, in *Modern Techniques in Neuroscience Research*, Windhorst, U. and Johansson, H., Eds., Springer-Verlag, Berlin, 1999, chap. 34.

13. Masino, S.A. et al., Characterization of functional organization within rat barrel cortex using intrinsic signal optical imaging through a thinned skull, *Proc. Natl. Acad. Sci. U.S.A.*, 90, 9998, 1993.
14. Prakash, N. et al., Malformation of the functional organization of somatosensory cortex in adult ephrin-A5 knock-out mice revealed by *in vivo* functional imaging, *J. Neurosci.*, 20, 5841, 2000.
15. Armstrong–James, M., The nature and plasticity of sensory processing within adult rat barrel cortex, in *Cerebr. Cortex*, Jones, E.G. and Diamond, I.T., Eds., Plenum Press, New York, 1995, chap. 8.
16. Simons, D.J., Neuronal integration in the somatosensory whisker/barrel cortex, in *Cerebr. Cortex,* Jones, E.G. and Diamond, I.T., Eds., Plenum Press, New York, 1995, chap. 6.
17. Cohen, L.B., Changes in neuron structure during action potential propagation and synaptic transmission, *Physiol. Rev.,* 53, 373, 1973.
18. Malonek, D. and Grinvald, A., Interactions between electrical activity and cortical microcirculation revealed by imaging spectroscopy: implications for functional brain mapping, *Science,* 272, 551, 1996.
19. Malonek, D. et al., Vascular imprints of neuronal activity: relationships between the dynamics of cortical blood flow, oxygenation, and volume changes following sensory stimulation, *Proc. Natl. Acad. Sci. U.S.A.,* 94, 14826, 1997.
20. Sengpiel, F., Stawinski, P., and Bonhoeffer, T., Influence of experience on orientation maps in cat visual cortex, *Nat. Neurosci.,* 2, 727, 1999.
21. Dragoi, V., Sharma, J., and Sur, M., Adaptation-induced plasticity of orientation tuning in adult visual cortex, *Neuron*, 28, 287, 2000.
22. Peterson, B.E., Goldreich, D., and Merzenich, M.M., Optical imaging and electrophysiology of rat barrel cortex. I. Responses to small single-vibrissa deflections, *Cerebr. Cortex,* 8, 173, 1998.
23. Polley, D.B., Chen–Bee, C.H., and Frostig, R.D., Two directions of plasticity in the sensory-deprived adult cortex, *Neuron,* 24, 623, 1999.
24. Brett–Green, B., Chen–Bee, C.H., and Frostig, R.D., Comparing the functional representations of central and border whiskers in rat primary somatosensory cortex, *J. Neurosci.,* in press, 2001.
25. Chen–Bee, C.H. et al., Areal extent quantification of functional representations using intrinsic signal optical imaging, *J. Neurosci. Methods,* 68, 27, 1996.
26. Chen–Bee, C.H. et al., Visualizing and quantifying evoked cortical activity assessed with intrinsic signal imaging, *J. Neurosci. Methods,* 97, 157, 2000.
27. Hand, P.J., Plasticity of the rat cortical barrel system, in *Institute of Neurological Sciences Symposium in Neurobiology*, Morrison, A.R. and Strick, P.L., Eds., Academic Press, Philadelphia, 1982, 49.
28. Kossut, M. and Hand, P., Early development of changes in cortical representation of C3 vibrissa following neonatal denervation of surrounding vibrissa receptors: a 2-deoxyglucose study in the rat, *Neurosci. Lett.,* 46, 7, 1984.
29. Fox, K., A critical period for experience-dependent synaptic plasticity in rat barrel cortex, *J. Neurosci.*, 12, 1826, 1992.
30. Milani, H., Steiner, H., and Huston, J.P., Analysis of recovery from behavioral asymmetries induced by unilateral removal of vibrissae in the rat, *Behav. Neurosci.,* 103, 1067, 1989.
31. Jenkins, W.M. et al., Functional reorganization of primary somatosensory cortex in adult owl monkeys after behaviorally controlled tactile stimulation, *J. Neurophysiol.,* 63, 82, 1990.

32. Recanzone, G.H. et al., Topographic reorganization of the hand representation in cortical area 3b owl monkeys trained in a frequency-discrimination task, *J. Neurophysiol.*, 67, 1031, 1992.
33. Wang, X. et al., Remodeling of hand representation in adult cortex determined by timing of tactile stimulation, *Nature,* 378, 71, 1995.
34. Xerri, C. et al., Experience-induced plasticity of cutaneous maps in the primary somatosensory cortex of adult monkeys and rats, *J. Physiol. Paris*, 90, 277, 1996.
35. Prakash, N., Cohen–Cory, S., and Frostig, R.D., Rapid and opposite effects of BDNF and NGF on the functional organization of the adult cortex *in vivo, Nature,* 381, 702, 1996.
36. Penschuck, S. et al., *In vivo* modulation of a cortical functional representation shortly after topical cholinergic agent application, submitted, 2002.

3 Analysis Methods for Optical Imaging

Lawrence Sirovich and Ehud Kaplan

CONTENTS

3.1 INTRODUCTION

Imaging methods have become widespread in the biomedical sciences as well as in other sciences and technologies. As we learn to extract the full measure of information latent in images, their role can only increase. Various imaging methods, such as functional magnetic resonance imaging (fMRI), optical imaging, positron emission tomography (PET), magneto-encephalography (MEG) scans, and so forth present similar challenges. Optical imaging of the brain, whether it relies on intrinsic signals or on voltage-sensitive dyes, is an extreme example of these challenges. Of

0-8493-2389-4/02/$0.00+$1.50
© 2002 by CRC Press LLC

the various imaging methods mentioned, it gives rise to some of the most difficult problems because of the typically low signal-to-noise ratio. It is therefore with little loss of generality that in this review we emphasize approaches for the analysis of data obtained in optical imaging experiments.

First we introduce some of the problems and challenges inherent in the analysis of such data sets, and then we focus on mathematical methods that have proven fruitful in such analysis. Although we often make reference to experiments in visual physiology, the methods are general and the discussion applies to other types of imaging data as well as to other signals.

Analysis methods of optical images involve varying degrees of mathematical treatment. In the interest of completeness, some effort has been made to include the appropriate mathematical underpinnings while still communicating with readers who are less interested in mathematical content. Thus, we have made an attempt to explain the methods with a minimal amount of formal mathematics. Mathematical background has been placed in section appendices (which may be skipped by the uninterested reader). Following mathematical tradition, such sections are indicated by the symbol ☡.

3.1.1 The Small Signal Problem

The term *optical imaging* usually describes the acquisition of images with light in the visible or near-visible range.[17] In biological applications, this imaging can sometimes take advantage of relatively large signal-to-noise ratios (SNRs) found, for example, in calcium imaging. Under these circumstances there is little need for extensive image processing. However, it is more common for optical imaging in biological applications to deal with a very small SNR, no better than 1 part in 1000. This occurs because the optical signal is usually very small, and the noise, which arises from unwanted signals from several possible sources (irrelevant physiological activity, instrumentation or digitization noise, shot noise, movement artifacts, etc.) is usually large.[1] This situation demands that sophisticated analysis techniques that are sensitive and robust be applied to the data. In this chapter we shall review several methods, some in common use and some recently developed, that can be used to extract the desired information from a data set of images. The examples used in this chapter come from our own work, with an emphasis on optical imaging of the mammalian visual cortex.

3.1.2 Single Condition Analysis

A common approach to image analysis is differential imaging (Section 3.2.3), in which two experimental conditions are contrasted. Such conditions are of two general types: 1) an experimental and a control condition, in which the experimental stimulus influences some parts of the image, whereas the control situation exerts a smaller (or no) influence; and 2) two complementary conditions, in which each experimental stimulus influences nonoverlapping parts of the image. In the context of imaging the visual cortex, an example of the first type might be to contrast cortical images obtained during visual stimulation with images obtained when the animal viewed a blank or dark screen in order to visualize regions excited by visual stimuli. An example of the second type might be contrasting images obtained during stimulation with vertical gratings with those obtained during stimulation with horizontal gratings in an attempt to visualize orientation-selective domains.

In some cases, it is straightforward to choose two appropriate conditions for either of the two types described. In other cases it is much more difficult, and sometimes impossible, to determine the appropriate control condition or the appropriate complementary conditions. For instance, in studying the representation of color in the cortex, any choice of such conditions depends strongly on the particular theory of color perception that one is entertaining at the moment. Therefore, in many cases there is a strong motivation to develop and use analysis methods that circumvent the difficulties associated with differential imaging. Ideally, one would like to apply the experimental condition and directly observe its impact on the imaged target without need for differential imaging. Such images are sometimes referred to as *single condition* images or maps.

3.1.3 What is the Appropriate Control Image?

In trying to determine the appropriate control stimulus for differential imaging, several investigators have chosen to contrast responses obtained during stimulation with the relevant stimulus against the average of all responses to all the stimuli used in the study ("cocktail blank response"). This procedure works well when the responses to all stimuli have nearly the same amplitude, as is the case in studies of orientation selectivity in the visual cortex. However, when the response amplitude varies with the stimulus parameter under investigation, as is the case when one studies representation of spatial or temporal frequency in the visual cortex, the cocktail blank response is dominated by the strong responses in the cocktail (i.e., the responses in the middle range to spatial or temporal frequency stimuli). In the absence of a proper weighting of the data, this results in a distortion of the contrasted responses (see, for example, the discussions in Everson et al.[12] and Issa et al.[20]).

A far better control is obtained if the stimulus-driven response is contrasted with the blank response, that is, when the stimulus contains none of the investigated parameters. In the context of visual physiology, such data might be the cortical responses obtained while the animal views a blank screen, which has no orientation, contrast, or spatial frequency but maintains the average luminance and adaptation state of the system under study. It is therefore important to include in the stimulus set instances of a blank stimulus. The responses to such a stimulus at various times during the experiment can also be used to characterize the noise (signals unrelated to any stimulus parameter) inherent in the physiological system under investigation.

3.1.4 Large Data Sets

A common problem with imaging data, including data from optical imaging, is that digitized images require large storage. An image might contain 640×480 pixels, each digitized at 12 bits. Often such images are obtained at a high rate; sometimes (for example, in studies that employ voltage-sensitive dyes) the frame rate can exceed 1000 frames/sec. Even at the more moderate frame rates used in optical imaging of intrinsic signals from the brain, the data rates are formidable: at 30 frames/sec, the data rate is $640 \times 480 \times 12 \times 30 \times 3600 = 49{,}766{,}400{,}000$ bits per h, or over 49 gigabytes/h. The sheer size of such data sets presents technical problems of storage and processing, even with today's rapidly advancing technology. In addition, sets of this size present conceptual problems of dimensional reduction — one has to decide whether the data should all be retained and what might be lost if the data were compressed, filtered, or truncated. At another level is the challenge of rendering such extremely large data sets into a human-sized and comprehensible format. Such issues motivate the treatment that follows.

3.2 ANALYSIS METHODS

This section reviews various methods used to analyze data from optical imaging experiments. Although the treatment is somewhat more mathematical than is usually found in biological texts, the hesitation that some readers might experience can be overcome by use of the tutorial and Matlab package that can be found at http://camelot.mssm.edu/imaging.html. Much of the material presented here corresponds to the tutorial and to the algorithms available there.

In any discussion of image analysis, there are compelling arguments for representing the data in terms of its principal components analysis (PCA). Such methods have already found their way into optical imaging,[8,9,11,12,27,30,31,34] MRI imaging,[14] and PET scans.[13] We therefore present here a brief introduction to this subject. More details can be found in Stewart and Sun,[35] Glaser and Ruchkin,[16] and other references cited there. Following this general discussion of PCA, we introduce several specific methods for the analysis of optical imaging data.

3.2.1 Principal Components Analysis

Before we can get on with the analysis of optical imaging data, we require a framework and a language for the specification and treatment of images. This will be accomplished by a mathematical structure that goes under a variety of names, e.g., Karhunen-Loève, singular value decomposition, factor analysis, principal components analysis (PCA), and empirical eigenfunctions. The last two terminologies will be used in this presentation.

For purposes of exposition, we will illustrate the mathematical apparatus with the problem of face recognition. Consider an ensemble of pictures of human faces. Six exemplars are shown in Figure 3.1 (from a collection of 280 located at http://camelot.mssm.edu/imaging.html).

We formally can represent any ensemble of images by

$$f = f(t, \mathbf{x}) \tag{3.1}$$

where t is the index of the image (say, the time stamp of the image), $\mathbf{x}$ is the pixel location, and f is the gray level of image t at location $\mathbf{x}$.

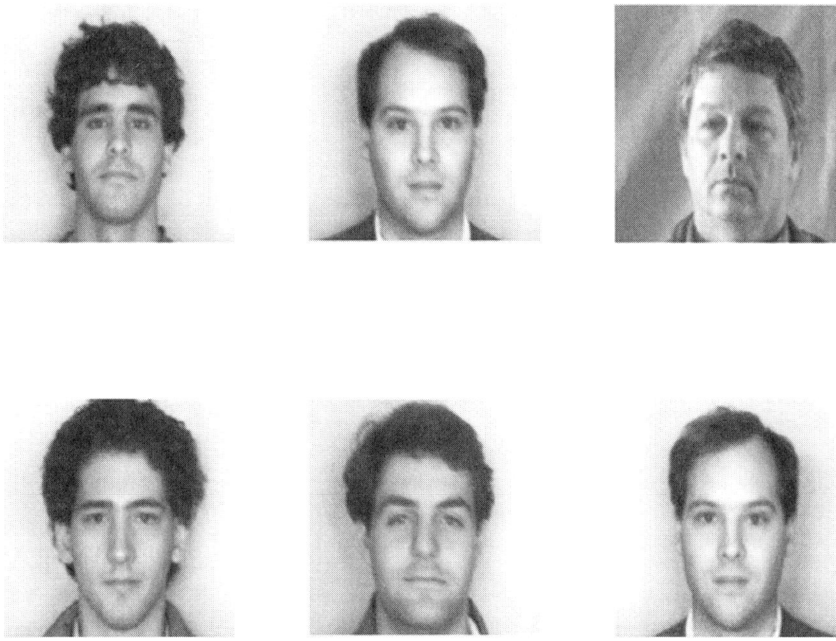

FIGURE 3.1 Six faces chosen from a collection of 280 faces. Note that the sixth face is the mirror image of the second.

Each image in Figure 3.1 contains roughly $100 \times 100 = O(10^4)$ pixels, and therefore each face of the ensemble is a point in a vector space of $O(10^4)$ dimensions; the point location in this space tells us the gray level at each pixel location of the image of the face. One refers to this vector space as the state space of the ensemble of faces. The images acquired in an optical imaging experiment will also be represented in the form of Equation 3.1. Typically, optical images can have more than $300 \times 300 = O(10^5)$ pixels, and more than $O(10^3)$ images are acquired in the course of an experiment.

We may ask whether there is a better space than the raw pixel state space in which to represent the faces. What is meant by "better" will emerge from our discussion, but for the moment we seek a more compact way of treating an ensemble of faces. We will answer the question, "Can we use fewer than the $O(10^4)$ parameters required in the pixel space and, if so, what is the smallest number of required parameters?" This challenge is referred to as the Rogue's Gallery problem.*

As it now stands, specification of a face requires specification of gray scales at $O(10^4)$ locations or, as stated earlier, a point in a space of $O(10^4)$ dimensions. In this view each pixel of an image is treated independently, and correlations that can exist between the gray scale values of neighboring pixels are ignored. To remedy this we can try to represent an image by a superposition of smooth functions.

* This problem was introduced by Sirovich and Kirby in 1987[29a] (see also Kirby and Sirovich, 1990[21a]) and was intended originally as a toy problem to model more complicated data analysis problems. Much of the current face-recognition technology has its origin in these two papers.

In the same spirit as one expands a function in a Fourier series, one may expand any ensemble of images $f(t, \mathbf{x})$ in a series,

$$f(t,\mathbf{x}) = \sum_n a_n(t)\mu_n\varphi_n(\mathbf{x}), \tag{3.2}$$

where the $\{\varphi_n(x)\}$ play the role of, but are not generally, sinusoids. In the mathematical appendage to this section, we show how to construct functions from the data such that any finite approximating sum of Equation 3.2 is optimal in the sense that the error is always minimal. Moreover, like sinusoids, these functions are uncorrelated with one another, so that with proper normalization, they fulfill the relation

$$(\varphi_n, \varphi_m)_\mathbf{x} \overset{def}{=} \sum_\mathbf{x} \varphi_n(\mathbf{x})\varphi_m(\mathbf{x}) = \delta_{nm}, \tag{3.3}$$

and are said to be orthonormal. The combination $a_n(t)\mu n$ gives the required Fourier coefficients for the image of index t. However, unlike the generic Fourier case, the coefficients $\{a_n(t)\}$, which change with each image, are also uncorrelated,

$$(a_n, a_m)_t = \delta_{mn}. \tag{3.4}$$

The coefficients $\{\mu_n\}$ naturally arise in the form $\lambda_n = \mu_n^2$ as characteristic values (eigenvalues) of equations for both $a_n(t)$ and $\varphi_n(\mathbf{x})$. Eigenfunctions correspond to λ_n, a terminology that we will use throughout this text. It should be observed that μ_n in Equation 3.2 measures the importance of the nth eigenfunction in the sum in Equation 3.2.

Rogue's Gallery Problem

We illustrate the formalism by returning to the Rogue's Gallery problem. The images of this data set have been prepared so that eye location has been placed in a fixed position, lighting has been normalized, and so forth. As is clear from Figure 3.1, 100×100 pixels furnish an adequately resolved image of a face.

The number of different faces in the ensemble is actually 140, but this number has been doubled by a procedure worth noting. We denote the set of faces by

$$f_n(x, y),\ n = 1,\ \ldots,\ 280. \tag{3.5}$$

Here y denotes a vertical variable and x denotes the horizontal variable measured from the midline of the face. The original ensemble has been doubled by including $f_n(-x, y)$, the mirror image of each face. Note that the sixth image in Figure 3.1 is the mirror image of the second face.

Obviously, each image generated in this way is an admissible face and one might believe that, in a large enough population of faces, something like a mirror-image face might eventually be realized. Aside from doubling our original ensemble, this maneuver has the interesting and useful consequence of making every eigenfunction either even in the midline

$$\varphi_n(x, y) = \varphi_n(-x, y). \tag{3.6}$$

or odd in the midline

$$\varphi_n(-x, y) = -\varphi_n(x, y). \tag{3.7}$$

Symmetry considerations, as were used above with faces, can also be used to extend cortical data.[31]

The first six eigenfunctions, arranged in descending order of λ_n, are shown in Figure 3.2; the first five are even and the sixth is odd.

The range of values of the eigenvalues is of some interest. As mentioned earlier, $\lambda_n = \mu_n^2$ indicates the importance of a_n and φ_n in reconstructing the image. In Figure 3.3 we plot λ_n vs. n for the Rogue's Gallery problem. It is reasonable to assume that eigenvalues with large index numbers correspond to noise. This is illustrated in Figure 3.4, which shows the last three eigenfunctions. Features of a face are seen in each of these, which is evidence that the number of faces is small. Thus, if we extrapolate back to see where the curve in Figure 3.3 departs from the noise portion of the curve, we estimate that about 100 eigenfunctions are significant for this set.

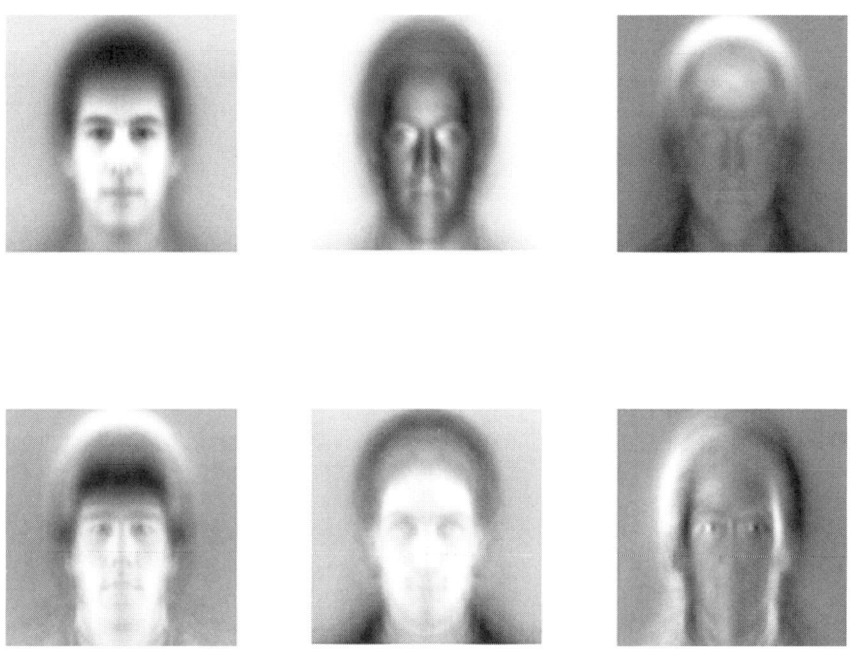

FIGURE 3.2 The first six eigenfaces arranged in descending order of eigenvalues, λ_n, for the ensemble specified in Figure 3.1.

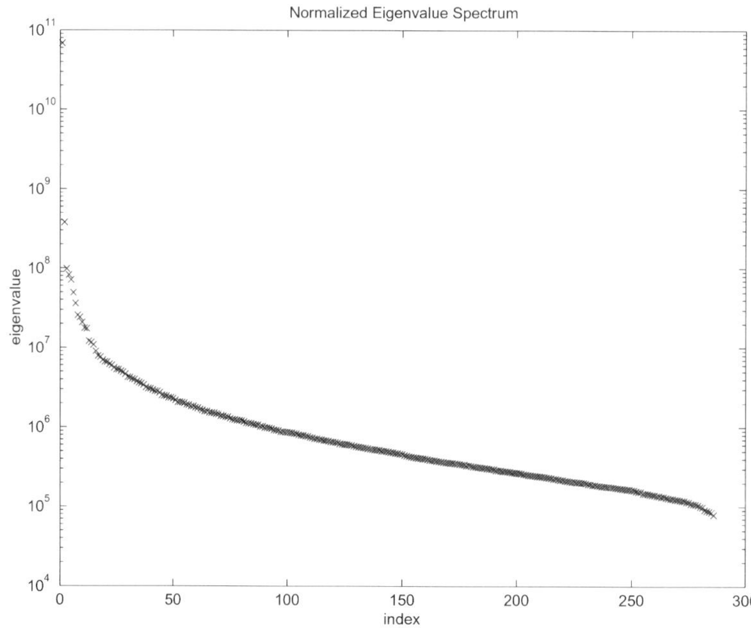

FIGURE 3.3 The eigenvalues as a function of their indexs, for the ensemble specified in Figure 3.1.

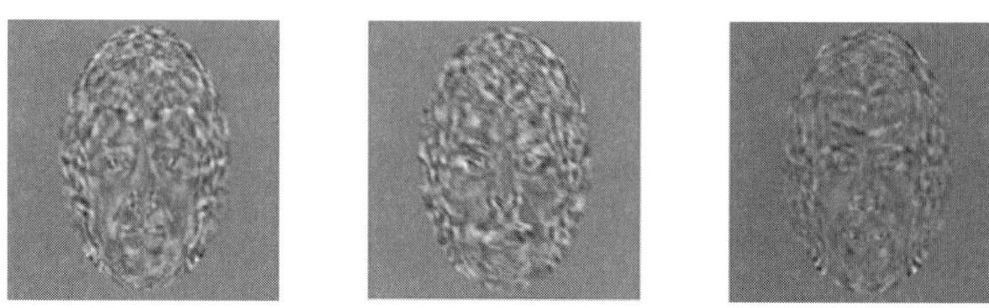

FIGURE 3.4 Last three of 280 eigenfaces. Examples of noisy eigenfaces, but the smallness of the population is apparent from the presence of facial features.

Each face of the ensemble can be assembled from an admixture of about 100 eigenfaces, 6 of which appear in Figure 3.2. Thus, in this instance, we have accomplished the sought-after reduction from $O(10^4)$ to $O(10^2)$ parameters.

Mathematical Appendix

We observe that the expansion of Equation 3.2, i.e., a sum of products of functions, each of which comes from an orthogonal set, cannot usually be assumed to exist and might be regarded as remarkable. If Equations 3.3 and 3.4 are applied to Equation 3.2, we obtain

$$a_n(t) = (\varphi_n, f)_{\mathbf{x}} / \mu_n \tag{3.8}$$

and

$$\varphi_n(\mathbf{x}) = (a_n, f)_t / \mu_n \tag{3.9}$$

If Equation 3.8 is substituted back into Equation 3.9, we obtain

$$K\varphi_n \stackrel{def}{=} \sum_{\mathbf{y}} K(\mathbf{x}, \mathbf{y})\varphi_n(\mathbf{y}) = \lambda_n \varphi_n(\mathbf{x}) \tag{3.10}$$

where

$$K(\mathbf{x}, \mathbf{y}) = (f(t, \mathbf{x})f(t, \mathbf{y}))_t = K(\mathbf{y}, \mathbf{x}), \tag{3.11}$$

is the spatial covariance and $\lambda_n = \mu_n^2$. Thus, we have shown that the assertion of Equation 3.2 leads to the eigenfunction framework of Equation 3.10.

Instead of dealing with Equation 3.10, we can substitute Equation 3.9 back into Equation 3.8 in the above to obtain

$$Ca_n \stackrel{def}{=} \sum_{s=1}^{T} C(t, s)a_n(s) = \lambda_n a_n(t) \tag{3.12}$$

where

$$C = C(t, s) = (f(t, \mathbf{x}), f(s, \mathbf{x}))_{\mathbf{x}} = C(s, t) \tag{3.13}$$

which, in a rigorous sense, is proportional to the covariance among pictures as introduced in Equation 3.1. As is clear from Equation 3.12, the order of C is no more than T, not P, an issue that we shall take up shortly.

Since Equations 3.11 and 3.13 lead to symmetric operators, standard theorems from analysis show that $\{a_n\}$ and $\{\varphi_n\}$ exist, and each can be taken to be an orthornormal system, Equations 3.3 and 3.4. Thus, the assertion that we can describe the data in the form shown in Equation 3.2 has been proven. We note that only one of the two eigenfunction problems (Equation 3.10 or 3.12) needs be solved, for once one set of eigenfunctions is determined, the complementary set is determined from Equation 3.8 or 3.9, whichever is appropriate.

The presentation just given was chosen mainly for its conciseness, and as a result we have not focused on the optimality of the representation. We now state without proof that, for the class of all orthonormal sets $\{\varphi_n'(\mathbf{x})\}$ and coefficients $\{\alpha_n(t)\}$ such that

$$f(t, \mathbf{x}) = \sum_{n=1}^{N} \alpha_n(t)\varphi_n'(\mathbf{x}) + \varepsilon_N(t, \mathbf{x}), \tag{3.14}$$

we make the smallest error $\varepsilon_N(t, \mathbf{x})$ for any N if $\varphi'_n = \varphi_n$ and $\alpha_n = \mu_n a_n$, so that

$$(\alpha_n, \alpha_m)_t = \lambda_n \delta_{nm} \tag{3.15}$$

and

$$\lambda_n = \left\| (\varphi_n, f)^2_{\mathbf{x}} \right\|_t = \mu_n^2. \tag{3.16}$$

Thus, a representation in terms of empirical eigenfunctions provides an optimal compression of the data.

Remarks

We have shown that for the data set $f(t, \mathbf{x})$ of T images and $P >> T$ pixels, there are at most T nonzero eigenvalues (all of which are nonnegative). However, if we solve in pixel space, there would appear to be P eigenvalues, which is a much larger number than T. The apparent discrepancy is easily resolved. In pixel space, each image $f(t, \mathbf{x})$, t fixed, can be thought of as a vector, and thus the entire database is composed of T vectors. Clearly there must be $P - T$ vectors orthogonal to our database and therefore K has a large nullspace; i.e., $\lambda_n = 0$, for $n > T$, has this degeneracy. From these deliberations, Equation 3.12 rather than 3.10 is the problem to solve. If we return to the calculation of the $\{\varphi_n\}$ in this instance, the values are given by Equation 3.9. This has the plausible but interesting interpretation that every eigenfunction is an admixture of images from the ensemble. For this reason the procedure just outlined is referred to as the *method of snapshots*.[29]

Another issue of interest concerns the time-like index t. A little reflection reveals that if the sequence of images is shuffled, the eigenpictures $\varphi_n(\mathbf{x})$ remain the same. This invariance of $\{\varphi_n\}$ under shuffling is, of course, not true for the $a_n(t)$, each of which, in fact, undergoes a predictable shuffled time course under the transformation; the values of t are permuted by the shuffle. For the Rogue's Gallery problem, the order in which we acquire pictures is of no consequence; i.e., Equation 3.1 is not a true time series. However, in a typical optical imaging study, Equation 3.1 is a true time series, since it reflects the dynamics underlying the phenomena of interest. In such cases the dynamical evolution of the modal coefficients $\{a_n(t)\}$ may be as important as the features of the modes $\{\varphi_n\}$. This will be taken up in Section 3.2.8.

A tutorial that illustrates the concepts of this section can be found at http://www.camelot.edu/imaging.html.

3.2.2 Optical Imaging Data

We now review a variety of techniques for the analysis of optical images and extraction of the information latent in them.

The deliberations of the previous section apply generally to databases of many types and in particular to those obtained in optical imaging studies. As already mentioned, empirical eigenfunctions compress data in an optimal fashion. To illustrate this concept, we note that it is not unusual to acquire $O(10^4)$ images in the course of an experiment. Nevertheless, experience has taught us that no more than $O(10^2)$ eigenfunctions are needed to represent any of these data sets. As might be expected, this reduces and simplifies virtually all calculations.

We continue to use the notation of the previous section. An optical image, acquired at time t, is denoted by

$$f = f(t, \mathbf{x}), \tag{3.17}$$

where f is the reflectance, measured as a gray level at pixel location $\mathbf{x}$, and captured by a camera, usually a charged coupled device (CCD). For conceptual reasons, the signal in a typical image can be usefully divided into two parts

$$f(t, \mathbf{x}) = B(t, \mathbf{x}) + s(t, \mathbf{x}). \tag{3.18}$$

$B(t, \mathbf{x})$ represents the background activity, which includes effects of heartbeat, respiration, and motion, and $s(t, \mathbf{x})$ represents the (sought-after) specific response signal evoked by a sensory stimulus. Any nonspecific response* that might be elicited by the stimulus is also included in $B(t, \mathbf{x})$. Note that we do not explicitly distinguish the noise coming from instruments and tissue but regard it as included in the background B. Since we are interested only in the signal, we will refer (somewhat imprecisely) to B as *noise*. In such terms, the signal-to-noise ratio for optical imaging of intrinsic signals is typically smaller than $O(10^{-3})$. We next review several methods that have been developed to extract this small signal from the noisy environment in which it is embedded.

3.2.3 Differential Imaging

We start our discussion with a continuous-time caricature of the signal extraction problem confronting us. Suppose that the background signal is denoted by $B(t, \mathbf{x})$, and for $t \geq 0$ a stimulus elicits a steady response $p(\mathbf{x})$. The image record is then given by

$$f'(t, \mathbf{x}) = B(t, \mathbf{x}) + H(t)p(\mathbf{x}), \tag{3.19}$$

where $H(t) = 1$ for $t > 0$ and $H(t) = 0$ for $t < 0$. Suppose the mean background is denoted by

$$\bar{B}(\mathbf{x}) = \lim_{T\uparrow\infty} \frac{1}{2T} \int_{-T}^{T} B(t, \mathbf{x})\, dt. \tag{3.20}$$

* A nonspecific response is a response evoked by the stimulus but not dependent on the specific nature of the stimulus.

Then, if $b(t, \mathbf{x}) = B(t, \mathbf{x}) - \bar{B}$, we may define the mean subtracted response as

$$f(t,\ x) = b(t,\ \mathbf{x}) + \frac{1}{2} sgn(t) p(\mathbf{x}), \tag{3.21}$$

where

$$sgn(t) = \begin{cases} 1, & t > 0 \\ -1, & t < 0 \end{cases}. \tag{3.22}$$

The unwanted background is now easily eliminated, since

$$\lim_{T \uparrow \infty} \frac{1}{T} \int_{-T}^{T} sgn(t) f(t,\ \mathbf{x})\, dt = p(\mathbf{x}). \tag{3.23}$$

This suggests that we might seek to remove the unwanted background signal by a subtraction procedure. As discussed in the introductory sections, this may be done through the use of a pair of stimuli that evoke appropriate complementary responses. For example, in the visual cortex, ocular dominance columns are visualized by complementary right eye vs. left eye stimulation, and orientation-selective patches are visualized by complementary θ vs. $(\theta + \pi/2)$ oriented stimuli. On occasion the absence of an explicit stimulus, which results in the so-called blank response, is used as a complementary, or control, image. Also, on occasion a *cocktail image,* made up of responses to all the stimuli used in the experiment, is the comparison image. The last can have some deleterious effects, as already mentioned in Section 3.1.3.

If the desired signal is $s(t,\ \mathbf{x})$ we denote the complementary or comparison stimulus by $s_c(t,\ x)$, and might arrange the experiment so that

$$f(t,\ x) = B(t,\ x) + \begin{cases} s(t,\ \mathbf{x}) & t > 0 \\ s_c(t,\ \mathbf{x}) & t < 0 \end{cases}. \tag{3.24}$$

If the stimulus has no effect on the background $B(t,\ \mathbf{x})$, then it is reasonable to believe that the activity map, $A(\mathbf{x})$, is given by

$$A(\mathbf{x}) = \lim_{T \uparrow \infty} \frac{1}{2T} \int_{-T}^{T} f(t,\ \mathbf{x})\, sgn(t)\, dt = \frac{\bar{s}(\mathbf{x}) - \bar{s}_c(\mathbf{x})}{2}. \tag{3.25}$$

(Average quantities are denoted by a bar, $\langle s(\mathbf{x}, t) \rangle_t = \bar{s}(\mathbf{x})$.) The right-hand side of Equation 3.25 gives the desired activity map, such as ocular dominance stripes, orientation patches, etc. This is the method of differential imaging, which has been used extensively in the field of optical imaging.[4,5,20,28] Since the amount of data is always finite, this procedure becomes problematical, as we now illustrate within the framework of the above caricature Equation 3.19 or 3.21. As mentioned in the introductory discussion, the signal-to-noise ratio is small:

$$\frac{p, \bar{s}}{\bar{B}} \approx O(10^{-4}) \tag{3.26}$$

Also, in practice it is found that the standard deviation of the background fluctuations

$$\sigma_B = \sqrt{\left\langle (B - \bar{B})^2 \right\rangle_t}. \tag{3.27}$$

can be as large as

$$\frac{\sigma_B}{\bar{B}} = O(10^{-2}). \tag{3.28}$$

From these estimates we have

$$\frac{1}{2T}\int_{-T}^{T} sgn(t)B(t, \mathbf{x})\, dt \approx \frac{\sigma_B}{2\sqrt{N}} \tag{3.29}$$

if N is the number of images in the interval $t \in (0, T)$. Therefore, the ratio of background error to signal intensity can be estimated by

$$\frac{\frac{1}{2T}\int_{-T}^{T} sgn(t)B(t, \mathbf{x})dt}{\bar{s}} \approx \frac{\sigma_B}{2\bar{B}\sqrt{N}} \cdot \frac{\bar{B}}{\bar{s}} \approx \frac{10^{-2}}{2\sqrt{N}} 10^4. \tag{3.30}$$

Thus, even if $N = 2500$, this ratio is close to one. As a result, activity maps that are latent in a given data set often cannot be extracted by differential imaging.

Instead of differencing, one may also take a ratio of images to extract the signal, which, because of the very small signal,[7] is tantamount to differencing.

3.2.4 Activity (Preference) Maps

Before discussing possible improvements to differential methods, it is useful to review the manner in which activity maps are derived. We illustrate this process by examples from the functional architecture of the visual cortex.

3.2.4.1 Ocular Dominance

The extraction of ocular dominance activity maps represents a rather simple example of the determination of preferential maps of the cortex. The required complementary image sets are the responses from a (rich enough) class of responses elicited by right eye vs. left eye stimulus sets. The data are arranged so that one eye is stimulated for $t < 0$ and the other eye for $t > 0$. Then the familiar ocular dominance stripes, which we denote by $D(\mathbf{x})$, emerge from consideration of

$$D(\mathbf{x}) = \frac{1}{2T}\int_{-T}^{T} f(t, \mathbf{x})\, sgn(t)\, dt. \tag{3.31}$$

3.2.4.2 Orientation and Direction

The determination of the response map to oriented stimuli is more subtle. In preparation for this discussion, we introduce a slight extension in notation. When needed, we will write

$$f(t, \mathbf{x}) = f(t, \mathbf{x}\!:\!\theta(t)) \tag{3.32}$$

where $\theta(t)$ indicates the orientation of the stimulus at time t. We then write

$$f(\theta, \mathbf{x}) = \langle f(t, \mathbf{x} : \theta'(\mathrm{t}))\rangle_{\theta'(t)=\theta}, \tag{3.33}$$

where the average is taken over all times, for which the orientation is θ. It has been known for some time[19] that many cortical neurons respond maximally to an oriented bar or edge and respond poorly (or not at all) to the same stimulus rotated by $\pi/2$. Based on these observations, Blasdel and Salama[5] used optical imaging to visualize orientation-specific patches

$$p_O(\theta, \mathbf{x}) = f(\theta, \mathbf{x}) - f\left(\theta + \frac{\pi}{2}, \mathbf{x}\right). \tag{3.34}$$

In effect the timestamp t in Equation 3.33 has been replaced by an equivalent index, e.g., orientation, θ.

Similarly, we might visualize direction-specific patches by constructing

$$p_D(\theta, \mathbf{x}) = f(\theta, \mathbf{x}) - f(\theta + \pi, \mathbf{x}). \tag{3.35}$$

(See Sirovich et al.[31] and Shmuel and Grinvald.[28])

The maps $p_O(\theta, \mathbf{x})$ and $p_D(\theta, \mathbf{x})$ display how well each pixel responds to the stimulus inclined at θ. With a full collection of such maps, one can construct orientation and direction preference maps.* To construct such maps, the orientation or direction that produces the maximal response at each pixel is determined, and a color-coded rendition of these preferences is displayed over an area of cortical tissue. This produces a map of the preferences at each location. An additional consideration

* In this abbreviated presentation we neglect the fine points dealing with matters such as time course (Sec. 3.2.8) and spatial frequency.[20]

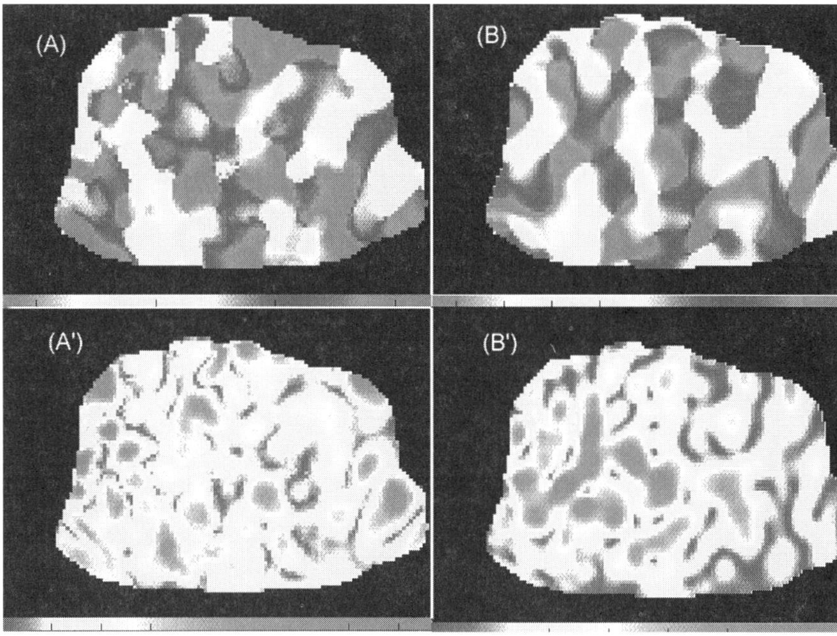

FIGURE 3.5 (See Color Figure 3.5.) Orientation and direction from areas 17/18 in the cat cortex. Tissue size is 5 × 7 mm. Direction preference (A) and response strength (A′) at each pixel. The upper color bar indicates a variation between $-\pi$ and π. The lower color bar refers to an order of magnitude change in amplitude. Orientation preference (B) and response amplitude (B′). The upper color bar indicates a variation between $-\pi/2$ and $\pi/2$. The lower color bar refers to an order of magnitude change in relative response amplitude.

is the relative strength of the preference map. Examples of direction and orientation preference maps, with their corresponding strengths, are shown in Color Figure 3.5.*

Mathematical Appendix

The constructions described previously may be put into a mathematical framework by revisiting the eigenfunction analysis. For this purpose we represent the orientation data, Equation 3.33, in the form

$$f(\theta, \mathbf{x}) = \sum_n \mu_n a_n(\theta)\phi_n(\mathbf{x}), \tag{3.36}$$

where $\{a_n\}$ and $\{\phi_n(\mathbf{x})\}$ are orthonormal sets. (Observe that t is replaced by θ.) Then, from the mathematical formalism presented in Section 3.2.1, the equation for the coefficients, $a_n(\theta)$, is given by

* Color figures follow page 112.

$$\sum_{\theta'} C(\theta,\ \theta')a_n(\theta') = \lambda_n a_n(\theta) \tag{3.37}$$

where $\lambda_n = \mu_n^2$ and

$$C(\theta,\ \theta') = (f(\theta,\ \mathbf{x}), f(\theta',\ \mathbf{x}))_{\mathbf{x}}. \tag{3.38}$$

As argued in Sirovich et al.,[31] if all orientations are democratically and evenly represented in the cortex, and if the piece of cortical tissue considered is sufficiently large to realize this, it then follows that

$$C(\theta,\ \theta') = C(\theta - \theta') = C(\theta' - \theta). \tag{3.39}$$

In this case Equation 3.37 is easily seen to be solved by the orthonormal functions

$$a_n = e^{in\theta} / \sqrt{2\pi}. \tag{3.40}$$

The eigenvalues are then given by

$$\lambda_n = \sum_{\theta} C(\theta) \cos n\theta, \tag{3.41}$$

since $C(\theta)$, Equation 3.39, is an even function of its argument. The corresponding spatial eigenfunctions are then given by Equation 3.9, which in this case is

$$\phi_n(\mathbf{x}) = \sum_{\theta} e^{-in\theta} f(\theta,\ \mathbf{x}) / \sqrt{2\pi\lambda_n}. \tag{3.42}$$

Observe that

$$\phi_1(\mathbf{x}) = \sum_{0 \le \theta \le \pi} e^{-i\theta}\{f(\theta,\ x) - f(\theta + \pi,\ \mathbf{x})\} / \sqrt{2\pi\lambda_1}. \tag{3.43}$$

evidently represents the activity map of directionally sensitive cells. Since roughly 25% of the neurons in the primate visual cortex exhibit this property,[10,18] λ_1 is relatively small, which is borne out in calculations. It should be noted that in the present instance, we follow an indexing based on Equation 3.40 and not on the magnitude of λ_n.

Except for a small error due to directionally sensitive cells, the orientation preference map is given by

$$\phi_2 = 2\sum_{0\le\theta<\pi} e^{-2i\theta}\hat{f}(\theta,\ \mathbf{x})\,/\,\sqrt{2\pi\lambda_2}\,, \tag{3.44}$$

where

$$\hat{f}(\theta,\ \mathbf{x}) = \frac{1}{2}(f(\theta,\ \mathbf{x}) + f(\theta+\pi,\ \mathbf{x})), \tag{3.45}$$

and in such terms

$$\phi_2(\mathbf{x}) = 2\sum_{0<\theta<\frac{\pi}{2}} e^{-2i\theta}\left\{\hat{f}(\theta,\ \mathbf{x}) - \hat{f}\left(\theta+\frac{\pi}{2},\ \mathbf{x}\right)\right\}/\sqrt{2\pi\lambda_2}\,. \tag{3.46}$$

The quantity in the curly brackets in Equation 3.46 is the orientation patch, Equation 3.34, that is obtained by differential imaging. It is important to note that ϕ_n is composed of two functions, $Re(\phi_n)$ and $Im(\phi_n)$. Thus, direction and orientation require two functions each for their descriptions.

Color Figure 3.5A shows the phase and Color Figure 3.5A′ shows the amplitude of ϕ_1 and, similarly, Color Figure 3.5B shows the phase and Color Figure 3.5B′, the amplitude of ϕ_2. The amplitude indicates signal strength and the phase indicates the preferred angle for direction or orientation, as the case may be.

To see the latter, suppose that at each pixel location, $f(\theta,\ \mathbf{x})$ has a relatively sharp maximum in orientation, θ, which we denote by $\theta^0(\mathbf{x})$. Then, for example, Equation 3.43 can be approximately evaluated to give

$$\phi_1(\mathbf{x}) \propto e^{i\theta^0(\mathbf{x})}\{f(\theta^0(\mathbf{x}),\mathbf{x}) - f(\theta^0(\mathbf{x})+\pi,\mathbf{x})\}, \tag{3.47}$$

where the constant of proportionality, which is real, depends only on the sharpness of tuning. The important feature of Equation 3.47 is that its phase is θ^0, the preferred orientation. A similar result follows for $\phi_2(\mathbf{x})$. Both phase maps exhibit pinwheels, a property dictated by topological considerations related to the smooth and continuous representation of orientation preference across the cortical surface. A little thought indicates that $\phi_3(\mathbf{x})$ in Equation 3.42 can be regarded as a refinement of the directionality determined by $\phi_1(\mathbf{x})$, and $\phi_4(\mathbf{x})$ refines the orientation as determined by $\phi_2(\mathrm{x})$.

3.2.5 Truncated Differences

We now take up the first of several methods that attempt to improve on simple differencing.

Consider the case modeled by Equation 3.24 for $-T < t < T$, i.e., a finite image record length. Denote by $\{\varphi_n\}$ the set of empirical eigenfunctions generated by this record. Then, from the general considerations of Section 3.2.1, we can express the record of images in terms of its empirical eigenfunctions,

$$f(t,\ \mathbf{x}) = \sum_n \alpha_n(t)\varphi_n(\mathbf{x}), \tag{3.48}$$

where

$$\alpha_n(t) = (\varphi_n,\ f)_{\mathbf{x}}, \tag{3.49}$$

and

$$(\alpha_n,\ \alpha_m)_t = \lambda_n \delta_{nm}. \tag{3.50}$$

In such terms the approximate activity map, Equation 3.25, is given by

$$A(\mathbf{x}) = \frac{1}{2T}\int_{-T}^{T} sgn(t)f(t,\ \mathbf{x})dt = \sum_n \Delta_n\varphi_n(\mathbf{x}) \tag{3.51}$$

where

$$\Delta_n = \frac{1}{2T}\int_{-T}^{T} \alpha_n(t)\, sgn(t)dt. \tag{3.52}$$

In the limit $T \uparrow \infty$, Equation 3.51 should become $(s(\mathbf{x}) - s_c(\mathbf{x}))/2$, as given by Equation 3.25.

The object of the present analysis is to determine, by *objective* means, the significance of the coefficients in Equation 3.51, in order to reduce errors arising from a finite sample of records. We can expect fluctuations, unrelated to the stimulus, from background biological conditions as well as from genuine noise in the imaging system. Reason and experience indicate that the former appears primarily in low index modes and the latter in high index modes. In essence, we seek a bandpass filter, which, when applied to Equation 3.51, will remove the undesirable contributions. As a criterion for retaining a mode, we can reasonably require that the coefficients $\alpha_n(t)$ be significantly correlated with the sequential ordering of the stimulus, $sgn(t)$. Thus we consider the statistical significance of the correlation

$$r_n = \frac{(\alpha_n,\ sgn)_t}{\|\alpha_n\|_t \|sgn\|_t} = \frac{1}{2}\frac{\Delta_n}{\sqrt{\lambda_n}}. \tag{3.53}$$

Under the reasonable assumption that the probability distribution for Δ_n, $\rho(\Delta_n)$ is Gaussian, the probability that the measured correlation is due to chance is given by

$$P_n = 1 - erf\left(\frac{\sqrt{N}}{2} r_n\right). \tag{3.54}$$

(See Gabbay et al.[15] for details.) Thus, the confidence level that a given component is correlated with the appearance of a stimulus is given by

$$\bar{P}_n = 1 - P_n = erf\left(\frac{\sqrt{N}}{2} r_n\right). \tag{3.55}$$

As shown in Gabbay et al.,[15] this can be used to select a band of statistically significant modes, which we write as $L \leq n \leq H$. Thus, under this procedure Equation 5.51 is replaced by

$$A(\mathbf{x}) \approx \sum_{n=L}^{H} \Delta_n \varphi_n(\mathbf{x}). \tag{3.56}$$

It should be noted that these deliberations suggest that $\Delta_n \to 0$ or is negligible as $T \uparrow \infty$ for $n < L$ and $n > H$.

Following Gabbay et al.,[15] we illustrate this with a model of the form of Equation 3.21. The artificial response pattern $p(\mathbf{x})$ to a stimulus is the checkerboard shown in Figure 3.6a. This is added digitally to $O(10^3)$ frames of optical images of macaque visual cortex acquired under unstimulated conditions (Figure 3.6b). The reference, or complementary set, is again a set of $O(10^3)$ images of unstimulated macaque visual cortex (Figure 3.6c). Figures 3.6b and 3.6c are visually indistinguishable, since the reflectance of the dark squares of the checkerboard is only 0.5×10^{-3} compared to the average reflectance of the background.

The stimulated and unstimulated image records are then concatenated, say, as $t < 0$ and $t > 0$, respectively. Next the empirical eigenfunctions $\{\varphi_n\}$ of the full record are determined, and the decomposition (Equation 3.48) is performed. The correlation coefficients, r_n, are shown in Figure 3.7a and the corresponding confidence levels, $\bar{P}_n$, are shown in Figure 3.7b. The criterion for retention of a mode is taken as $\bar{P}_n > 0.99$.

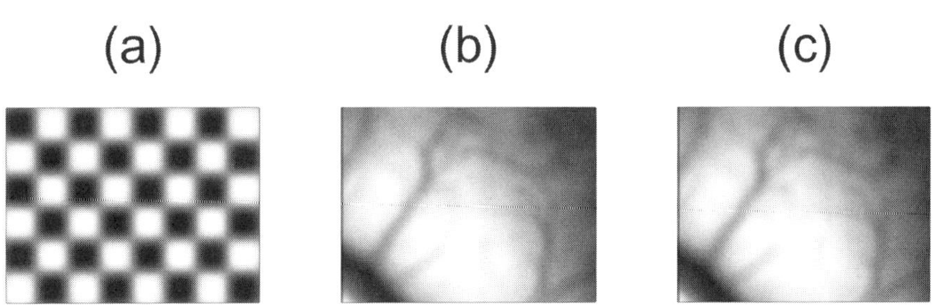

FIGURE 3.6 The test set: (a) test pattern; (b) a member of the stimulated set; (c) a member of the reference set. The cortical tissue measures 5 × 7 mm.

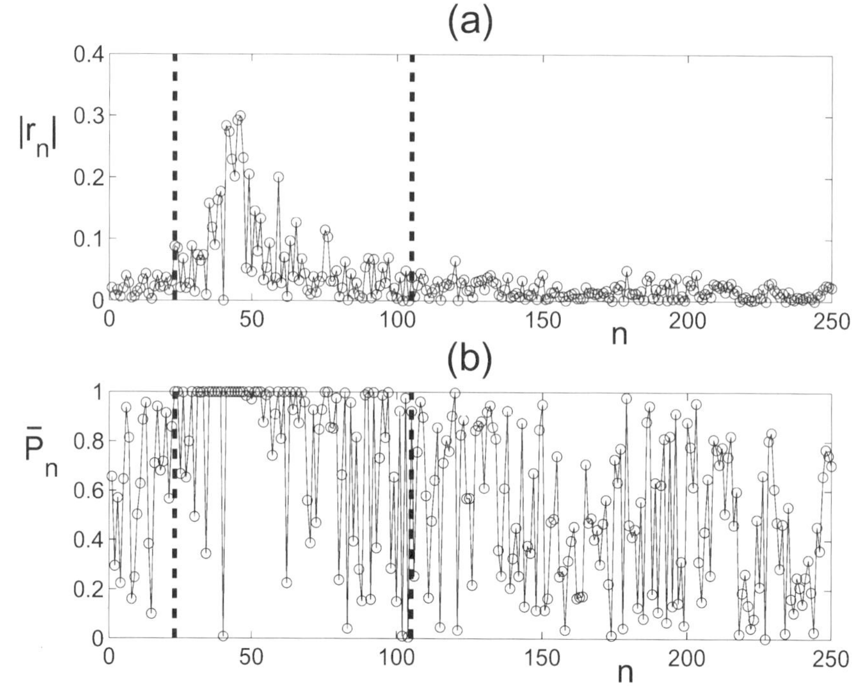

FIGURE 3.7 Magnitude of the correlations (a) and associated confidence levels vs. mode number (b). Dashed lines indicate truncation points.

Thus, as indicated in Figure 3.7b, the band of modes between the two vertical dashed lines is extracted from the record. The resulting image is shown in Figure 3.8a and contrasted with the standard difference map of Figure 3.8b. For further discussion of this method, the reader is referred to Gabbay et al.[15] Clearly, this method furnishes a significant improvement over the common practice of differential imaging, by anchoring the procedure in accepted criteria of statistical significance. A MatLab software package based on this procedure can be found at http://www.camelot.edu/imaging.

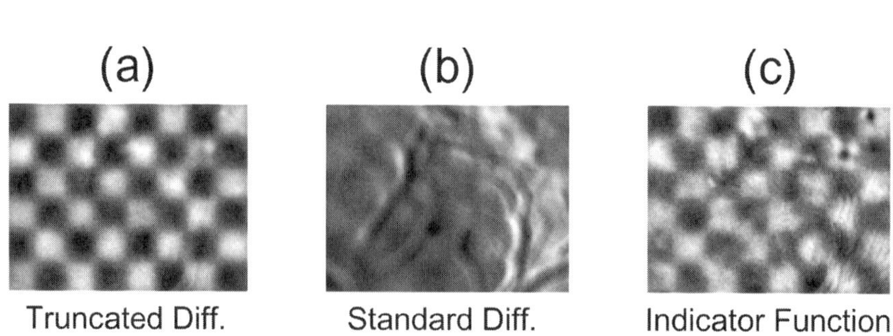

FIGURE 3.8 Activity maps: (a) truncated differences; (b) standard differences; and (c) the indicator function for the test data.

3.2.6 Indicator Functions

We next consider an approach for extracting activity maps with a different point of view. This method seeks an image (the indicator function) most closely correlated with the appearance of a stimulus while $t > 0$ and anticorrelated (or uncorrelated) with the appearance of the complementary stimulus (or with the absence of a stimulus) while $t < 0$. More specifically, we would like to find an indicator function, $\Phi(\mathbf{x})$ such that

$$(f, \Phi(\mathbf{x}))_{\mathbf{x}} = \begin{cases} 1, & t > 0 \\ -1, & t < 0 \end{cases}. \tag{3.57}$$

Stated in such terms, the problem is overdetermined. Instead, in the mathematical appendix we attempt to satisfy Equation 3.57 in a least-squares sense. Using the same notation as the previous section, the result of the calculation is explicitly given by

$$\Phi(\mathbf{x}) = \sum_n \frac{\Delta_n \varphi_n(\mathbf{x})}{\lambda_n}. \tag{3.58}$$

An immediate concern with Equation 3.58 is that $\lambda_n \downarrow 0$ as $n \uparrow \infty$, which, if ignored, will give more weight to the noisy eigenfunctions, i.e., those of large index. This issue was given considerable attention in Everson et al.[11] and treated there from several perspectives. One of these is related to the discussion of statistical significance that appeared in the previous section. For present purposes we carry over part of the analysis from the last section to the issue at hand. Namely, we truncate the right-hand side of Equation 3.58 at $n = H$, when the confidence level discussed earlier (Equation 3.55) falls below 99%. Thus, we take

$$\Phi = \sum_{n=1}^{H} \frac{\Delta_n \varphi_n(\mathbf{x})}{\lambda_n}, \tag{3.59}$$

which, in contrast to Equation 3.56, has the effect of a smooth low pass filter. Since the principal eigenvalues, λ_n, for low index n are relatively large, this serves to attenuate the background eigenfunctions associated with low index. A software package for the calculation of the indicator function can be found at http://camelot.mssm.edu/imaging.html.

Mathematical Appendix

As stated earlier, Equation 3.57 is overdetermined. Instead, we try to fulfill this equation in the least-squares sense by minimizing the quantity

$$C(\Phi) = \frac{1}{2T}\int_{-T}^{T} [(f,\ \Phi)_{\mathbf{x}} - sgn(t)]^2\, dt. \tag{3.60}$$

The minimization requires that

$$\frac{\delta C}{\delta \Phi}(\Phi) = 0 \tag{3.61}$$

and, by standard arguments from the calculus of variations, this leads to the following equation for $\Phi(\mathbf{x})$:

$$\sum_{\mathbf{y}} K(\mathbf{x},\ \mathbf{y})\Phi(\mathbf{y}) = \frac{1}{2T}\int_{-T}^{T} f(t,\ \mathbf{x})\, sgn(t) dt \tag{3.62}$$

K is just the covariance as calculated from the full set, $f(t,\ \mathbf{x})$, viz.,

$$K(\mathbf{x},\ \mathbf{y}) = \frac{1}{2T}\int_{-T}^{T} f(t,\ \mathbf{x}) f(t,\ \mathbf{y}) dt. \tag{3.63}$$

For later purposes note that, in the limit of $T \uparrow \infty$, the right-hand side of Equation 3.62 is given by

$$\lim_{T\uparrow\infty} \frac{1}{2T}\int_{-T}^{T} f(t,\ \mathbf{x})\, sgn(t) dt = \frac{\bar{s}(\mathbf{x}) - \bar{s}_c(\mathbf{x})}{2}. \tag{3.64}$$

To solve Equation 3.62, we again denote the eigenfunctions of K by $\{\varphi_n\}$, and the right-hand side of this equation is given by Equation 3.64, so that Equation 3.62 in these terms is expressed by

$$\sum_{\mathbf{y}}\left(\sum_{n} \lambda_n \varphi_n(\mathbf{x})\varphi_n(\mathbf{y})\right)\Phi(\mathbf{y}) = \sum_{n} \Delta_n \varphi_n(\mathbf{x}). \tag{3.65}$$

Equation 3.65 is easily solved and gives Equation 3.58.

The form of the indicator function and its derivation appear to be quite different from the steps leading to the truncated difference described. Since Equation 3.25 approaches the correct form as $T \uparrow \infty$, we can be concerned about the credentials of Equation 3.59 under the same limit. To quiet such concerns, we return to Equation 3.62 and model the situation by taking

$$f(t,\ \mathbf{x}) = B(t,\ \mathbf{x}) + \begin{cases} \frac{1}{2} v(t)(\bar{s}(\mathbf{x}) - \bar{s}_c(\mathbf{x})),\ t > 0 \\ \frac{1}{2} v(t)(\bar{s}_c(\mathbf{x}) - \bar{s}(\mathbf{x})),\ t < 0 \end{cases}, \tag{3.66}$$

where $v(t) > 0$ is random with mean unity. Under this hypothesis it follows that Equation 3.66 approaches Equation 3.64. In the limit of $T \uparrow \infty$, the covariance, Equation 3.63, is given by

$$K(\mathbf{x}, \mathbf{y}) = \lim_{T\uparrow\infty} \left(\frac{1}{2T} \int_{-T}^{T} B(t, \mathbf{x})B(t, \mathbf{y})dt \right) - \frac{1}{4}(\bar{s}(\mathbf{x}) - \bar{s}_c(\mathbf{x}))(\bar{s}(\mathbf{y}) - \bar{s}_c(\mathbf{y})). \quad (3.67)$$

It therefore follows that, for $T \uparrow \infty$, Equation 3.62 becomes

$$\sum_{\mathbf{y}} \left\{ K_B(\mathbf{x}, \mathbf{y}) - \frac{1}{4}(\bar{s}(\mathbf{x}) - \bar{s}_c(\mathbf{x}))(\bar{s}(\mathbf{y}) - \bar{s}_c(\mathbf{y})) \right\} \Phi(\mathbf{y}) = \frac{(\bar{s}(\mathbf{x}) - \bar{s}_c(\mathbf{y}))}{2}, \quad (3.68)$$

where K_B is the first term on the right-hand side of Equation 3.67. It is clear from this equation that if

$$\sum_{\mathbf{y}} K_B(\mathbf{x}, \mathbf{y})(\bar{s}(\mathbf{y}) - \bar{s}_c(\mathbf{y})) \approx 0, \quad (3.69)$$

i.e., it is zero or small, then the indicator function method yields the correct limit, e.g., that

$$\Phi(\mathbf{x}) \propto (s(x) - s_c(\mathbf{x})). \quad (3.70)$$

To show that Equation 3.69 is indeed the case, we observe that, for $T \uparrow \infty$

$$B \sim B^\circ(\mathbf{x}) + \eta(t, \mathbf{x}) \quad (3.71)$$

where $B^\circ = \langle B \rangle_t$ is primarily a display of the vascularity of the piece of tissue and η is noise. Thus,

$$K_B = B^\circ(\mathbf{x})B^\circ(\mathbf{y}) + \langle \eta(t, \mathbf{x})\eta(t, \mathbf{y}) \rangle_t. \quad (3.72)$$

On the grounds of spatial scale and location, we expect $(B^\circ, \bar{s} - \bar{s}_c)_{\mathbf{x}} \approx 0$. If we assume that the irrelevant contribution from instrument noise and other sources is also small, then, since its spatial scale is small, we expect its action on $s(\mathbf{x}) - s_c(\mathbf{x})$ to be small, as well. Thus, it is reasonable to believe that Equation 3.69 and hence Equation 3.70 hold as $T \uparrow \infty$.

When the indicator function analysis is applied to the test pattern of the previous section, it results in Figure 3.8c. While it is seen to do better than standard differences, in this instance it does not do as well as the truncated difference method of the previous section, as shown in Figure 3.8a. As will be seen in the next section, in other instances the indicator function method performs better than the truncated difference approach and no generally applicable comparison of methods can be made.

3.2.7 The Generalized Indicator Function

We next present a new and promising approach to image analysis, which is still under development and which we term *the generalized indicator method*.[37]

Our discussion so far has dealt with situations in which two stimulus conditions were designed to elicit two complementary responses (Section 3.1.2). Now we consider an ensemble of several stimuli and their corresponding responses. We write the image responses as

$$f_m = f_m(t, \mathbf{x}), \ m = 1, \ \ldots, \ M; \ t = 1, \ \ldots, \ T \tag{3.73}$$

where the subscript m identifies the M different stimulus conditions, for example, drifting gratings at M different orientations. For ease of exposition, we have made the (unnecessary) assumption that each stimulus condition f_m has exactly T images.

We first define

$$\bar{f}_m(\mathbf{x}) = \langle f_m(t, \mathbf{x}) \rangle_t \tag{3.74}$$

as the mth condition average and

$$\bar{f}(\mathbf{x}) = \langle \bar{f}_m(\mathbf{x}) \rangle_m, \tag{3.75}$$

as the average computed over all conditions. For sufficiently large T we expect that

$$\bar{f}_m = \bar{B}(\mathbf{x}) + \bar{s}_m(\mathbf{x}) \tag{3.76}$$

where, as before, $\bar{B}(\mathbf{x})$ is background and $\bar{s}_m(\mathbf{x})$ is the specific signal. For present purposes it is useful to define the cocktail

$$\bar{f} = \langle \bar{f}_m \rangle_m = \bar{B}(\mathbf{x}) + \bar{s} \tag{3.77}$$

where $\langle \bar{s}_m \rangle_m = \bar{s}$ and to write

$$f_m(t, \mathbf{x}) = \bar{B}(\mathbf{x}) + s_m(t, \mathbf{x}). \tag{3.78}$$

We can eliminate the background by forming

$$\bar{f}_m - \bar{f} = \bar{s}_m(\mathbf{x}) - \bar{s}(\mathbf{x}). \tag{3.79}$$

Thus, it is seen that the desired activity maps are contained in this expression. On this basis we can estimate the amount of signal contained in a test image $\Phi(\mathbf{x})$, which is of unit length

$$(\Phi(\mathbf{x}),\ \Phi(\mathbf{x}))_{\mathbf{x}} = \|\Phi\|_{\mathbf{x}}^{2} = 1. \tag{3.80}$$

We take this signal to be measured by

$$S(\Phi) = \left\langle (\Phi,\ \bar{f}_m - \bar{f})_{\mathbf{x}}^{2} \right\rangle_m = \left\langle (\Phi,\ \bar{s}_m - \bar{s})_{\mathbf{x}}^{2} \right\rangle_m. \tag{3.81}$$

In the same way, the noise contained in $\Phi(\mathbf{x})$ is

$$N(\Phi) = \left\langle (\Phi, f_m - \bar{f}_m)_{\mathbf{x}}^{2} \right\rangle_{t,m} = \left\langle (\Phi, s_m(t, \mathbf{x}) - \bar{s}_m(\mathbf{x}))_{\mathbf{x}}^{2} \right\rangle_{t,m}, \tag{3.82}$$

from which we can form the signal-to-noise ratio

$$F(\Phi) = \frac{S(\Phi)}{N(\Phi)}. \tag{3.83}$$

In order to determine Φ, we might try to maximize the signal-to-noise ratio as depicted by Equation 3.83. This has the form of a Rayleigh quotient,[3,6] and in the statistical literature this is related to the method of canonical variates.[6] For our purposes, this procedure proves to have serious shortcomings. The heart of the problem is that, within image space, the denominator can be made arbitrarily small, and thus inappropriate small signal "solutions" make $F(\Phi)$ unbounded. We point out that this was a key issue in considering truncated differences and the indicator function method. A more extensive discussion of this point can be found in Yokoo et al.[37]

Instead of the above approach, we attempt to maximize Equation 3.81 while minimizing Equation 3.82. To accomplish this, we define the criterion function

$$C(\Phi) = S(\Phi)_{\mathbf{x}} - \alpha N(\Phi) \tag{3.84}$$

which is to be maximized subject to the constraint Equation 3.80. The constant $\alpha > 0$ reflects the fact that, for the moment, we do not know the proper admixture of Equations 3.81 and 3.82. This problem is addressed in the following mathematical appendix.

Color Figure 3.9 contrasts the results of applying the three methods discussed here to the responses to oriented stimuli analyzed in Everson et al.[12] What is shown are the two functions that govern orientation preference, discussed in Section 3.2.4, e.g., *Re* ϕ_2 and *Im* ϕ_2. The signal-to-noise ratio, F, of thc map is indicated in each panel. Thus, we see from the lower panel that the generalized indicator has a significantly higher confidence level than the middle panel, which shows the result from the indicator function approach. In this instance, unlike the case depicted in Section 3.2.5, the indicator function does a better job than truncated differences. With these data, standard differencing does not produce anything worth showing.

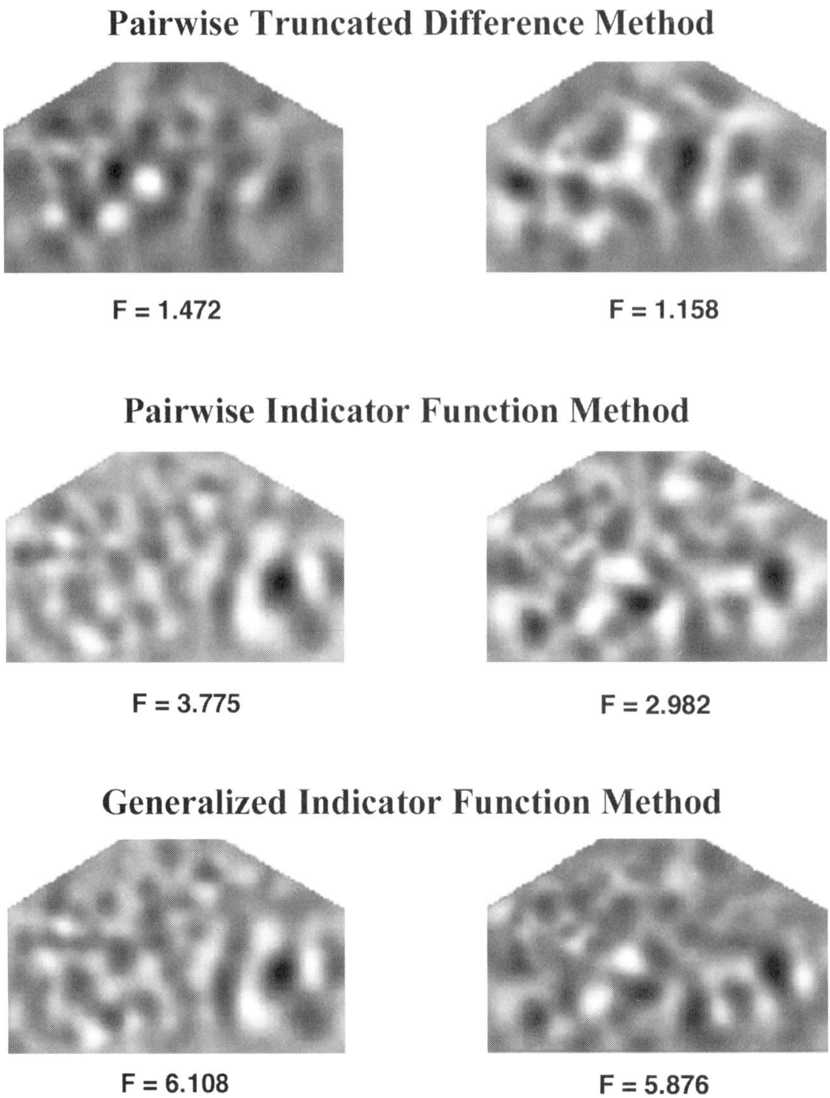

FIGURE 3.9 (See Color Figure 3.9.) Orientation eigenfunction pairs, each shown with its signal-to-noise ratio, *F*, statistic as calculated by truncated differences (upper row); indicator functions (middle row); generalized indicator functions (lower row). These figures are based on data from the cat cortex reported in Everson et al., *PNAS,* 95, 8334–8338, 1998.

A more thorough discussion of this new method can be found in Yokoo et al.,[37] http://camelot.mssm.edu/imaging.html.

Mathematical Appendix

If we define spatial covariances

$$K_S(\mathbf{x},\ \mathbf{y}) = \left\langle (f_m((\mathbf{x}) - \bar{f}(\mathbf{x}))(f_m(\mathbf{y}) - \bar{f}(\mathbf{y})) \right\rangle_m \tag{3.85}$$

and

$$K_N(\mathbf{x},\ \mathbf{y}) = \left\langle (f_m(t,\ \mathbf{x}) - \bar{f}_m(\mathbf{x}))(f_m(t,\ \mathbf{y}) - \bar{f}_m(\mathbf{y})) \right\rangle_{m,t} \tag{3.86}$$

we can write Equations 3.81 and 3.82 as

$$S(\Phi) = (\Phi,\ K_s\Phi)_{\mathbf{x}} \tag{3.87}$$

and

$$N(\Phi) = (\Phi,\ K_N\Phi)_{\mathbf{x}}. \tag{3.88}$$

It is interesting to note that

$$\left\langle f_m(t,\ \mathbf{x}) f_m(t,\ \mathbf{y}) \right\rangle_{m,t} = K(\mathbf{x},\ \mathbf{y}) = \bar{f}(\mathbf{x})\bar{f}(\mathbf{y}) + K_S(\mathbf{x},\ \mathbf{y}) + K_N(\mathbf{x},\ \mathbf{y}). \tag{3.89}$$

In such terms the maximization of Equation 3.84 subject to Equation 3.80 yields the eigenvalue problem

$$(K_S - \alpha K_N)\Phi = \gamma\Phi \tag{3.90}$$

for the generalized indicator function Φ.[37]

If we take the inner product of Equation 3.90 with Φ,we obtain

$$(\Phi,\ K_S\Phi) - \alpha(\Phi,\ K_S\Phi) = \gamma \tag{3.91}$$

and from this we can form the signal-to-noise ratio

$$F(\Phi) = \alpha + \frac{\gamma}{(\Phi,\ K_N\Phi)_{\mathbf{x}}} > \alpha, \quad \text{for all } \gamma > 0. \tag{3.92}$$

This suggests that α be taken such that

$$\alpha = F_c, \tag{3.93}$$

where F_c is an *a priori* chosen signal-to-noise ratio. In this way each eigenfunction of Equation 3.90 for $\gamma > 0$ is guaranteed to have an acceptable signal-to-noise ratio.[37]

3.2.8 Imaging Dynamics

Our discussion of preferential activity maps has emphasized spatial patterns of neuronal activity rather than the temporal aspects of the imaging data. In this section we remedy this situation and, in the process, introduce a formalism leading to an improvement in the overall quality of the extracted optical signal.[33]

We consider a number of sets of images $\{s_j(t, \mathbf{x})\}$, $j = 1, \ldots, P$, all of duration T and all in response to repeated presentations of the same stimulus. (In practice, such response records are usually interspersed randomly among responses to other stimuli in order to avoid conditioning of the imaged tissue.) All these records are then concatenated into a single record

$$f = f(t, \mathbf{x}) \tag{3.94}$$

such that

$$f(t' + (j-1)T, \mathbf{x}) = s_j(t, \mathbf{x});\ 0 \le t' \le T. \tag{3.95}$$

If our experimental preparation is stable over time, then, except for noise, f is "T-periodic." In this respect, the ordering of the concatenated different responses to the stimulus should be inconsequential. The record Equation 3.94 can be resolved into its empirical eigenfunctions such that

$$f(t, \mathbf{x}) = \bar{f}(\mathbf{x}) + \sum_n \alpha_n(t)\varphi_n(\mathbf{x}) \tag{3.96}$$

where

$$\bar{f}(\mathbf{x}) = \langle f(t, \mathbf{x}) \rangle_t. \tag{3.97}$$

The eigenfunctions $\{\varphi_n(x)\}$ satisfy the orthonormality condition (Equation 3.3) and $\alpha_n(t)$, the orthogonality condition (Equation 3.15). In the present context, it is also useful to Fourier-transform f in the time domain

$$f(\omega, \mathbf{x}) = \bar{f}(\mathbf{x}) + \sum_n \alpha_n(\omega)\varphi_n(\mathbf{x}), \tag{3.98}$$

where

$$\alpha_n(\omega) = \int dt\, e^{-i\omega t}\alpha_n(t). \tag{3.99}$$

The orthogonality condition carries over to the transformed functions

$$\int \alpha_n(\omega)\alpha_m(\omega)d\omega = 2\pi\lambda_n\delta_{nm}. \tag{3.100}$$

According to our fundamental assumption that f should be T-periodic, it follows that for each $\alpha_n(\omega)$ we should filter out all frequencies except those that are integer multiples of

$$\omega_o = \frac{2\pi}{T}. \tag{3.101}$$

If we denote the filtered $\alpha_n(\omega)$ by $\hat{\alpha}(\omega)$ and its transform by $\hat{\alpha}(t)$, by then the cleansed data record is given by

$$\hat{f}(t, \mathbf{x}) \approx \bar{f}(\mathbf{x}) + \sum_n \hat{\alpha}_n(t)\varphi_n(\mathbf{x}). \tag{3.102}$$

It should be noted that we do not expect the $\{\hat{\alpha}_n\}$ to be orthogonal, and a repeated PCA on $\hat{f}(t,\mathbf{x})$ can be performed as a next step.

These deliberations assume a method for the accurate determination of Fourier spectra. For this purpose we have used the powerful multitaper methods introduced in recent times by Thomson,[36] Mitra and Pesaran,[24] and by Sornborger et al.[33] (See also Percival and Walden[25] for an exposition of these methods.)

These dynamical techniques have been applied to optical images of mouse heart activity as visualized by voltage-sensitive dyes.[32] Color Figure 3.10A (raw) shows the temporal spectral concentration vs. PCA mode number for a record of activity, a film clip of 10 heartbeats, with about 100 frames per beat. The methodology just described led to the cleansed spectral concentration shown in Color Figure 3.10A (signal). As can be seen from the third panel, Color Figure 3.10A (noise), the spectral noise content of the set of images, contains a large part of the complete spectrum, Color Figure 3.10A (raw). The most dramatic effect of this analysis comes in viewing the cleansed temporal film clip with the artifactual effects removed. Phenomena such as wave propagation are then crisply observed. It has been estimated that the SNR has been improved by a factor of 50. For present purposes we compare two still images in Color Figure 3.10B, one from the raw film clip (untreated) and the other from the treated film clip (cleansed). Improvement is evident even at this single frame level. The graininess seen in the untreated image is gone in the stronger cleansed signal.

As a last example, we consider orientation preference data from the cat visual cortex, V1, shown in Figure 3.11, in which a single condition orientation patch is considered. The panel on the left shows the result of applying the generalized indicator function method to the raw data comprising 20 records, each containing 128 frames. To the right we see the result of the generalized indicator function method applied to the cleansed temporal records. The enhancement of the image speaks for itself.

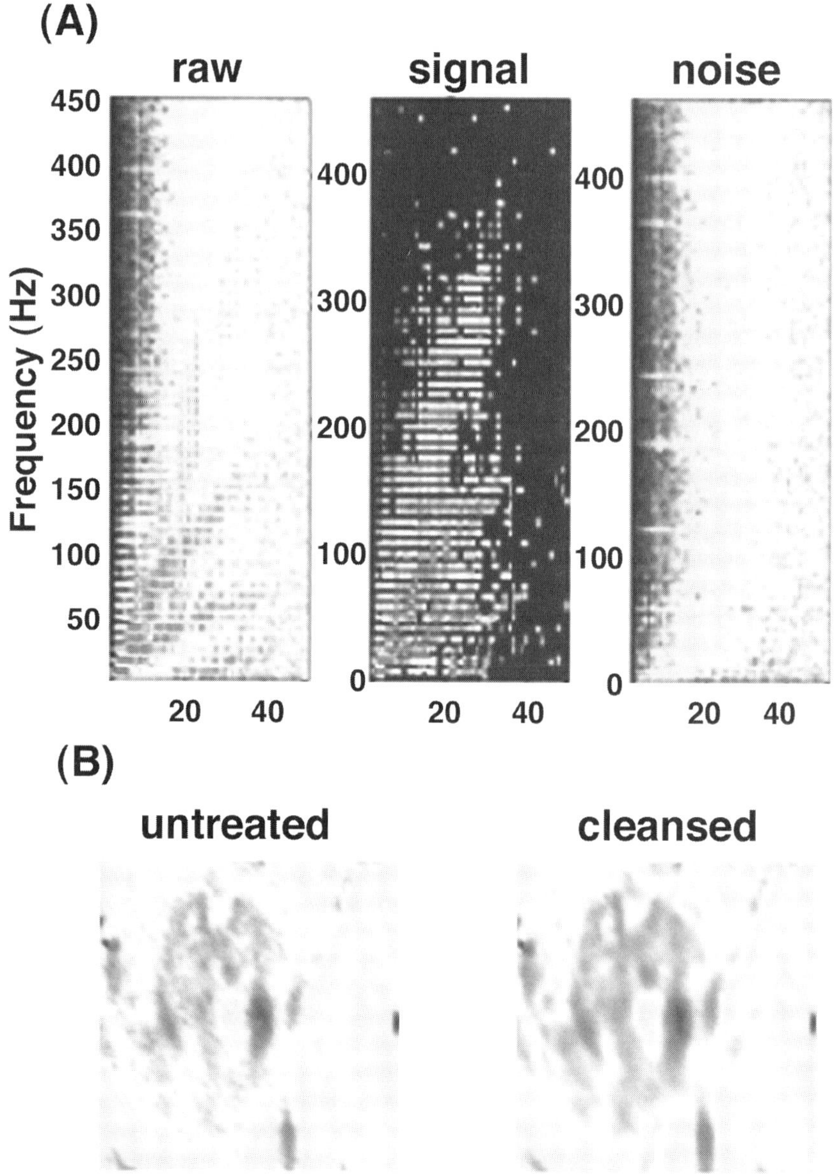

FIGURE 3.10 (See Color Figure 3.10.) Analysis of fluorescence images of a voltage-sensitive dye from the *in vitro* mouse heart, as a function of time. The raw data analyzed were collected by Gregory Morley, Mount Sinai School of Medicine; image rate was 914 frames/sec. A: (raw) spectral concentration of data vs. PCA mode number; (signal) spectrum with noise removed; (noise) residual noise. B: mouse heart activity: (untreated) record; (cleansed) record.

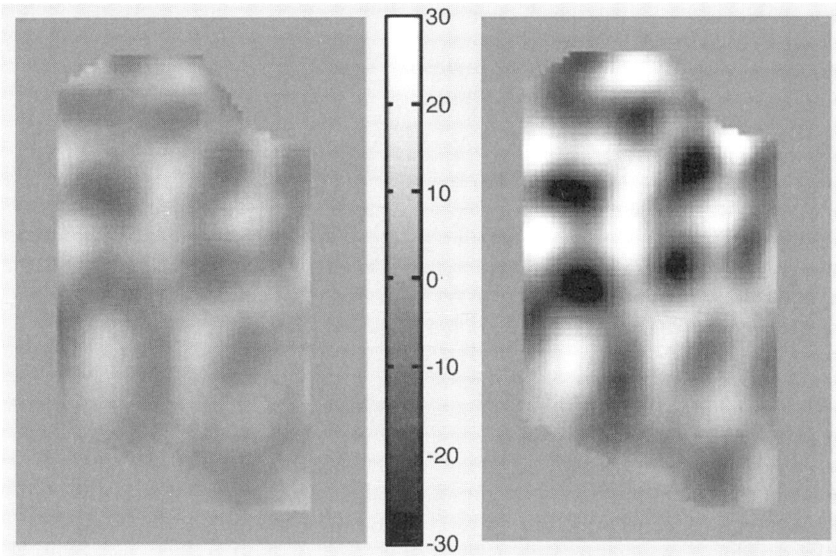

FIGURE 3.11 Orientation patches in the primary visual cortex of the cat: (a) untreated record; (b) cleansed record. Data collected by Cynthia Sailstad, Mount Sinai School of Medicine. The grayscale in the middle shows the relative response amplitude. Image is 4 × 3 mm.

3.3 OTHER METHODS

The challenges presented in the introduction have stimulated various groups to seek improved methods for the analysis of imaging data. Because of space limitations, we cannot provide a comprehensive review of those methods, but for completeness we briefly mention three approaches.

Wavelets. In dealing with time-varying optical signals, one may use an analysis method based on wavelets. For example, Carmona et al.[9] have used that approach, demonstrating its utility in reducing the impact of the vasculature on imaged data from the brain. In addition, after cleaning the data of vasculature artifacts, their method (by using PCA) was able to follow the evolution of reflectance changes in the activated regions of the brain. Other groups have used wavelets to analyze images from positron emission tomography (for example, Millet et al.[23]).

Independent Components Analysis (ICA). A technique that has attracted much attention recently is independent components analysis (ICA), which is aimed at a "blind" separation of various sources, including imaging, mixed together during various types of data recording. (See Bell and Sejnowski[2] and a survey of ICA by Aapo Hyvärinen at http://www.cis.hut.fi/aapo/papers/NCS99web/NCS99web.html).

This approach utilizes not only the second-order correlations used by PCA but also correlations of higher order. Like PCA, ICA seeks a linear decomposition of the data but, importantly, it relaxes the requirement made by PCA that the separated components be orthogonal. On the other hand, ICA makes certain assumptions about the statistical nature of the various sources that combine to produce the recorded data. Some of these assumptions have not been verified or are known not to hold for optical imaging data. Several groups have used ICA for the analysis of imaging data, for example, McKeown and Sejnowski[22] and Jung et al.[21]

Extended Spatial Decorrelation (EDS). Recent work (Schiessle et al.[26] and Stetter et al.[34]) has applied the method of extended spatial correlation (EDS) to optical imaging data. This method makes more assumptions than ICA about the relationships among sources contributing to the recorded images, but it uses only second-order statistics, like PCA-based methods. EDS takes advantage of the spatial scattering of light within the cortical tissue, especially at long wavelengths, and does not require that the various sources to be separated have similar variance. The cited references compare the underlying assumptions and performance of EDS with those of PCA and ICA.

3.4 CONCLUSIONS

We have provided a survey of several methods that experimenters can use to analyze optical (and other) imaging data. Several of the methods reviewed here have been developed in our laboratory over the past few years, but we also briefly mentioned several other methods. Experience shows that there is no *one* perfect method for all situations and data types. In particular, the nature of the noise in the data, and knowledge of the recording system and the tissue being imaged, can help in guiding the user to the analysis method most appropriate for the data at hand. The constant evolution of imaging and computing technologies is certain to continue and challenge us to develop new analysis methods in the future. Information extraction from imaging is a powerful tool for scientific investigation; with new methods now on the horizon, our ability to mine this information will improve.

ACKNOWLEDGMENTS

Our work has been supported by NIH-NEI EY11276, NIH-NIMH MH50166, NIH-NEI EY12867 and Darpa MDA972-01-1-0028. Ehud Kaplan is the Jules and Doris Stein Research to Prevent Blindness professor at the Ophthalmology Department at Mt. Sinai. The authors are grateful to A. Casti, B.W. Knight, A. Sornborger, and T. Yokoo for their helpful comments, and to Cynthia Sailstad and Greg Morley for sharing their data.

REFERENCES

1. Arieli, A. et al., Dynamics of on-going activity: explanation of the large variabity in evoked responses, *Science*, 273, 1868–1871, 1996.
2. Bell, A. and Sejnowski, T., An information maximization approach to blind separation and blind deconvolution, *Neural Comput.*, 7, 1129–1159, 1995.
3. Bellman, R., *Introduction to Matrix Analysis*, McGraw-Hill, New York, 1960.
4. Blasdel, G., Orientation selectivity, preference, and continuity in monkey striate cortex, *J. Neurosci.*, 12, 3139–3161, 1992.

5. Blasdel, G. and Salama, G., Voltage sensitive dyes reveal a modular organization in monkey striate cortex, *Nature*, 321, 579–585, 1986.
6. Boch, R., *Multivariate Statistical Methods in Behavioral Research*, McGraw-Hill, New York, 1975.
7. Bonhoeffer, T. et al., Optical imaging of the layout of functional domains in area 17 and across the area 17/18 border in cat visual cortex, *J. Neurosci.*, 7, 1973–1988, 1995.
8. Cannestra, A. et al., The evolution of optical signals in human and rodent cortex, *NeuroImage*, 3, 202–208, 1996.
9. Carmona, R., Hwang, W., and Frostig, R., Wavelet analysis for brain function imaging, *IEEE Trans. Med. Imaging*, 14(3), 556–564, 1995.
10. DeValois, R., Yund, E., and Hepler, N., The orientation and direction selectivity of cells in macaque visual cortex, *VISRES*, 22, 531–544, 1982.
11. Everson, R., Knight, B., and Sirovich, L., Separating spatially distributed response to stimulation from background. I. Optical imaging, *Biol. Cybern.*, 77, 407–417, 1997.
12. Everson, R. et al., Representation of spatial frequency and orientation in the visual cortex, *PNAS*, 95, 8334–8338, 1998.
13. Friston, K. et al., Functional connectivity: the principal-component analysis of large (PET) data sets, *J. Cereb. Blood Flow Metab.*, 13, 5–14, 1993.
14. Friston, K. et al., Characterizing dynamic brain responses with fMRI: a multivariate approach, *Neuro Image*, 2, 166–172, 1995.
15. Gabbay, M. et al., A principal components-based method for the detection of neuronal activity maps: application to optical imaging, *NeuroImage*, 11, 313–325, 2000.
16. Glaser, E. and Ruchkin, D., *Principles of Neurophysiological Signal Analysis*, Academic Press, New York, 1976.
17. Grinvald, A. et al., *In-vivo* optical imaging of cortical architecture and dynamics, in *Modern Techniques in Neuroscience Research,* Windhorst, U. and Johansson, H., Eds., Springer-Verlag, Heidelberg, 1999.
18. Hawken, M., Parker, A., and Lund, J., Laminar organization and contrast sensitivity of direction-selective cells in the striate cortex of the old world monkey, *J. Neurosci.*, 8, 3541–3548, 1988.
19. Hubel, D. and Wiesel, T., Receptive fields, binocular interaction and functional architecture in the cat's visual cortex, *J. Physiol. Lond.*, 160, 106–154, 1962.
20. Issa, N., Trepel, C., and Stryker, M., Spatial frequency maps in cat visual cortex, *J. Neurosci.*, 20, 8504–8514, 2000.
21. Jung, T.-P. et al., Imaging brain dynamics using independent component analysis, *Proc. IEEE*, 89(7), 1107–1122, 2001.
21a. Kirby, M. and Sirovich, L., Application of Karhuen-Loève Procedure for the characterization of human faces, IEEE Transactions on Pattern Analysis and Machine Intelligence, 12(1), 103–108, 1990.
22. McKeown, M. and Sejnowski, T., Independent component analysis of fMRI data: Examining the assumptions, *Human Brain Mapping*, 6, 368–372, 1998.
23. Millet, P. et al., Wavelet analysis of dynamic PET data: application to the parametric imaging of benzodiazepine receptor concentration, *NeuroImage*, 11, 458–472, 2000.
24. Mitra, A. and Pesaran, B., Analysis of dynamic brain imaging data, *Biophys. J.*, 76, 691–708, 1999.
25. Percival, D. and Walden, A., *Spectral Analysis for Physical Applications*, Cambridge University Press, Cambridge, UK, 1993.
26. Schiessle, M. et al., Blind signal separation from optical imaging recordings with extended spatial decorrelation, *IEEE Trans. Biomed. Engin.*, 47(5), 573–577, 2000.

27. Senseman, D. and Robbins, K., Modal behavior of cortical neural networks during visual processing, *J. Neurosci.*, 19(RC3), 1–7, 1999.
28. Shmuel, A. and Grinvald, A., Functional organization for direction of motion and its relationship to orientation maps in cat area 18, *J. Neurosci.*, 16(21), 6945–6964, 1996.
29. Sirovich, L., Turbulence and the dynamics of coherent structures, parts i, ii, and iii, *Q. Appl. Math.*, XLV(3), 561–590, 1987.
29a. Sirovich, L. and Kirby, M., Low-dimensional procedure for the characterization of human faces, *J. Opt. Soc. Am.,* 4, 519–524, 1987.
30. Sirovich, L. and Everson, R., Management and analysis of large scientific datasets, *Int. J. Supercomputer Appl.*, 6, 50–68, 1992.
31. Sirovich, L. et al., Modeling the functional organization of the visual cortex, *Physica D*, 96, 355–366, 1996.
32. Sornborger, A. et al., A method for denoising periodic multivariate signals: Application to voltage dye imaging data of the mouse heart, *IEEE Trans. Med. Imag.*, submitted, 2001.
33. Sornborger, A. et al., Spatio-temporal analysis of optical imaging data, *NeuroImage,* submitted, 2001.
34. Stetter, M. et al., Principal component analysis and blind separation of sources for optical imaging of intrinsic signals, *NeuroImage*, 11, 482–490, 2000.
35. Stewart, G. and Sun, J., *Matrix Perturbation Theory*, Academic Press, San Diego, 1990.
36. Thomson, D., Spectrum estimation and harmonic analysis, *Proc. IEEE*, 70, 1055–1096, 1982.
37. Yokoo, T., Knight, B., and Sirovich, L., An optimization approach to signal extraction from noisy multivariate data, *NeuroImage*, 14, 1309–1326, 2001.

4 Optical Imaging of Neural Activity in Sleep–Waking States

Ronald M. Harper, David M. Rector, Gina R. Poe, Morten P. Kristensen, and Christopher A. Richard

CONTENTS

0-8493-2389-4/02/$0.00+$1.50
© 2002 by CRC Press LLC

4.1 INTRODUCTION

4.1.1 Advantages to Intrinsic Optical Imaging

Certain behaviors involve the integrated action of large numbers of neurons, often in different brain sites, which show different activity patterns over substantial periods of time. These behaviors include sleep and waking states, homeostatic responses to everyday challenges, developmental changes, acquisition of learned tasks, and development of addictive processes or changes in neural response to pharmacologic agents. A need exists to examine activity of the large numbers of neurons simultaneously in brainstem, subcortical, and cortical brain areas associated with these behaviors; that examination often involves acquisition of signals from conscious, freely moving, and drug-free subjects.

Description of neural processes underlying induction and maintenance of sleep states or mediating processing of information during different states represents an extreme example of complexity of assessment of multiple neural interactions. However, expression of even the simplest actions requires the interaction of large numbers of nerve cells in multiple areas that typically involve both excitatory and inhibitory processes. Optical procedures provide significant advantages for long-term determination of activity changes in large numbers of neurons over several areas. Because sleep states provide an "experiment of nature" that substantially reorganizes neural responses to physiological challenges, do so repeatedly for prolonged periods of time, and recruit many brain areas for expression of the reorganization, principles of optical recording will be illustrated largely from examples acquired during sleep and waking conditions.

A principal classic procedure used to assess neural activity in freely behaving animals has been the closed-skull rigid[1] or flexible[2,3] microelectrode technique, which allows long-term monitoring of a few neurons; this procedure has recently been extended to allow collection of data from several-to-many microelectrode tips placed in close proximity to a brain site.[4,5] The procedure offers many advantages, including the capability for deep, subcortical placement and a certainty of timing and identification of action potentials. Long-term single neuron recording has been invaluable for differentiating actions of neurotransmitter groups during sleep states[6] and for describing integrated action of information exchange in such brain functions as spatial mapping in the hippocampus.[4] The technique possesses certain disadvantages, however, including the obvious number of electrodes vs. size of electrode bundle tradeoff, an inability to examine certain brain sites in a freely moving animal (e.g., very superficial brain areas located behind substantial soft tissue, such as the ventral medullary surface, VMS), and computationally intensive analyses for study of interactions between spike trains.[7]

4.1.2 Limitations of Other Imaging Methodologies

A number of strategies have been employed to visualize neural activity in intact animals. Visualization of regional activity with a several-mm^3 volume resolution can be achieved with functional magnetic resonance (fMRI) procedures — if the behaviors can be performed within the confines of an imaging device while the neural structures are immobile and if the behavioral tasks are compatible with the substantial audio noise associated with magnetic resonance scans. The procedure is valuable for screening the entire brain for behaviorally associated activity; however, only relatively slow temporal characteristics (currently, approximately 1/sec sampling rate) can be measured. Similarly, visualization of neural functional activity can be obtained, although with even more limited spatial (tens of mm^3) and temporal (multiple sec) resolution, using positron emission tomography (PET) procedures. However, ionizing radiation, a component of PET techniques, poses substantial risk vs. benefit issues with human subjects and limits repeated measures in both animal and human preparations. In addition, the cost of access to a cyclotron must be taken into consideration.

Determination of interactions between neurons in localized regions can be effectively visualized with calcium- or voltage-sensitive dyes at relatively high temporal rates. Such visualization has been exceptionally valuable for demonstrating the time course of integrated neural action in several systems. Dye procedures, however, have only finite usefulness for certain long-term studies: the dyes are often toxic, limiting multiple trials, and application to subcortical areas can be difficult. The potential for toxicity precludes examination of activity over long-term periods in a single subject, as might be required during sleep or developmental studies.

Dye-free imaging procedures[8] offer an alternate means to visualize metabolic activity over large brain areas in intact animals. These techniques provide high spatial (<10 μm) and temporal (<1 msec) resolution over long time periods (>4 weeks), with minimal perturbation of the tissue. The functional organization of columnar structures in visual cortex[8] or lamellar analysis of other neural structures can be imaged.

4.1.3 Biophysics of Intrinsic Optical Imaging

The capability to visualize neural activity with imaging procedures derives from a number of physical changes in cells associated with discharge. For over 100 years, cell firing has been known to be associated with changes in cell size.[9] When cell membranes are altered by discharge, imposed illumination will scatter differently and regionally modify returned light. Tasaki and Iwasa[10] demonstrated that discharge-related membrane movement occurs at action potential rates, a finding which suggests that capturing scattered and reflected light at very high rates offers the potential to follow cell firing visually.

Changes in light transmission through axon bundles occur at action potential rates.[11] When ionic flow is blocked in hippocampal slices, light transmission changes cease, suggesting that ionic inflow modifies cell swelling and, thus, light transmission.[12] Stimulation decreases the extracellular volume fraction without changing overall tissue volume, suggesting an increase in intracellular volume upon activation.[13] Manipulating the osmolarity of the environment surrounding hippocampal cells alters light reflectance *in vivo*,[14] supporting the concept that relative ionic flow into cells, and resultant cellular volume changes, are indicated by alterations in optic properties. All of these changes in membrane characteristics that occur with discharge result without interactions with added dyes responsive to calcium, voltage, or other changes. The light characteristics changes result from light action on membrane properties alone.

Other changes accompanying cell discharge can be visualized, such as increased blood flow and volume that accompany firing; in addition, changes in the relative ratio of oxyhemoglobin and deoxyhemoglobin occur with changing cell metabolic demands. Hemodynamic changes occur more slowly than membrane changes associated with cell discharge, and tissue illumination with wavelengths specific to vascular aspects can also reveal lasting functional organization of neural elements.[15]

The pioneering studies for visualizing regional neural activity took advantage of localized functional topography of visual cortex to reveal, among other aspects, ocular dominance columns. A very sensitive, wide-dynamic-range, slow-scan camera was used to capture reflected light from illuminated cortical areas through a skull window.[8] Later procedures, using longer wavelengths, allowed assessment of activity through thinned skull, reducing contributions from brain movement associated with open skull preparations and improving surgical logistics.[15,16]

The initial procedures, however, restricted observations to cortical surface structures and relied on reflected light passing through a lens-coupled system remote from the neural tissue. A lens induces substantial light loss, and detected reflectance changes from a distant cortical surface are small, because of the large amount of background light. Typical signals from a lens-coupled system are a fraction of a percent reflectance change, requiring considerable signal averaging to ensure an adequate signal-to-noise ratio and expensive, highly sensitive cameras, cooled to assist operation at low noise. Such signal averaging reduces the temporal resolution of the device, a significant concern for detection of rapid neural events. Initially, highly sensitive cameras had relatively slow scanning rates, but that timing aspect has improved.

4.2 CHRONIC IMAGING PROCEDURES

4.2.1 Components and Properties

We developed an optical device sufficiently small to allow placement of the camera within body tissue near the brain of freely moving animals for weeks at a time.[17,18] The procedure has been used to assess regional neural activity over areas as large as 24 mm^2 in surface sites and allows subcortical placement of smaller probes. The device uses a coherent image conduit, typically 1.0 to 3.2 mm in diameter, encompassing approximately 10,000 5- to 12-μm optic fibers; tapered probes as large as a 6×4 mm exposed area have been used successfully on the ventral medullary surface of cats. One end of the conduit is placed directly on neural tissue or on dura over the neural tissue; the other end is attached to a charge-coupled device (CCD).

Development of CCDs has been extremely rapid, with improved resolution, dynamic range, sensitivity, scanning speed properties, and noise characteristics, making possible rapid image acquisition at costs not conceivable a few years ago. Current CCD devices measure only a few millimeters on a side and contain in excess of 400,000 pixels; we have used the Texas Instruments (Texas Instruments, Dallas, TX) TC211 series and, more recently, custom-developed rapidly scanning devices from Dalsa Instruments (Dalsa Life Sciences, Tucson, AZ).

The conduit is surrounded by fibers that carry illumination from narrow bandwidth light-emitting diodes (LEDs) to the neural surface; the light is scattered by tissue structures that are altered by cell firing, and scattered light is returned by way of the image conduit to the CCD. Since the optic fibers rest directly on the tissue surface, light loss from dispersion is minimized; the illumination and light capture are analogous to that of dark-field microscopic procedures. In addition to light loss through lens coupling (proportional to lens properties), remote viewing systems have a disadvantage of light loss proportional to 1/r (r = distance from cortical surface). The optic conduit avoids the scatter loss from lens distance from the neural surface and gains the advantage of increased light capture and enhanced signal-to-noise ratio through direct tissue contact. This gain in signal to noise can be used to considerable advantage for imaging; the need for exceptional dynamic range of capture cameras is reduced because signals are larger, reducing equipment cost or camera cooling needs; the necessity for image averaging to determine signal from noise is diminished, thereby improving temporal resolution.

The spatial resolution of the device depends on optic fiber diameters constituting the conduit and pixel dimensions of the CCD elements. We use optic conduits with individual fiber diameters down to 5 μm. Our earlier CCDs[18] used 25-μm pixel elements; however, CCDs with 10-μm elements are available. The output of the CCD camera is sent to pixel-scanning circuitry. In early versions, video circuitry operating at 50 to 60 fields/sec was incorporated with the CCD chip; later, higher scan rate CCD arrays were implemented, providing an effective frame rate of 1 KHz.[19] The extremely rapid scan rates allow examination of neural events taking place in the time domain of action potentials.

Early versions of the cameras used an 8-bit digitizer, a relatively coarse resolution for low light level signals, and required careful sampling of the video signal to optimize the dynamic range. The video black level was set so that the darkest pixel was 10% above a digitized value of zero. Such judicious manipulation provided an effective 10-bit dynamic range; however, such adjustments now are not necessary with newer digitizers that have wider dynamic range (e.g., 12 to 14 bits). The enhanced signal-to-noise ratio offered by the optic conduit and direct tissue contact provides significant advantages in reducing requirements for very wide dynamic range in the cameras.

Although camera cooling would obviously reduce electronic noise associated with acquisition and was used in earlier models,[17] such cooling can be used only with camera placements external to the body in conjunction with an optic conduit and is precluded from internal siting. However, with current cameras, noise is sufficiently low at body temperatures for image acquisition. The recording ensemble forms an extremely small unit that can be mounted on the head for recording from dorsal brain structures or buried behind the esophagus and trachea within the neck to acquire images of the ventral brain surface (Figure 4.1).

Stereotactic procedures are used to place the probe at the appropriate site and the skull is sealed with dental acrylic; as is the case with microelectrodes, closed skull preparations minimize brain movements relative to the probe. Recordings can be collected for 3 to 6 weeks before significant tissue deterioration occurs at the probe tip. Since light loss through the conduit is negligible, the optic fibers can be any

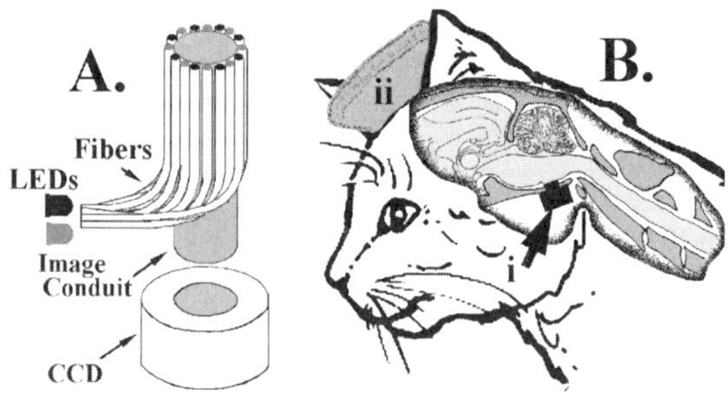

FIGURE 4.1 A. Schematic drawing of imaging device. A coherent image conduit, composed of optic fibers, is directly attached to a CCD chip. Illumination fibers from red and green LEDs (shown here as black and gray representing red and green, respectively; different wavelengths can be electronically switched in sequence) are arranged around the bundle. Electrophysiological recording electrodes or dialysis probes can be attached along the probe. Video and electrophysiological signals are transferred externally via flexible wires and headpiece connectors. B. Probe placement on VMS. An amplifier sits on the CCD array chip attached to the optic fiber bundle (i, arrow); the array chip also contains scanning circuitry. Cables are passed to a dental cement headpiece, ii. (From Rector, D.M. et al., *Am. J. Physiol.*, 278, R1090, 2000.)

appropriate length to reach any brain area. Moreover, since the fibers are embedded in glass, the probe can be curved following momentary heating so that the tip can record from a surface angle other than horizontal, such as in the medial hypothalamus.[20]

4.2.2 Illumination Wavelengths

We use a relatively long wavelength (660 nm) to visualize changes associated with cell discharge; comparably long wavelengths have been used in blood-free preparations to image synaptic potentials[12] or cortical columnar structures.[21] The signal changes presumably result from alterations in cellular swelling and extracellular space. A shorter-wavelength illumination (560 nm) is closer to absorption peaks for oxyhemoglobin or deoxyhemoglobin; absorption for hemoglobin values differs by a factor of 500 over values at 660 nm and was used to assess vascular contributions simultaneously. Measures from the shorter wavelengths consistently show pattern differences in recordings from ventral medullary sites during sleep.[22]

4.3 IMAGE CHANGES WITH BEHAVIOR

4.3.1 Spontaneous Activity during Sleep

The capability to record over long periods from structures that are difficult to access can be seen in Figure 4.2, which shows a trace representing the overall average of pixels from the probe placed on the VMS of a cat across a several-hour recording period. The figure demonstrates the remarkable decline in ventral medullary surface activity during rapid-eye-movement (REM) sleep,[22] a finding also appearing in another species, the goat, using similar techniques.[23] Since the ventral medullary

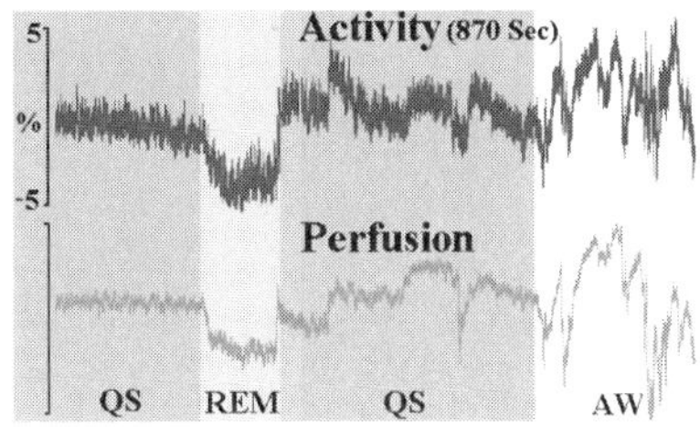

FIGURE 4.2 Moment-to-moment values of successive changes in activity from the intermediate area ventral medullary surface of an adult cat during the course of a transition from quiet to REM sleep and to waking. Light reflectance and scattering from two illumination wavelengths, 660 nm (red, top trace) and 560 nm (green, lower trace) are shown, representing activity from cellular activity and perfusion, respectively. Both activity and the hemodynamic signals substantially declined during REM sleep. (From Richard, C.A. et al., *Am. J. Physiol.*, 277, R1239, 1999.)

surface is a critical structure in control of blood pressure and breathing, investigators who examine state-related disorders of breathing and cardiovascular function are particularly concerned with the capability to monitor activity of such a strategic area during sleep. However, the area is located deep under the soft tissue of the neck and is virtually inaccessible by either dorsal or ventral approaches from conventional microelectrode techniques in a chronic preparation, thus requiring assessment only in anesthetized preparations.

A typical sleep recording requires 3 to 6 h of acquisition; for species in which REM sleep is a minor component of total sleep time, even longer recordings may be necessary; for example, 6 to 12 h of recording are necessary to obtain adequate representation of REM sleep in the goat.[23] Such recording times require stress-free conditions for natural sleep and, thus, must particularly avoid animal restraint. Even if placement of microelectrodes in a chronic preparation were possible in the VMS over such a time period, retaining continuous contact with single neurons so near the surface of the brain is not a realistic option. For these reasons, the capability to place a miniaturized imaging device on a freely behaving animal is especially beneficial.

Figure 4.2 also demonstrates that hemodynamic aspects measured by the optic probe often differ from activity measures assessed by longer wavelength illumination. Overall, the trends are comparable, but close inspection reveals significant timing and extent of change differences between the two signals.

4.3.2 Conscious Preparations vs. Anesthesia

The use of anesthetics is so common in studies of neural function that it is easy to overlook the dramatic effect of such anesthetic agents on the measures under study. In addition to local transmitter changes in the brain region examined, influences from more rostral or other brain areas may be dramatically affected in the anesthetized state. The concern is illustrated in Figure 4.3, which shows the goat rostral VMS response to an intravenous administration of a blood pressure challenge (15 μg phenylephrine). After an initial small activity decline, activity rises modestly during waking, but under halothane anesthesia, activity declines and remains lower for a

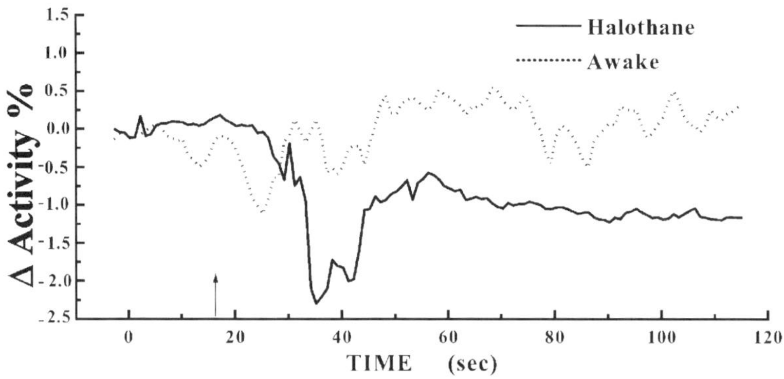

FIGURE 4.3 Traces of percentage change in overall activity from a region in the rostral ventral medullary surface of one goat during waking (dashed line) or under Halothane anesthesia (solid line) in response to a 15-μg/kg phenylephrine intravenous administration (onset at arrow) as a pressor challenge. The trace during anesthesia differs substantially from the awake response. (Derived from Harper, R.M. et al., *Am. J. Physiol.,* 270, R182, 1996.)

substantial period of time.[24] Obviously, description of activity of the VMS to blood pressure challenges must consider whether the challenge was delivered in a "natural" or anesthetized state.

4.3.3 Physiological Challenges during Sleep

Because optic procedures can examine activity even in subcortical areas in the conscious or sleeping animal, the technique offers other benefits for assessment of challenges to vital functions during natural conditions.[25] For respiratory control investigators, the technique allows an unprecedented view of activity in respiratory-related brain areas to increased CO_2 exposure during sleep (Figure 4.4), an aspect of clinical interest because of the failure of some patients to enhance breathing with diaphragmatic and upper airway muscles to elevated CO_2, leading to disordered breathing during sleep. Activity within rostral and intermediate VMS areas was found to decline to elevated CO_2 challenges in both quiet sleep and REM states, as well as in waking conditions, with a smaller decline in quiet sleep compared to waking and intermediate values in REM sleep.[26]

The rates of declines and topography of activation and deactivation, as well as the latency to nadir and recovery, differed between states. Similarly, investigators who study neural responses to blood pressure manipulation can examine their hypotheses under natural conditions. Application of sodium nitroprusside lowers blood pressure and enhances ventilation, and the optical procedures show that, under anesthesia, a narrow column of activated cells emerges on the rostral VMS during such a challenge.[27] Because anesthesia affects neural functioning so dramatically, it is essential to evaluate cardiovascular life-threatening challenges in the intact, unanesthetized state. In the awake animal, activity on the VMS closely tracks blood pressure lowering during blood loss, with activity diminishing as the animal enters the first stage of shock and elevates blood pressure in an attempt to compensate for

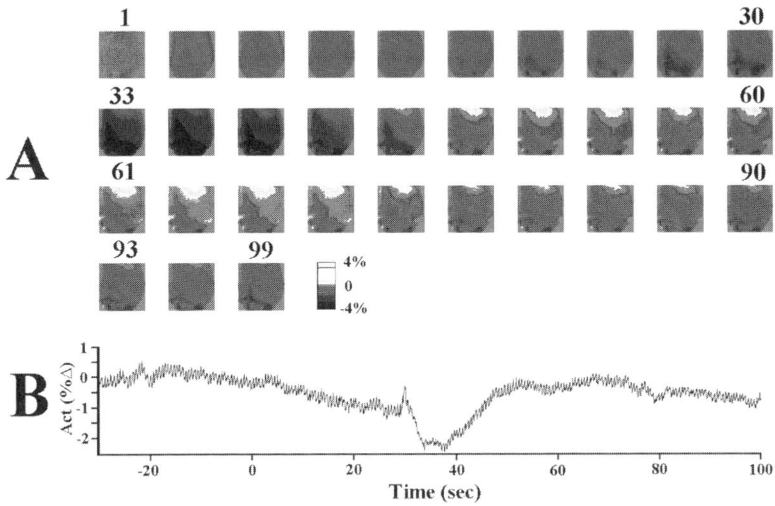

FIGURE 4.4 A. Topographical changes following a CO_2 challenge on the VMS in REM sleep plotted with the averaged VMS activity trace. B. Each image in A is an average over 3 sec, subtracted from baseline values and gray scale coded for magnitude of change, and is 3 mm in diameter. The images are gray scale coded so that lighter scales represent increased activity and darker tones represent decreased activity. The numbers above the images correspond to time, in sec, along the axis in B. (From Richard, C.A. et al., *Neurosci. Lett.,* 286, 175, 2000.)

onset of blood withdrawal (Figure 4.5).[28] Activity then increases in the second stage of shock, as blood pressure collapses, and recovers as the animal initiates somatic and respiratory efforts to restore cardiovascular tone.

Since baseline activity on the VMS declines during REM sleep, a subject may possibly not be able to recover if a profound loss of blood pressure occurs during that state; the need to activate the VMS during low blood pressure may not be accommodated because spontaneous VMS activity has reached too low a level during REM sleep. That scenario remains a realistic concern, since some clinical conditions are associated with substantial blood pressure control issues during sleep, such as sudden infant death syndrome. In fact, we found that the VMS is far more responsive during REM sleep, as if the structure were "released" from dampening control during that state.[29]

The optical procedure allows description of regional changes, the temporal pattern of changes, and repeated measures across the same animal under different state conditions, a circumstance not possible by dye imaging procedures.

4.3.4 Regional and Columnar Findings

Figure 4.6 illustrates two of the more significant aspects of the chronic imaging technique: capability to visualize activity over a large area and, after a transition to another state, to visualize regions of activation and inactivation that can provide significant insights into function. That capability shows, for example, that, without

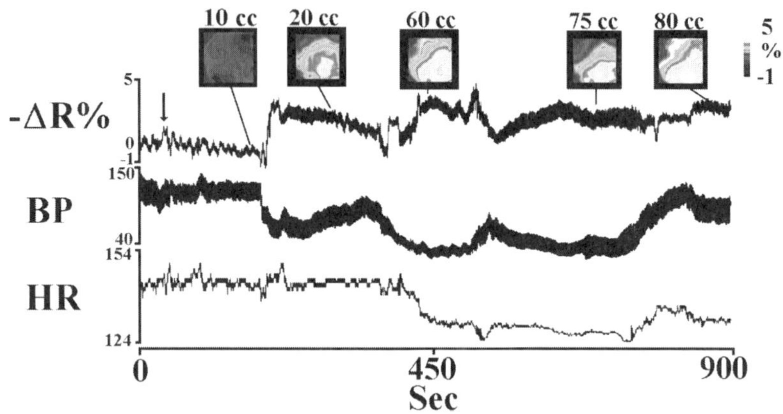

FIGURE 4.5 Traces of percent activity change (inverse of reflectance –ΔR%) on the intermediate area of the VMS, blood pressure (BP), and heart rate (HR) from a cat undergoing blood loss. The preparation shows response characteristics of two stages of shock, an initial compensatory phase, with a slight rise in blood pressure accompanied by a decline in VMS activity (darker image at 10-cc blood withdrawal), and the second stage which is concomitant with a large drop in blood pressure and a rise in VMS activity (lighter areas, 20-cc blood withdrawal). As the preparation makes efforts to restore blood pressure, activity on the VMS waxes and wanes. (From Harper, R.M. et al., *Neuroscience,* 94(2), 579, 1999.)

changing conditions except for state, remarkable reorganization of active and inactive areas occurs. In Color Figure 4.7,* images were collected from the dorsal hippocampus during waking and REM states; these images demonstrate curved bands of coordinated activity in the hippocampus, occurring preferentially during REM sleep, with an absence of such organization during waking. The reorganization of activity within the hippocampus during REM sleep, as assessed by cell discharge and a shift to highly synchronous rhythmical slow electroencephalographic activity,[30] develops columns of functional organization that differ from the relatively homogenous pattern of activity found in waking.[31] It would be possible to determine this columnar organization with microelectrode recording, but only with prodigious recording efforts.

4.3.5 Usefulness for Pharmacologic Studies

Imaging procedures are exceptionally valuable for examining pharmacologic action on neural structures of freely behaving animals, since the technique offers wide-area visualization of activity and long-term recording potential. Such an application is shown in Color Figure 4.8, which plots activity from two probes, one on the suprasylvian cortex and another on the dorsal hippocampus during a baseline period and following IV administration of 3.5 mg/kg of cocaine to conscious cats.[32] The images illustrate bands of enhanced and unchanged activity on the hippocampus, similar to the activation patterns found in the earlier study during REM sleep (Color Figure 4.7), and a rather muted response in the neocortex, with no pronounced

* Color figures follow page 112.

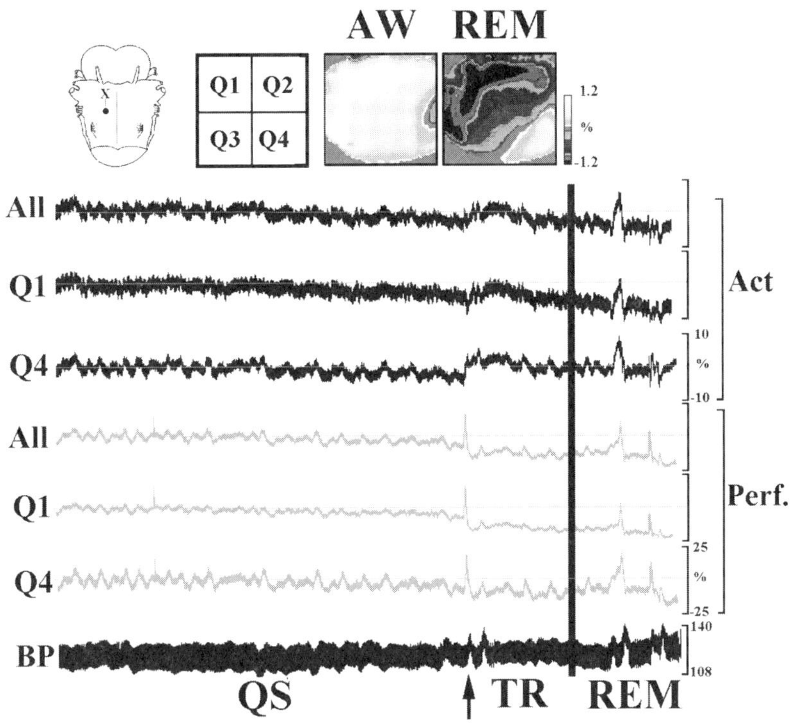

FIGURE 4.6 Traces of regional activity (Act) and hemodynamic activity (Perf) from quadrants 1 (Q1) and 4 (Q4) and overall activity from 4 quadrants (All) across QS, transition (TR; at arrow), and REM sleep states. Gray scale representation of activity in images of all four quadrants from a 5-sec epoch of AW and REM, subtracted from an epoch of QS, is shown above the traces. Regional cellular activation emerged between states, with one area significantly activated while another was significantly deactivated or unchanged. The traces indicate that the decline in activity during the REM state differs among quadrants, with Q4 showing much less decline over Q1. The probe is at X on the VMS outline. The spike during REM sleep indicates a phasic period, with restoration of VMS activity and perfusion. (From Richard, C.A. et al., *Am. J. Physiol.*, 277, R1239, 1999.)

columns of activation in the cortical area. The images were collected every 3 sec over a 90-min period. The technique is particularly useful in demonstrating highly organized functional structure in the hippocampus, contrasted with the relatively unstructured organization in the overlying cortex.

Activation of brain regions by pharmacologic agents has been examined by 2-deoxyglucose or other functional anatomic procedures.[33] Such procedures, however, can sample only one point in time and require sacrifice of the animal at that time, while imaging procedures allow moment-by-moment assessment of activity. Recordings can obviously be repeated on subsequent trials; activity can be monitored between trials and, if appropriate brain regions are sampled and repeated drug exposures are implemented, neural activity associated with addictive processes can be assessed.

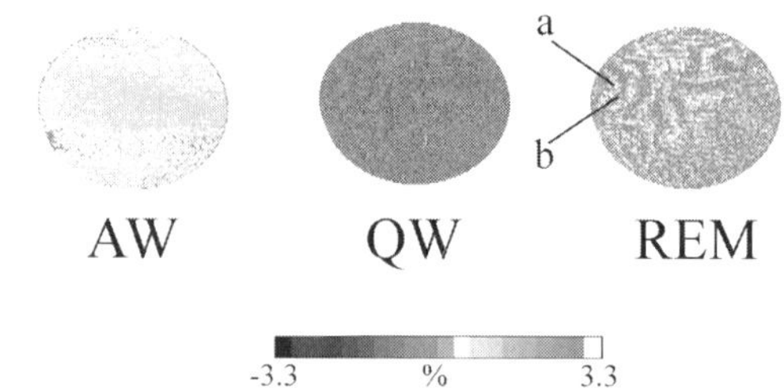

FIGURE 4.7 (See Color Figure 4.7.) Representative averaged (n = 40) images of reflectance, collected from the dorsal hippocampus of a cat at 3-sec intervals, during waking (AW), quiet waking (QW), and rapid eye movement (REM) sleep. The hippocampal image is 1.6 mm in diameter. Note the curved columns of enhanced (a) or diminished (b) activity occurring preferentially during REM sleep and the relative absence of such organization in AW. The color bar in the lower portion of the figure is the pseudocolor scale used to illustrate changes in activity. Yellow to red pixels indicate areas of increased activity (decreased reflectance), with white representing an activity increase of 3.3%. (Derived from Poe, G.R. et al., *J. Neurosci.,* 14(5), 2933, 1994.)

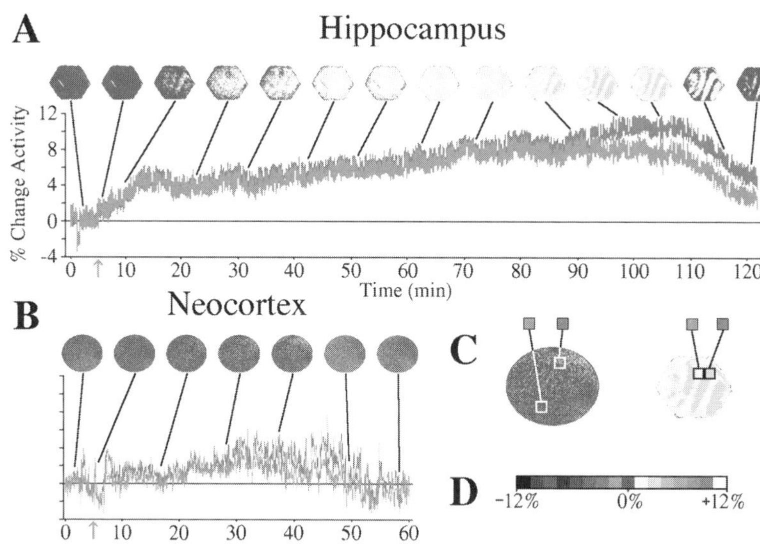

FIGURE 4.8 (See Color Figure 4.8.) Representative images of activity changes from the hippocampus (A) of a cat and from the neocortex (B) of another cat before and after (arrows) administration of 3.5 mg/kg of cocaine; traces of enhanced (red) and diminished (blue) activity are shown below the images. Each image is an average of 40 frames collected at 3-sec intervals and subtracted from a baseline pre-cocaine image. The images are pseudocolored, using the scale in D. In the hippocampal placement, banding of activity occurs (C, right), while the cortical site (C, left) shows no such organization and only modest activation. Images are derived from a 1 × 0.5 mm section from within the probe area. (From Rector, D.M. et al., *Neuroscience,* 54(3), 633, 1993.)

4.4 SOURCES OF REFLECTANCE CHANGES

4.4.1 CYTOARCHITECTONICS

The extent of scattered light and reflectance changes substantially depends on structural organization and type of stimulus. Studies which include examination of "spontaneous" and stimulus-evoked responses of the hippocampus and suprasylvian cortex uniformly show more marked reflectance changes in the hippocampus.[31] Although the biophysical sources of reflectance differences between different areas are unknown, several possibilities exist. The structure of the dorsal hippocampus consists of a highly organized layering of cells that differs from the less patterned arrangement in cortical regions; hippocampal cells in these layers are also more densely packed than cells in other brain areas.[13]

4.4.2 GLIA

A portion of the slower components of the imaging response likely derives from glial sources;[34] hippocampal and cortical regions may differ in structural aspects of glia. Those aspects likely will modify overall signal return and must be considered in any analyses.

4.5 POTENTIAL DEVELOPMENTS

4.5.1 ADVANCES IN CCD TECHNOLOGY

Chronic optical recording procedures have rapidly developed as technology has improved. Improvements have derived principally from advanced photocollection chips, with new CMOS technology offering great promise. The significant advances in computer-related image acquisition capabilities have been important; faster processors, availability of large mass storage devices, and improved framegrabbing and image-processing hardware have also been significant. CCD chips that allow selection and rapid scanning of individual elements are now available at very low cost, provide excellent dynamic range (70 dB), and are available in much smaller physical configurations than earlier chips. The new devices allow capture of 100×100 pixel images at >3 kHz rates, imposing substantial needs for data transfer and storage. However, the rapid image capture allows visualization of much faster neural changes than was available in the past.

4.5.2 DEPTH IMAGING

A current limitation of the technique is that light-scattering sources cannot be differentiated easily by depth. Illumination can reach to 0.5 mm below the tissue surface, as determined by control studies using the probe in a semiopaque fluid medium. However, a number of possible extensions to the procedure have the potential to illuminate tissue at defined depths or to minimize return from all but specific depths and, thus, to allow reconstruction of light-scattering images in the third dimension. Those issues are being explored.

Very short wavelength light can be used to visualize other physical aspects of cell discharge, such as protein conformation changes. However, short wavelengths require special consideration in selection of optical bandwidth characteristics of the imaging conduit and responsivity of the CCD elements; typically, optic fiber and CCD elements have reduced transmission and response sensitivity, respectively, at the short wavelengths necessary for imaging metabolic processes. Neurons can also be labeled prior to imaging, using dyes with particular absorption properties to visualize various activation parameters or to follow developmental processes in young animals.

4.6 SUMMARY

Neuroscientists have long needed techniques that could dynamically visualize activity over large areas for very long time periods comparable to that which *c-fos* or 2-deoxyglucose functional anatomic procedures provide in a one-time fashion. Although classic functional anatomic techniques have greatly assisted interpretation of neural interactions, such techniques visualize activity at only one instant in time rather than dynamically. Chronic optical techniques provide a means to visualize different aspects of activity associated with cell discharge; these aspects include changes in blood perfusion, cell membrane movement, protein activity, and glial changes. Physiologic processes occur at different rates, and these temporal characteristics may assist in separating different components of the collected signals. Reflectance imaging can record the spatial distribution of activity at very high rates, even as high as cell discharge rates, and thus can visualize high-resolution spatial and temporal characteristics of cell responses during unrestrained behavior in drug-free animals.

ACKNOWLEDGMENTS

Supported by HD-22506 and HL-22418. A number of investigators, Drs. G. Aljadeff, J. Carroll, X-W. Dong, H. Forster, D. Gozal, and P. Ohtake, also participated in these studies.

REFERENCES

1. Evarts, E.V., Temporal patterns of discharge of pyramidal tract neurons during sleep and waking in the monkey, *J. Neurophysiol.,* 27, 152, 1964.
2. Strumwasser, F., Long-term recording from single neurons in brain of unrestrained mammals, *Science,* 127, 469, 1958.
3. Harper, R.M. and McGinty, D.J., A technique for recording single neurons from unrestrained animals, in *Brain Unit Activity and Behavior*, Phillips, M.I., Ed., Charles C. Thomas, Springfield, IL, 1973, 80.
4. O'Keefe, J. and Recce, M.L., Phase relationship between hippocampal place units and the EEG theta rhythm, *Hippocampus,* 3, 317, 1993.

5. Wilson, M.A. and McNaughton, B.L., Dynamics of the hippocampal ensemble code for space, *Science,* 261(5124), 1005, 1993.
6. McGinty, D.J. and Harper, R.M., Dorsal raphe neurons: depression of firing during sleep in cats, *Brain Res.*, 101, 569, 1976.
7. Lindsey, B.G., Shannon, R., and Gerstein, G.L., Gravitational representation of simultaneously recorded brainstem respiratory neuron spike trains, *Brain Res.*, 483, 373, 1989.
8. Grinvald, A. et al., Functional architecture of cortex revealed by optical imaging of intrinsic signals, *Nature,* 324, 361, 1986.
9. Mann, G., Histological changes induced in sympathetic, motor, and sensory nerve cells by functional activity, *J. Anat. Physiol.*, 29, 101, 1894.
10. Tasaki, I. and Iwasa, K., Rapid pressure changes and surface displacements in the squid giant axon associated with production of action potentials, *Jpn. J. Physiol.,* 32, 505, 1982.
11. Cohen, L.B., Changes in neuron structure during action potential propagation and synaptic transmission, *Physiol. Rev.,* 53, 373, 1973.
12. MacVicar, B.A. and Hochman, D., Imaging of synaptically evoked intrinsic optical signals in hippocampal slices, *J. Neurosci.,* 11, 1458, 1991.
13. McBain, C.J., Traynelis, S.F., and Dingledine, R., Regional variation of extracellular space in the hippocampus, *Science*, 249, 674, 1990.
14. Poe, G.R. et al., Concurrent reflectance imaging and microdialysis in the freely behaving cat, *J. Neurosci. Meth.,* 65, 143, 1996.
15. Frostig, R.D. et al., Cortical functional architecture and local coupling between neuronal activity and the microcirculation revealed by *in vivo* high-resolution optical imaging of intrinsic signals, *Proc. Natl. Acad. Sci. USA*, 87, 6082, 1990.
16. Masino, S. et. al., Characterization of functional organization within rat barrel cortex using intrinsic signal optical imaging through a thinned skull, *Proc. Natl. Acad. Sci. USA,* 90, 9998, 1993.
17. Rector, D. and Harper, R.M., Imaging of hippocampal neural activity in freely behaving animals, *Behav. Brain Res.,* 42, 143, 1991.
18. Rector, D.M. et al., A miniature CCD video camera for high-sensitivity measurements in freely behaving animals, *J. Neurosci. Meth.,* 78, 85, 1997.
19. Rector, D.M. et al., Scattered light imaging *in-vivo* tracks fast and slow processes of neurophysiological activation, *NeuroImage*, 14, 977, 2001.
20. Kristensen, M.P. et al., State dependent cellular activity patterns of the cat paraventricular hypothalamus measured by reflectance imaging, *Brain Res.,* 727, 107, 1996.
21. Holthoff, K. and Witte, O.W., Intrinsic optical signals in rat neocortical slices measured with near-infrared dark-field microscopy reveal changes in extracellular space, *J. Neurosci.*, 16(8), 2740, 1996.
22. Richard, C.A. et al., Optical imaging of the ventral medullary surface across sleep–wake states, *Am. J. Physiol.,* 277, R1239, 1999.
23. Rector, D.M. et al., Ventral medullary surface activity during sleep, waking and anesthetic states in the goat, *Am. J. Physiol.,* 267, R1154, 1994.
24. Harper, R.M. et al., Imaging of VMS activity during blood pressure challenges in awake and anesthetized goats, *Am. J. Physiol.,* 270, R182, 1996.
25. Forster, H.V. et al., Ventral medullary surface activity during hypoxia in awake and anesthetized goats, *Respir. Physiol.,* 103, 45, 1996.
26. Richard, C.A. et al., Changes in ventral medullary light reflectance during hypercapnia in awake and sleeping cats, *Neurosci. Lett.,* 286, 175, 2000.

27. Harper, R.M. et al., Pressor-induced responses of the cat ventral medullary surface, *Am. J. Physiol.,* 268, R324, 1995.
28. Harper, R.M., Richard, C.A., and Rector, D.M., Physiological and ventral medullary surface activity during hypovolemia, *Neuroscience,* 94(2), 579, 1999.
29. Rector, D.M. et al., Sleep states alter ventral medullary surface responses to blood pressure challenges, *Am. J. Physiol.,* 278, R1090, 2000.
30. Harper, R.M., Frequency changes in hippocampal electrical activity during movement and tonic immobility, *Physiol. Behav.,* 7, 55, 1971.
31. Poe, G.R., Rector, D.M., and Harper, R.M., Hippocampal reflected optical patterns during sleep and waking states in the freely behaving cat, *J. Neurosci.,* 14(5), 2933, 1994.
32. Rector, D.M., Poe, G.R., and Harper, R.M., Imaging of hippocampal and neocortical neural activity following intravenous cocaine administration in freely behaving animals, *Neuroscience*, 54(3), 633, 1993.
33. Clow, D.W. and Hammer, R.P., Cocaine abstinence following chronic treatment alters cerebral metabolism in dopaminergic reward regions. Bromocriptine enhances recovery, *Neuropsychopharmacology,* 4(1), 71, 1991.
34. Andrew, R.D. and MacVicar, B.A., Imaging cell volume changes and neuronal excitation in the hippocampal slice, *Neuroscience*, 62, 371, 1994.

5 *In Vivo* Observations of Rapid Scattered-Light Changes Associated with Electrical Events

David M. Rector, Ronald M. Harper, and John S. George

CONTENTS

0-8493-2389-4/02/$0.00+$1.50
© 2002 by CRC Press LLC

5.1 INTRODUCTION

In the last decade, it has become increasingly important to record simultaneously from large neural populations to assess how neurons within a network interact with each other to perform complex tasks. Theoretical arguments and experimental observations have suggested that correlated firing across many individual neurons may encode relationships within the data stream,[1] and recent studies have found significant information in the synchrony between local neural populations.[2,3] Such work has demonstrated that it is important to know not only when a neuron fires but also how such discharge occurs in relation to activity in other cells.

Most procedures for assessing activity of many neurons involve the use of multiple-electrode arrays. The complexity of these arrays has grown rapidly from a few electrodes to 100 or more electrodes in close proximity. Although electrode arrays provide excellent temporal resolution of neural activity, spatial resolution and sampling density are limited and invasive electrodes have the potential to damage tissue. Moreover, single-unit recordings by microelectrode arrays can be biased by preferential sampling of large neurons.

Several investigators have used changes in reflected light from cortical surface to assess neural activity patterns based on changes in blood flow and oxygenation. Sensitive optical techniques have demonstrated spatial organization of visual cortex columnar structures in a fashion that complements electrophysiological recording.[4-6] Spatial patterns of sensory activation in human temporal cortex[7] and rodent sensory cortex[8,9] have also been visualized. Such methods are most useful for revealing static or slowly evolving functional organization, since hemodynamic responses occur on time scales of hundreds of milliseconds to seconds.

Although optical measurements can temporally resolve the submillisecond dynamics of action potentials,[10,11] such measurements have typically employed single-channel detectors for speed and sensitivity. Even when fast signals are enhanced through the use of voltage-sensitive dyes, most investigators have used limited photodiode arrays consisting of a few dozen detectors. We have recently demonstrated the feasibility of imaging fast optical signals associated with neural activity by using solid-state imagers such as CCDs. To date, our measurements have been limited to evoked activity of neural populations acting in synchrony; optical changes from a single neuron are too minute to detect within a matrix of several thousand other neurons and tissue with present optical and imaging technology.

Our current studies involve tissue illumination with light of specific wavelengths while scattered light is collected by a coherent fiber optic image conduit or a gradient index (GRIN) lens and conveyed to a charged coupled device (CCD) camera. The present technology allows continuous long-term measurements from a two-dimensional tissue surface, with image capture rates up to 500 Hz. Some versions of our imager allow recording from deep brain structures in freely behaving animals without disrupting normal behavior. Such techniques are essential to assess the role of the brain in spontaneous-state-related and motor behaviors.

We have performed a number of physiological experiments to study the nature of light-scattering changes *in vivo* and to investigate brain functioning in acute preparations and in freely behaving animals. Our early studies focused primarily on relatively slow detection of optical signals and correlation of these changes with manipulations of physiological activation state. Our initial image-acquisition system was slow, capable of 1- to 30-Hz frame rates, which limited the time resolution of the recordings. Even though fast signals were presumably present, it was difficult to determine the time course of these signals. Indeed, in a stimulus and response paradigm, we usually observed a small deflection of the optical signal in the next frame after the stimulus. However, the signal eventually grew larger over time, suggesting the presence of slow signals as well. More recent studies have explored fast dynamics of the optical signals and have begun to explore the physiological basis of the responses.

Images of optical changes from the dorsal hippocampus and ventral medulla showed substantial regional patterns in response to physiologic or state manipulations corresponding to neural activation of these structures. These experiments demonstrated consistent changes after repeated stimulation or state alterations. The procedure allowed assessment of activity components difficult to measure or inaccessible with standard microelectrode techniques in freely behaving animals. Relationships between light-scattering changes and neural activation were established by analyzing reflectance changes during synchronous "spontaneous" and evoked electrical activity, pharmacologically induced activity, and spontaneous state changes.[12-14] Synchronous activation produced detectable light-scattering changes at similar frequencies to those observed in concurrent electroencephalographic recording.

We believe that a large portion of the signals obtained during these studies is due to changes in light scattering with some absorbance component. Since the wavelength of illumination typically used is 660 nm or longer, absorbance by hemoglobin is low. Indeed, through the use of spectral component modeling, Malonek and Grinvald[15] claim that the contribution of light scattering to the mapping components was larger than 70% at longer wavelengths. Additionally, the use of illumination around the perimeter of the image probe forces the light to enter the tissue and be scattered before returning to the detector. Thus, scattering events play a more prominent role in changes that we see, especially since the vasculature (the locus of the dominant absorbance changes) is located primarily on the surface of the tissue.

Earlier reports of fast optical signals associated with neural activity described scattering changes and birefringence changes.[10,11] Since we did not use polarized light in these experiments, we expect no contribution from birefringence. The best evidence for scattering would be direct measurement of a change in the angular distribution. Demonstration of a flat spectral dependence across a polarity change in the Hgb oxygenation difference spectrum (e.g., 780 to 820 nm) would also be a strong indication.

Many issues in neural interaction require temporal resolution in excess of those at electroencephalographic changes; our recent efforts have been directed toward assessment of faster neural changes by optical means. We found that Schaeffer's collateral stimulation, *in vivo,* activated hippocampal cell populations and produced light-scattering changes concomitant with evoked electrical responses. These fast optical changes have been imaged and further characterized in recent studies of the rat dorsal medulla.

Recordings from isolated crayfish nerve have been useful in optimizing scattered-light changes resulting from volleys of action potentials. These signals can be seen in single trials and show optical changes that occur on the submillisecond time scale, comparable to ionic flux across the neural membrane. Such studies with isolated nerve are ongoing in our laboratory to investigate the biophysical mechanisms of fast scattered-light changes associated with membrane potentials.

5.2 MECHANISMS OF OPTICAL RESPONSES

A number of classes of optical changes have been observed during neural activation. Absorption, refraction, birefringence, transmission, and reflectance processes can be used to measure changes in scattered light during neural discharge. The largest proportion of studies examines changes in blood volume and oxygenation associated with neural activation. Because oxygenated hemoglobin molecules primarily absorb green light, hemoglobin concentration will influence green light reflectance; red light, which is absorbed less efficiently, will be modified to a lesser extent (500 to 1000 times lower as determined from standard hemoglobin absorbance curves[16]) by this mechanism. Hemoglobin also mediates absorption changes in the near infrared, in principle allowing deeper measurements with diffuse light techniques.

5.2.1 Scattering and Birefringence

Light is transmitted through neural tissue with varying degrees of scattering. A process that increases light transmittance will result in decreased reflected light. Increased transmission (for example, by reduced back scattering) occurs when light-scattering particles become more dilute in a medium, as may be the case with cell swelling. Another potential mechanism involves reflectance changes. A reflective surface that alters its orientation will change the angle at which it reflects light, resulting in more or less light return. However, a systematic characterization of mechanisms that influence light scattering in response to neural activity remains to be accomplished.

Birefringence involves a change in the rotation of polarized light — a process associated with conformational changes in protein structure — and changes in the spacing of closely packed membranes. Many optical changes during tissue activation result from changes in polarized light.[10,11] Large protein molecules change conformation in response to membrane potential differences. For example, the voltage-activated sodium channel undergoes a dramatic twist in response to depolarization; the degree of this twist has been indexed by polarized light rotation.[17,18] Strong birefringence is also observed in multilayer lipid vesicles. It is likely that changes in the microscopic architecture or packing of cellular membranes or intracellular organelles might also contribute to a birefringence signal.

5.2.2 Absorption

Important molecules for tissue light absorption include hemoglobin, water, cytochromes, amino acids, lipids, nucleic acids, and sugars. Absorption measurements have been used to assess quantitative concentrations of oxygenated and deoxygenated hemoglobin. Additionally, the local blood volume of neural tissue increases with activation.[6,15,19-23] Typically, red light is used to visualize hemodynamic changes because tissue is more transparent to red light. Green light (515 to 530 nm) absorption can also be applied for determining local activation and is particularly useful for visualizing vasculature.

A disadvantage of measuring hemoglobin concentration changes is that the spatial resolution is limited by microvasculature and the temporal resolution is limited by vessel diameter fluctuation rates and hemoglobin deoxygenation. Cytochromes capture blue to ultraviolet light in conjunction with increased metabolism and rapidly alter their optical properties with cellular activation. Such absorption changes can be measured with single cell resolution; however, the relatively low efficiency of blue light detection requires high, and potentially damaging, illumination levels.

5.2.3 Cellular Swelling

Hill[24,25] reported decreased opacity of squid giant axon with action potential generation. After offset adjustment to background illumination transmitted through a nerve fiber, relative changes as large as 10% of residual background were observed with repeated stimulation. Increased axon volume in conjunction with opacity changes suggested that increased transmission resulted from dilution of intracellular contents. MacVicar and Hochman[26] examined white light transmission through a blood-free, *in vitro* hippocampal slice following electrical stimulation of Schaeffer's collaterals and found increased local broad spectrum light transmission through the tissue. When anion channels were blocked with furosemide, postsynaptic potentials were elicited, but optical changes were not observed, suggesting that transient changes in optical properties of the slice may result from volume changes associated with cellular swelling.

Within *in vivo* preparations, component analysis of spectral data reveals a wavelength-independent process that suggests light-scattering changes with neural activation.[15] Intracellular and secondary extracellular volume changes provide a basis for relating increased neural discharge to diminished reflectance via the cellular swelling hypothesis.[27] Volume changes modify tissue density, thereby changing the refractive index and reflective properties of tissue. Changes in neural activity in the CA1 region of the hippocampal slice are inversely related to changes in the extracellular volume fraction.[28] Additionally, increased synchronous firing accompanies shrinkage of extracellular space, even when synaptic transmission is blocked.[29-32]

Lipton provided evidence that activation of a blood-free cortical slice preparation produced decreased light reflected from the tissue.[33] Grinvald and colleagues[4] observed 0.1% absolute light reflectance decreases in 665- to 750-nm reflectance of the cat visual cortex in response to visual stimulation. In these experiments, functional maps of the visual cortex were obtained by illuminating the cortical surface and imaging reflectance with a high-sensitivity, slow-scan CCD camera.

5.2.4 Temporal Resolution of Light-Scattering Changes

Early studies using reflected light to map visual cortex activation patterns often required stimulation for many seconds to several minutes before changes could be measured.[4,34] The optical response time course depends largely on illumination wavelength and sensitivity to various physical mechanisms for optical changes. Optical changes occurring rapidly enough to follow the action potential typically require voltage-sensitive dyes or polarized light with high-sensitivity photodiodes.[10] Our studies suggest that, with sufficient sensitivity, light-scattering changes associated with neural activation can be measured and imaged.[35,36] With sufficient temporal resolution, it may be possible to make sophisticated measurements of time-correlated network behaviors within neural populations. However, achieving sufficient temporal resolution for resolving neurophysiological processes remains a major challenge for imaging studies.

5.2.5 Cell Types Contributing to Optical Responses

The cellular origin of light-scattering changes has generated much controversy. Hill[24,25] showed that increased giant squid axon volume corresponds to increased transmitted light. However, neural tissue contains glial cells, which constitute a major proportion of tissue volume. Glial cells also swell or shrink with changes in extracellular ion concentration; dendrites swell in response to activation.[37] Thus, at least three possible volume-related contributors to light-scattering changes exist: increases in neuronal cell body, dendrite size, and glial size. Cohen[10] suggested that compact clusters of small axons, such as in the crab nerve bundle, produce larger changes in reflectance than a single large axon, such as the squid giant axon. Our own experience suggests that densely packed tissues, such as gray matter, provide larger fractional signals than isolated nerves.

Since proteins and lipids constitute a large percentage of dry cell weight, these components probably contribute most of the scattering changes. A structure that possesses more scattering elements (such as dense dendrite or axon clusters, mitochondria, or other intracellular organelles) would be expected to exhibit a greater change in scattering. When measuring light scattering from brain tissue *in vivo*, both small processes and large cell body diameters contribute to the local changes observed with the probe; however, high-density components may contribute more to changes than large cell bodies. Indeed, MacVicar and Hochman[26] found a larger degree of light-scattering changes in the dendritic layers of the hippocampus; these also contain many small interneurons. Glial cells may also contribute to the spatially organized changes associated with neuronal discharge because they typically respond to osmotic changes in the local microenvironment. However, the time course for glial swelling is probably significantly slower than neural swelling rates.

5.2.6 Peripheral vs. Direct Illumination

Reflectance studies that use a lens-coupled system illuminate the tissue surface and detect light from a distance. High light levels are required, and most of the illumination detected by the camera consists of specular reflection from the tissue surface or scattering from superficial tissue. Only a small proportion of the light detected in this configuration interacts with tissue beneath the surface of the cortex. Most investigators employ high-sensitivity, cooled camera equipment with slow scan rates (1 to 10 fps) for detection of such small changes.

We developed another method of detecting scattered light using fiber optic image conduit or a gradient index lens. The image conduit is in direct contact with the tissue and the camera. Light is delivered to the tissue around the perimeter of the probe as in "dark-field" illumination. Thus, before entering the image conduit fibers, light must be scattered by objects below the tissue surface. The use of fiber optics in reflectance imaging greatly increases the detection sensitivity for light-scattering changes over lens-coupled devices. Because of improved collection efficiency, this technique does not require high illumination levels, which allows use of miniature LEDs. The geometry of light paths from illumination fibers to detection fibers is such that a large portion of reflected light is perturbed by cells that change their scattering properties with activation. Such procedures greatly enhance the contrast of tissue activity changes.

Imaging systems incorporating a gradient index lens are also placed in contact with the tissue surface, thus eliminating a number of optical artifacts that may arise from mechanical motion of the tissue surface. These systems may incorporate optics that allow focusing into tissue beyond the end of the imaging probe. This feature, along with optical geometry, facilitates advanced imaging strategies such as confocal microscopy. However, the GRIN lens is much less efficient at capturing and transmitting light than fiber optic image conduit, so considerably higher illumination levels are required.

5.3 METHODS

Our imaging and data acquisition system is optimized for spatial and temporal resolution and sensitivity for detecting small changes in scattered light. The optical system utilizes perimeter illumination to mimic dark-field microscopy to optimize contrast for scattered-light changes. Components directly in contact with the tissue reduce mechanical artifacts from blood pressure and respiration cycles. The optical system also provides the ability to form confocal images for better depth discrimination. The imaging system consists of one of several high-speed CCD sensors with good signal to noise (66 dB or better), coupled with custom scanning and digitizing circuitry.

5.3.1 Imager Design

Fiber optic perimeter illumination with image conduit or a GRIN lens provides an efficient arrangement for imaging. The device uses a straight piece of image conduit with illumination fibers arranged around the image conduit perimeter[38] (Figure 5.1). The arrangement precludes direct illumination; however, our studies showed that dark-field illumination doubled the field depth of view (to 0.600 mm) over direct illumination methods[39] and increased the magnitude of light-scattering signals.[36,40]

Optical techniques using coherent fiber optic image conduit[38,39,41] formed good images of the tissue surface. Also, since illumination surrounded the imaged area, dark-field methods eliminated specular reflectance and provided scattering information from deeper tissue. However, light from deeper structures was out of focus and the image was blurred. Because cells of interest are frequently located several

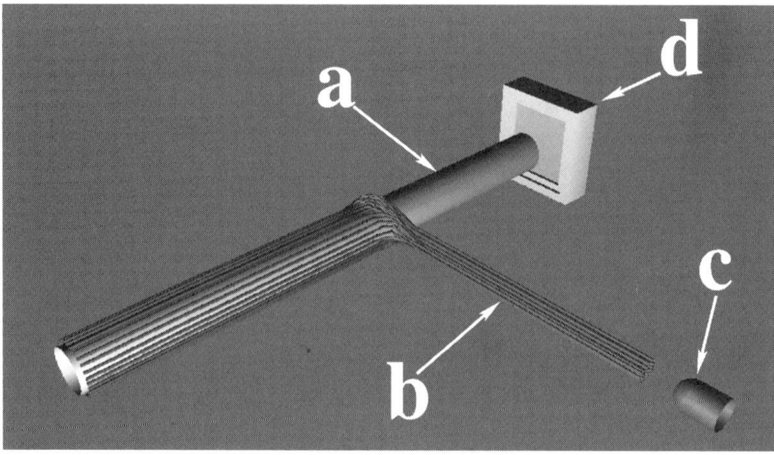

FIGURE 5.1 A three-dimensional rendering of the image conduit probe illustrates a coherent fiber image conduit (a) with plastic illumination fibers (b) around the perimeter of the image conduit and provides dark-field illumination from a light source (c), typically a light-emitting diode. Back-scattered light from the neural tissue is collected by the image conduit and an image from the neural surface is transmitted to a CCD camera (d).

hundred microns below the surface, it is desirable to focus below the surface to measure deep structures accurately.

We have recently developed an advanced imaging system (Figure 5.2) utilizing a gradient index lens, which serves as a relay lens from tissue and provides firm tissue contact to minimize movement and specular reflectance. A microscope objective projects an image onto a high-speed, miniature CCD camera.[36] The microscope objective and camera assembly are adjusted independently of the GRIN lens, allowing focus depth changes without disturbing the probe and tissue interface. Fiber optics around the probe perimeter provide diffuse illumination from a 780-nm laser, or 660- and 560-nm light-emitting diodes. The device has the potential for miniaturization to be chronically implanted in freely behaving animals. Related prototype devices developed can generate confocal and spectral images.

5.3.2 Depth-of-View Measurements

We tested the depth of field of the image probes by imaging a black suture (0.2-mm diameter) through successive tissue depths, measuring changes in intensity and sharpness of the suture image as a function of depth.[36] Figure 5.3 illustrates average suture images and intensity profiles across tissue thicknesses for 660- and 780-nm illumination. Plotted intensity profiles show a change in apparent shape as the suture image becomes blurred through increasing tissue thickness. Figure 5.3 illustrates two parameters of the model used to characterize the suture blurring. The amplitude and width parameters estimated from fits to the Gaussian used to model the blurring are plotted as a function of tissue depth. The suture disappeared from view (determined by amplitude parameter <0.1 and steep width parameter curve) after 0.720 mm using 660-nm illumination and after 0.900 mm using 780-nm illumination. Figure

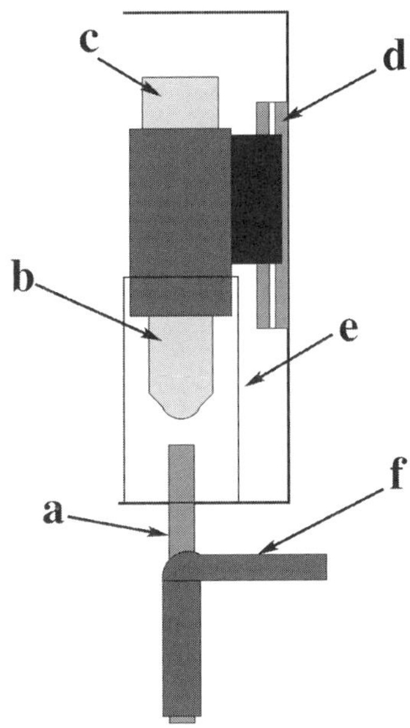

FIGURE 5.2 A schematic illustration of a focusing endoscope shows the gradient index lens (a) which projects an image from the neural surface through a microscope objective (b) onto a CCD camera (c). The point of focus is adjusted by a hydraulically positioned linear bearing stage (d) which slides the microscope objective or camera assembly through a centering tube (e). Illumination is provided through flexible plastic optical fibers (f) placed around the perimeter of the gradient index lens.

5.3 also compares data from the image conduit version of the probe analyzed in the same manner. Using 660-nm reflectance mode illumination and a macro lens, the suture is obscured after 0.300-mm tissue thickness. Surround illumination (dark-field) and image conduit increased visibility to 0.600 mm.

5.3.3 Digitizing Hardware

We built a custom device that digitizes video and multiple analog signals simultaneously at 12-bit resolution.[42,43] A computer system displays intensity histograms as well as raw or averaged images and dynamic average pixel intensity over the image or specified subregion, along with a strip-chart style display of physiological data. In our experience, continuous, real-time feedback of physiological and image data is extremely useful for consistent experimental success.

To increase sensitivity for detection of very small changes in light intensity, we employ techniques for increasing the effective dynamic range of the video-digitizing system. One method involves building video amplifiers that allow judicious selection of black level and gain. The black level was first set so that a digital value of 16 represented the dimmest pixel in the image, and a value of 240 represented the brightest pixel in the image. This procedure increases the ability to detect intensity changes by a factor of 6.7. One disadvantage is that it requires an extra calibration

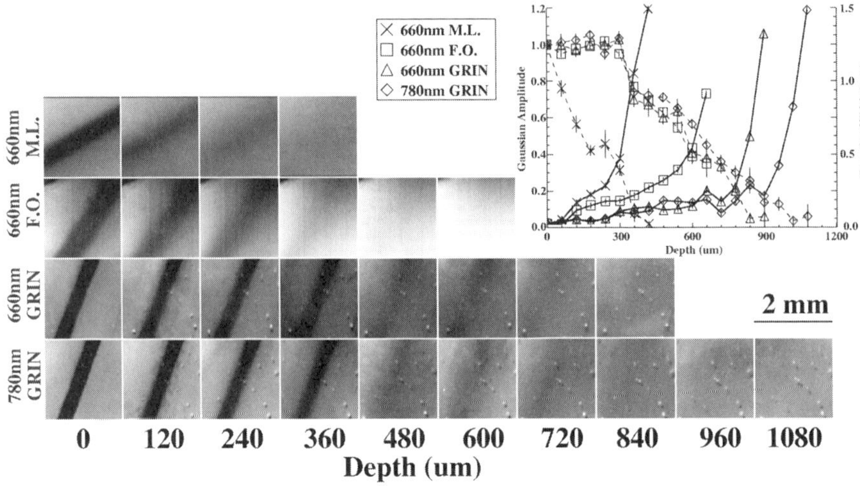

FIGURE 5.3 A black suture, 200 μm in diameter, was placed on a block of rat cortical tissue and imaged with four methods. Images were gathered though successive tissue thicknesses placed over the suture. For each depth measurement, the focus of the microscope lens was adjusted to bring the plane of focus onto the suture. Four sets of images were collected using 660- or 780-nm illumination with a remote macrolens (M.L.) with remote illumination, a fiber optic image conduit with perimeter illumination (F.O.), or a gradient index (GRIN) lens with perimeter illumination. Intensity profiles (representing the black suture cross section at each tissue thickness) were modeled by convolving the profile with a Gaussian curve, fitting amplitude and width parameters and plotted across depth. The thick lines represent Gaussian amplitude as a function of tissue thickness for four illumination types. The thin dashed lines represent the Gaussian width; vertical lines represent the SEM.

to recover absolute intensity values. A second, and more difficult, method for increasing dynamic range involves digitizing a reference video frame, storing that frame in memory, and using a differential video amplifier to subtract the reference frame from incoming frames. Such a procedure introduces many potential noise sources, although commercial systems based on this strategy are available.

The current digitizing system is built on a PCI interface card for high-speed data transfers. It contains 8 million bytes of double buffered memory for synchronous acquisition and download capability and uses 7-ns programmable logic devices (PLDs) to execute the various control functions for digitizing and CCD control. The acquisition hardware gains flexibility through field programmable arrays (FPGAs) and embedded digital signal processors (DSPs). The PCI interface allows 132 MB/sec burst and 66 MB/sec continuous data streaming to the host computer for archive, online display, and online analysis.[43]

This system has performed well for rapid and flexible video and electrophysiological acquisition. Data are stored in the on-board memory in a standard file format IFFPHYS, an internationally recognized format for physiological data. The PCI card plugs into any standard Intel motherboard. Simple drivers were written for a UNIX operating system in order to control the card and download buffered data. We have

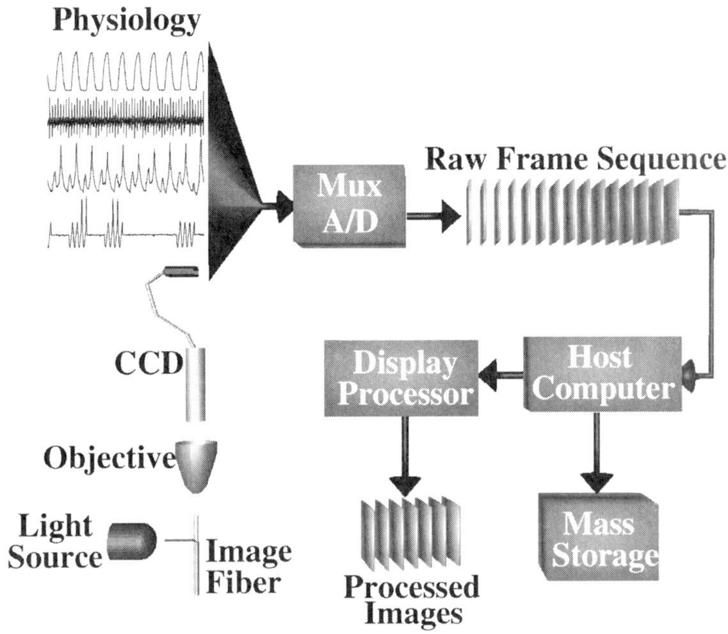

FIGURE 5.4 A diagram of system components shows a light source projecting light into flexible fiber optics mounted around the perimeter of a coherent fiber optic image conduit or gradient index lens. The imaging device was placed in contact with the neural structure of interest, which collected back-scattered light and formed an image through a microscope objective onto a CCD camera. A change in the focal plane was achieved by moving the CCD camera and objective relative to the image fiber, which remained fixed in position. Video frames from the CCD camera and 16 channels of electrophysiology were multiplexed and digitized continuously during the experiment. A Pentium host computer collected the raw frame sequences and streamed them continuously onto a mass storage device. The host computer also searched image sequences for stimulus trigger events, and epochs of 90 frames were averaged and transferred to a display processor via a fast ethernet connection for online analysis and visualization.

recently undertaken a redesign of this system to employ newer and higher performance technology.

In addition to real-time display during acquisition, our system performs analyses that recall a series of image frames in sequence, define image groups of interest, and produce averaged group images. To create animated sequences, the software performs sequential, time-triggered subtraction, ratio, and calibration calculations (Figure 5.4). Analysis includes frame-by-frame standard deviation and FFT computations.

5.4 BRAIN IMAGING IN INTACT ANIMALS

We have performed numerous studies of light-scattering changes within freely behaving animals, as well as on several acute preparations. In all studies, we have observed changes in back-scattered light corresponding to increased neural activity. Our procedures employ the image probe system described earlier to acquire data continuously to mass storage and to provide real-time display image processing.

5.4.1 Correlations with Evoked Potentials

Acquisition of scattered light at 2.4 kHz from a single point on the dorsal hippocampal surface with a high-sensitivity photodiode revealed a strong inverse correlation between evoked potentials and light-scattering changes in the hippocampus.[35] Stimulation produced a complex electrical response with at least two components: an early population spike within 10 msec of the stimulus artifact and a population postsynaptic potential within 50 msec. The light-scattering signal showed an inverse relationship to the electrical signal with at least three components: 1) a large, slow response corresponded to the population postsynaptic potential, 2) a very slow and long lasting response lasted several seconds and may result from vascular processes, 3) a small, fast response, the onset of which preceded the early population spike by 5 msec (Color Figure 5.5B).* The amplitude of the evoked light-scattering change was dependent on stimulus position corresponding to a laminar spatial representation of contralateral projections of the Schaeffer's collaterals.[44] When the photodiode was replaced with a CCD camera and individual images were captured at the peak of the evoked response, a topographical pattern of activation could be recorded with the camera when different regions of the dorsal hippocampus were stimulated in an 8×2 mm square region on the contralateral side. Some stimulating positions produced strong optical responses with significant spatial structure (Color Figure 5.5C), and other positions produced little response (Color Figure 5.5D).

5.4.2 Dorsal Medulla Recording from the Rat

We investigated fast optical responses to vagal nerve stimulation within the nucleus tractus solitarius of the rat dorsal medulla.[36,45] These experiments utilized the GRIN-based imager to record central neural responses to peripheral stimulation at very fast acquisition rates and at variable tissue depths. Vagal nerve stimulation elicited response patterns, detected as changes in back-scattered light. Regional responses disclosed four distinct time courses with components that paralleled fast electrical evoked responses as well as slower hemodynamic signals. Additionally, different response components could be optimally recorded at different depths.

Back-scattered light images were continuously digitized (100 fps) during vagal stimulation (400 mA, 0.1-msec pulse, 1- to 2-sec random interval, 400 pulses), together with blood pressure, EKG, tracheal pressure, and field potentials from a macrowire inserted under the probe. To determine changes in light scattering across time after the stimulus, images from each sequence were averaged on a pixel-by-pixel basis and divided by the average of the prestimulus baseline. Nerve shocks to the vagal bundle elicited an electrical population spike 30 msec after the stimulus artifact and a population-evoked potential with a peak 80 msec after the stimulus

* Color figures follow page 112.

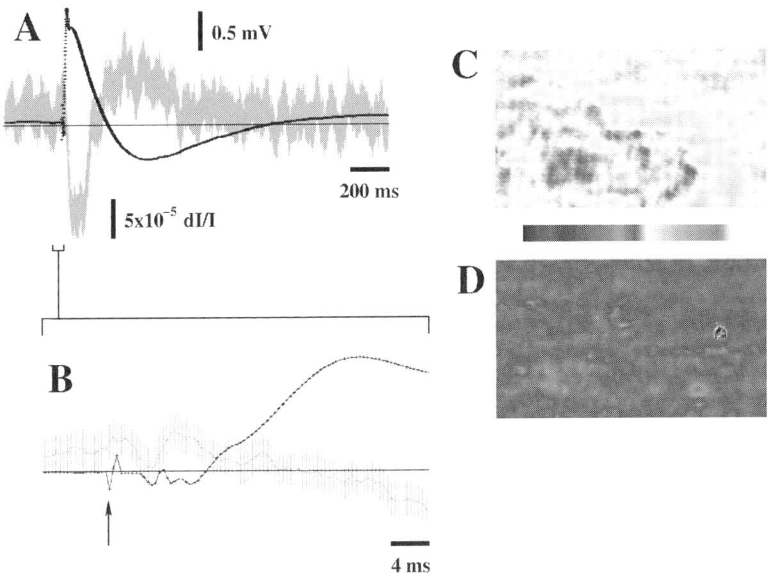

FIGURE 5.5 (See Color Figure 5.5.) A fiber optic probe with perimeter illumination was placed on the surface of the dorsal hippocampus of a cat. Back-scattered light was collected with a photodiode at 2.4 kHz or a CCD camera at 100 Hz during stimulation of the contralateral hippocampus. Stimulation elicited evoked electrical responses (dark lines, Panels A and B) and concurrent optical responses. Panel B shows an expanded view of the electrical and optical response around the stimulus artifact (arrow). Images of peak optical changes from the dorsal hippocampus surface reveal significant spatial patterns during stimulation of the contralateral structure that projects directly with a laminar organization to the area under the image probe (Panel C). Stimulation 1 mm away from the optimal site shows little optical response (Panel D). Images are pseudocolored such that decreases in light intensity (increased activation) are colored with warm colors (yellow to red), increases in light intensity (decreased activation) are colored with cool colors (blue to purple), and no significant change from baseline is colored green.

artifact (Figure 5.6, ERP). Concombinant imaging of back-scattered 780-nm light showed a significant temporal and spatial pattern of activation within the tissue, with four distinct components (Figure 5.6, 401VN): an early negative response (Figure 5.6, P30) lasting 20 msec; an intermediate positive response (Figure 5.6, N80), which peaked 80 msec after the stimulus; a late and long-lasting positive response (Figure 5.6, N300); and a slow negative response lasting at least 800 msec (Figure 5.6, P800). Different vagal stimulus–related signals across depth were also observed; the N80 signal increased in amplitude and the P300 signal decreased in amplitude in deeper tissue layers across all animals recorded ($p < .01$, $n = 9$, ANOVA).

Figure 5.7 shows the progression of the optical changes across a typical averaged image sequence. The optical response was modeled by four component processes illustrated in Figure 5.6 that paralleled four putative tissue mechanisms for optical changes. The P30 response was modeled after the timing of the population action potential as recorded by the macroelectrode placed under the imaged region; the

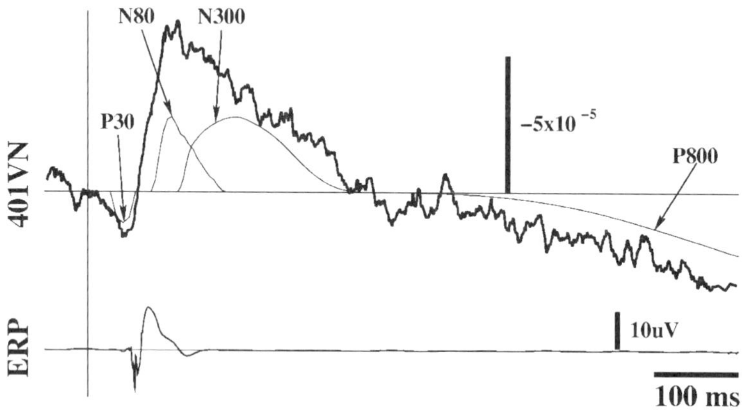

FIGURE 5.6 Average image intensity change of the NTS during stimulation of the vagus nerve (401VN, thick line) shows a typical optical response over a period of 800 msec after the stimulus onset (vertical line). The lower trace (ERP) represents the electrical evoked response as recorded by a macrowire placed under the imaged area of the NTS. Each trace represents an average of 2000 trials. The optical response was modeled using four component processes (thin lines under optical response). The P30 component corresponded to the population action potential, the N80 component corresponded to the population postsynaptic potentials, and the N300 and P800 components were modeled after hemodynamic responses.

N80 response was modeled after the electrical population postsynaptic potentials. The N300 component was modeled to match putative initial deoxygenation of the tissue due to the sudden metabolic demand, and the P800 response was modeled after a subsequent increase in oxygenated hemoglobin produced by an overcompensatory increase in blood flow to the activated region.

The image sequence was decomposed into component images of the four temporal components (basis functions). These images represent the relative contribution of one of the four components to the observed optical response. The two faster responses (Figure 5.7, P30, N80) were limited in their spatial distribution and may represent processes directly associated with neural activation. The two slower components (Figure 5.7, N300, P800) were more widespread and overlapped with regions showing distinct patterns in the blood pressure correlation images. The time course of the slower components matched the hemodynamic mapping responses reported in earlier studies,[6] while the faster responses probably represent direct neural response components not seen in previous studies. The spatial distribution of optical signals is comparable to early evoked potential results obtained by Humphrey (cat carotid sinus nerve (CSN)),[46] Seller and Illert (cat CSN),[47] and Gabriel and Seller (cat CSN and AN).[48]

5.5 FAST SCATTERED-LIGHT CHANGES IN ISOLATED NERVES

Large nerves from invertebrates, e.g., lobster leg nerve, crab nerve, and squid giant axon, provide an excellent means to evaluate scattered-light changes that parallel

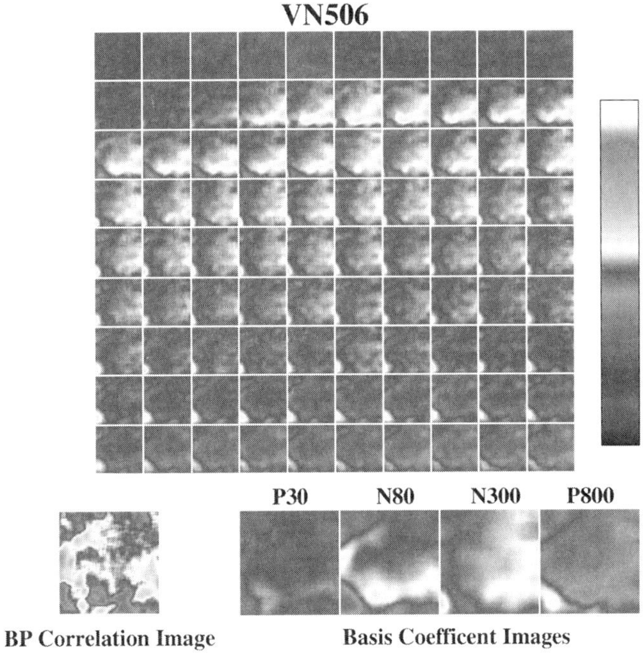

FIGURE 5.7 (See Color Figure 5.7.) Average image sequences synchronized to electrical stimulation of the vagus nerve show the spatial representation of evoked activity in the dorsal medulla (1000 trials). The top panel shows 90 frames at 10-msec intervals representing 100 msec before and 800 msec after a shock is delivered to the ipsilateral vagus nerve on frame 10. Images are pseudocolored so that warm colors represent percent reflectance decrease (to -5×10^{-4}) and cool colors represent percent reflectance increase (to 5×10^{-4}), while green represents no significant change from baseline. The lower left panel shows a correlation image of the blood pressure signal compared to an average image sequence triggered by the peak of the cardiac R-wave. Different regions of the stimulus-evoked sequence show unique response time courses that can be separated into four major components which model four physical processes associated with optical changes. Using four component temporal basis functions, four basis coefficient images were created on a pixel-by-pixel basis representing the fraction of the pixel intensity profile accounted for by each basis function (lower right panel). Since the P30 and P800 basis functions are opposite the N80 and N300 curves, the corresponding basis function images were inverted to parallel the reflectance change. The correlation and basis coefficient images are pseudocolored so that warm colors represent positive correlation or positive fractional coefficients, cool colors represent negative correlation or negative fractional coefficients, and green represents no correlation or a fractional coefficient of zero.

action potentials.[10,11] Crayfish claw nerves provide an alternative, although the nerve is ten times smaller than the lobster nerve, and total light intensity and scattered-light changes caused by action potentials are smaller than detected in other invertebrates. Earlier studies with isolated invertebrate nerves used very large stimuli that recruited large axon populations to produce a detectable response. We utilize more modest stimulus levels to produce stable responses, allowing systematic experimental manipulation and control.

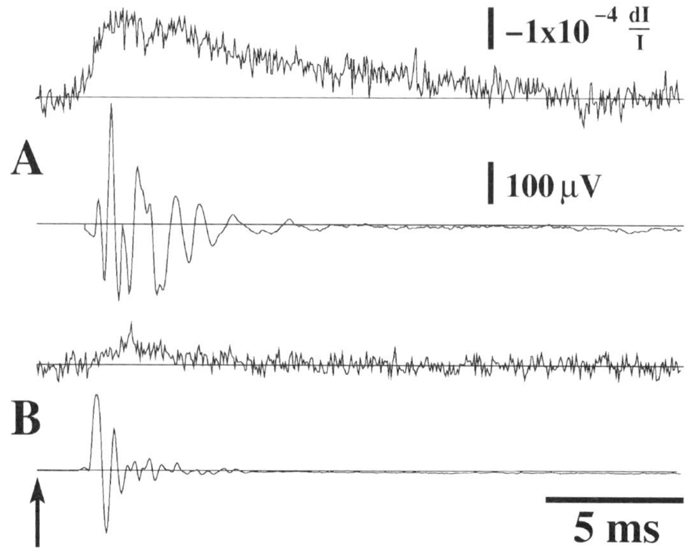

FIGURE 5.8 Stimulation of the crayfish claw nerve produces significant birefringence changes (upper panel A) with a rapid time course that parallels evoked action potentials (lower panel A). Ten-times-lower stimulation levels produce fewer action potentials with a corresponding lower optical response. However, the nerve survived much longer with lower stimulation levels.

We visualized scattered-light changes to single-action potential volleys in the crayfish claw nerve through improved optics and photodiode detection schemes. The first goal was to lower the stimulus required to generate an optical response to levels that allow eliciting of several hundred responses before the nerve shows signs of deterioration. Figure 5.8A shows a significant birefringence response to stimulus levels ten times lower than original studies; a nerve under these conditions will last for an hour or more while receiving stimulation at 1 Hz. Figure 5.8B was acquired using less than half the stimulus amplitude of trace 5.8A. Significantly fewer axons were recruited, as indicated by the evoked potential signal, yet the preparation still produced a detectable optical response. Photodiodes and amplification circuitry were optimized for low noise and high speed.

Responses show significantly more temporal structure than seen in earlier studies.[10] Simultaneous birefringence and large angle-scattering measurements assist in determining the biophysical mechanisms involved in the optical responses. The isolated nerve preparation will allow testing of hypotheses related to the biophysical basis of the fastest optical responses observed in brain tissue.

These and earlier studies of isolated nerves use single-channel, large area detectors for measuring optical changes with the generation of the action potential. As seen in Figure 5.8, the optical signal is blurred across time, appearing as a signal with a slower time course than the action potential volley. The slow components of the optical response could result from spatial integration of the neural response as different axons transmit action potentials down the nerve bundle at slightly different times or from a slow recovery of the swelling effect. We are currently configuring a linear photodiode array to measure the progression of the optical response at multiple points along the axis of the nerve, which will help to answer questions about the relationship between the temporal components of the optical response and individual action potentials. We are also combining an interferometer with an imaging detector to characterize micromechanical components of the neurophysiological response.

Evidence from our work and others[36,49] suggests that the fastest observed optical response may precede the arrival of the electrical signal. This intriguing possibility may arise by the action of membrane channels that initiate the flow of ions. However, our findings suggest that the fast response develops earlier than would be expected of such mechanisms. We hypothesize that the earliest response could be caused by a compression wave established at the stimulus site (and at sites of subsequent activation) that travels through the nerve at the speed of sound.

5.6 SUMMARY

Optical signals indicative of different aspects of neural signaling can now be imaged at very high temporal resolution. The signals can be derived from structures that remain in focus deep below the cortical surface and can be captured at relatively high signal-to-noise ratios by judicious illumination of activated structures to minimize extraneous scattering. Optical signals from various depths of the dorsal medulla parallel evoked electrical potentials elicited by stimulation of afferent nerves. Very rapid optical correlates of activity from invertebrate axons result in smaller optical signals than from central neural sites but have the potential to reveal insights into mechanisms of propagation that may not be apparent by electrophysiologic recording. Because near-infrared light can penetrate deeply into tissue, even passing through bone, it may be possible to perform imaging or optical tomography of cortical activation noninvasively,[9] perhaps in human subjects.[50,51]

ACKNOWLEDGMENTS

This study was funded in part by grants from NIH: HL22418 and MH60993, MH60263, LANL (LDRD), and the National Foundation for Functional Brain Imaging. David Rector was a Director's Fellow at LANL. MH60993 is a human brain project/neuroinformatics research project funded by the National Center for Research Resources, National Institute of Mental Health, National Institute on Drug Abuse, and the National Science Foundation.

REFERENCES

1. Fries, P. et al., Synchronization of oscillatory responses in visual cortex correlates with perception in interocular rivalry, *Proc. Nat. Acad. Sci. U.S.A.*, 94(23): 12699–12704, 1997.
2. Jones, M.S. and Barth, D.S., Spatiotemporal organization of fast (>200 Hz) electrical oscillations in rat Vibrissa/Barrel cortex, *J. Neurophysiol.*, 82(3):1599–1609, 1999.

3. Lindsey, B.G. et al., Repeated patterns of distributed synchrony in neuronal assemblies, *J. Neurophysiol.*, 78(3):1714–1719, 1997.
4. Grinvald, A. et al., Functional architecture of the cortex revealed by optical imaging of intrinsic signals, *Nature*, 324:361–364, 1986.
5. Ts'o, D.Y. et al., Functional organization in the primate visual cortex as revealed by optical imaging of intrinsic signals, *Science*, 249:417–420, 1990.
6. Frostig, R.D. et al., Cortical functional architecture and local coupling between neuronal activity and the microcirculation revealed by *in vivo* high-resolution optical imaging of intrinsic signals, *Proc. Natl. Acad. Sci. U.S.A.*, 87:6082–6086, 1990.
7. Haglund, M.M., Ojemann, G.A., and Hochman, D.W., Optical imaging of epileptiform and functional activity in human cerebral cortex, *Nature*, 358:668–671, 1992.
8. Narayan, S.M., Santori, E.M., and Toga, A.W., Mapping functional activity in rodent cortex using optical intrinsic signals, *Cerebr. Cort.*, 4(2):195–204, 1994.
9. Masino, S.A. et al., Characterization of functional organization within rat barrel cortex using intrinsic signal optical imaging through a thinned skull, *Proc. Natl. Acad. Sci. U.S.A.*, 90(21):9998–10002, 1993.
10. Cohen, L.B., Changes in neuron structure during action potential propagation and synaptic transmission, *Physiol. Rev.*, 53(2):373–413, 1973.
11. Tasaki, I. and Byrne, P.M., Rapid structural changes in nerve fibers evoked by electrical current pulses, *Biochem. Biophys. Res. Comm.*, 188(2):559–64, 1992.
12. Rector, D.M. et al., Imaging the dorsal hippocampus: light reflectance relationships to electroencephalographic patterns during sleep, *Brain Res.*, 696:151–160, 1995.
13. Rector, D.M., Poe, G.R., and Harper, R.M., Imaging of hippocampal and neocortical neural activity following intravenous cocaine administration in freely behaving animals, *Neuroscience,* 54(3):633–641, 1993.
14. Rector, D.M. et al., Imaging of the goat ventral medullary surface activity during sleep–waking states, *Am. J. Physiol.*, 267:R1154–R1160, 1994.
15. Malonek, D. and Grinvald, A. Interactions between electrical activity and cortical microcirculation revealed by imaging spectroscopy: implications for functional brain mapping, *Science*, 272(5261):551–554, 1996.
16. Perkampus, H.H., Ed., *UV Atlas of Organic Compounds,* Vol. V, New York: Plenum Press, 1971.
17. Landowne, D., Optical activity change with nerve impulses, *Biophys. J.,* 61:A109, 1992.
18. Landowne, D., Molecular motion underlying activation and inactivation of soduim channels in squid giant axons, *J. Membrane Biol.,* 88: 173–185, 1985.
19. Raichle, M.E. et al., Correlation between regional cerebral blood flow and oxidative metabolism, *Arch. Neurol.,* 33:523–526, 1976.
20. Baron, J.C. et al., Noninvasive measurement of blood flow, oxygen consumption, and glucose utilization in the same brain regions in man by positron emission tomography: concise communication, *J. Nucl. Med.*, 23(5):391–399, 1982.
21. Belliveau, J.W. et al., Functional mapping of the human visual cortex by magnetic resonance imaging, *Science*, 254:716–719, 1991.
22. Hayaishi, O., Oxyhemoglobin increases and deoxyhemoglobin decreases in the circulation of the brain of the rhesus monkey during REMS but not during slow wave sleep, in *Slow Wave Sleep: Its Measurement and Functional Significance*, Chase, M.H. and Roth, T., Eds., Los Angeles, CA: Brain Information Service/Brain Research Institute, University of California, Los Angeles, 1990.
23. Onoe, H. et al., REM sleep–associated hemoglobin oxygenation in the monkey forebrain studied using near-infrared spectrophotometry, *Neurosci. Lett.*, 129(2):209–13, 1991.

24. Hill, D.K., The effect of stimulation on the opacity of a crustacean nerve trunk and its relation to fibre diameter, *J. Physiol.*, 111:283–303, 1950.
25. Hill, D.K., The volume change resulting from stimulation of a giant nerve fibre, *J. Physiol.*, 111:304–327, 1950.
26. MacVicar, B.A. and Hochman, D., Imaging of synaptically evoked intrinsic optical signals in hippocampal slices, *J. Neurosci.*, 11(5):1458–69, 1991.
27. Ballanyi, K. and Grafe, P., Cell volume regulation in the nervous system, *Renal. Physiol. Biochem.*, 3–5:142–157, 1988.
28. McBain, C.J., Traynelis, S.F., and Dingledine, R., Regional variation of extracellular space in the hippocampus, *Science*, 249:674–677, 1990.
29. Dudek, F.E., Obenaus, A., and Tasker, J.G., Osmolality-induced changes in extracellular volume alter epileptiform bursts independent in chemical synapses in the rat: importance of non-synaptic mechanisms in hippocampal epileptogenesis, *Neurosci. Lett.*, 120:267–270, 1990.
30. Taylor, C.P. and Dudek, F.E., Excitation of hippocampal pyramidal cells by an electrical field effect, *J. Neurophysiol.*, 52(1):126–142, 1984a.
31. Taylor, C.P. and Dudek, F.E., Synchronization without active chemical synapses during hippocampal afterdischarges, J. *Neurophysiol.*, 52(1):143–155, 1984b.
32. Roper, S.N., Obenaus, A., and Dudek, F.E., Osmolality and nonsynaptic epileptiform bursts in rat CA1 and dentate gyrus, *Ann. Neurol.*, 31(1):81–85, 1992.
33. Lipton, P., Effects of membrane depolarization on light scattering by cerebral cortical slices, *J. Physiol.,* 231:365–383, 1973.
34. Blasdel, G.G., Visualization of neuronal activity in monkey striate cortex, *Annu. Rev. Physiol.*, 51:561–581, 1989.
35. Rector, D.M. et al., Light scattering changes follow evoked potentials from hippocampal schaeffer collateral stimulation, *J. Neurophysiol.*, 78:1707–1713, 1997.
36. Rector, D.M., Rogers, R.F., and George, J.S., A focusing image probe for assessing neural activity *in-vivo, J. Neurosci. Meth.*, 91:135–145, 1999.
37. Van Harreveld, A., Changes in the diameter of apical dendrites during spreading depression, *Am. J. Physiol.*, 192:457–463, 1958.
38. Rector, D.M., Poe, G.R., and Harper, R.M., Fiber optic imaging of subcortical neural tissue in freely behaving animals, in *Optical Imaging of Brain Function and Metabolism*, Dirnagl, U. et al., Eds., New York: Plenum Press, 81–86, 1993.
39. Rector, D.M. et al., A miniature CCD video camera for high-sensitivity light measurements in freely behaving animals, *J. Neurosci. Meth.*, 78:85–91, 1997.
40. Holthoff, K. and Witte, O.W., Intrinsic optical signals in rat neocortical slices measured with near-infrared dark-field microscopy reveal changes in extracellular space, *J. Neurosci.*, 16(8):2740–2749, 1996.
41. Rector, D.M. and Harper, R.M., Imaging of hippocampal neural activity in freely behaving animals, *Behav. Brain Res.*, 42:143–149, 1991.
42. Rector, D.M., Burk, P., and Harper, R.M., A data acquisition system for long-term monitoring of physiological and video signals, *Electroenceph. Clin. Neurophysiol.*, 87:380–384, 1993.
43. Rector, D.M. and George J.S., Continuous image and electrophysiological recording with real time processing and control, *Methods*, 25:151–163, 2001.
44. Tamamaki, N. and Nojyo, Y., Disposition of slab-like modules formed by axon branches originating from single CA1 pyamidal neurons in the rat hippocampus, *J. Comp. Neurol.,* 291:559–564, 1992.
45. Rector, D.M. et al., Scattered light imaging *in-vivo* tracks fast and slow processes of neurophysiological activation, *NeuroImage*, 14:977–994, 2001.

46. Humphrey, D.R., Neuronal activity in the medulla oblongata of cat evoked by stimulation of the carotid sinus nerve, in *Baroreceptors and Hypertension*, Kezdi, P., Ed., New York, Pergamon Press, 1967, 131–168.
47. Seller, H. and Illert, M., The localization of the first synapse in the carotid sinus baroreceptor reflex pathway and its alteration of afferent input, *Pfluegers Arch.*, 306:1–19, 1969.
48. Gabriel, M. and Seller, H., Interaction of baroreceptor afferents from carotid sinus and aorta at the nucleus tractus solitarii, *Pfluegers Arch.*, 318: 7–20, 1970.
49. Salzberg, B.M., Obaid, A.L., and Gainer, H., Large and rapid changes in light scattering accompany secretion by nerve terminals in the mammalian neurohypophysis, *J. Gen. Physiol.*, 86:395–411, 1985.
50. Gratton, G. et al., Noninvasive detection of fast signals from the cortex using frequency-domain optical methods, *Ann. N.Y. Acad. Sci.*, 820:286–299, 1997.
51. Steinbrink, J. et al., Somatosensory evoked fast optical intensity changes detected noninvasively in the adult human head, *Neurosci. Lett.*, 291(2):105–108, 2000.

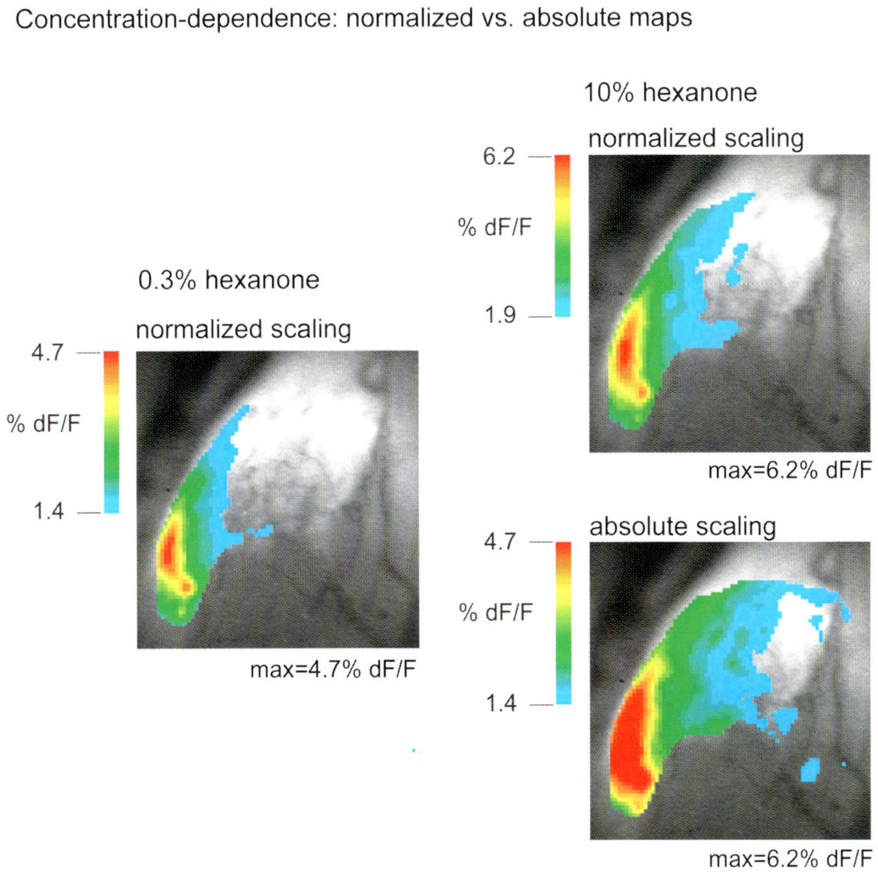

COLOR FIGURE 1.4 Normalized maps of receptor neuron input to the turtle olfactory bulb are concentration-invariant. The left panel shows a pseudocolor map of the response to a 0.3% dilution of saturated vapor of 2-hexanone. The map is normalized to the maximum signal amplitude for this trial. The right panels show pseudocolor maps of a response to a 10% dilution of 2-hexanone. The map on the top is normalized to its maximum signal amplitude. The map on the bottom (absolute scaling) shows the same data using the same scaling as for the 0.3% trial. The figure shows a concentration-dependent increase in the number of glomeruli activated above a given absolute level but shows a concentration invariance in the relative levels of input to all glomeruli activated by an odorant. 4X image magnification. The field of view is approximately 4 x 4 mm.

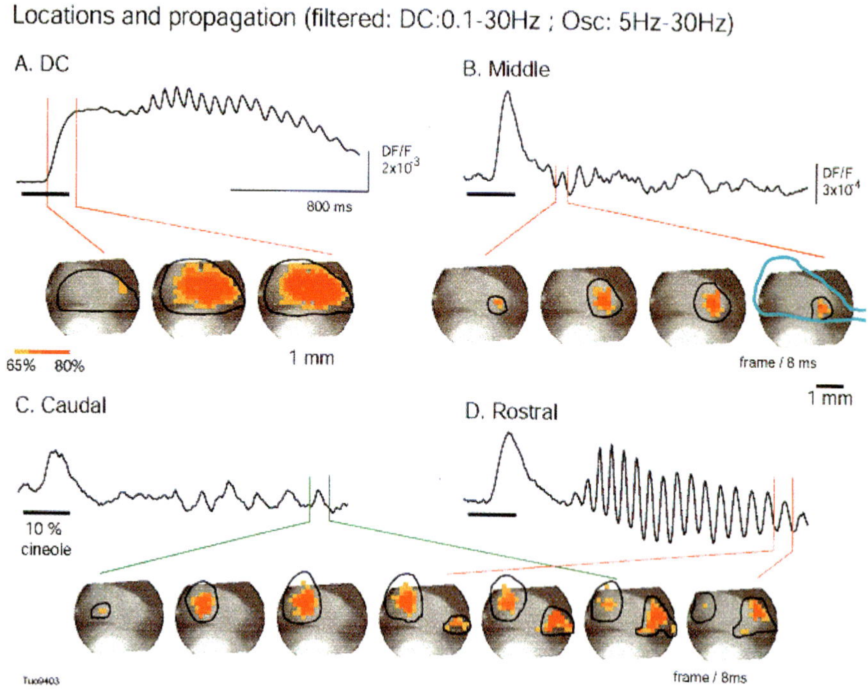

COLOR FIGURE 1.8 The locations and propagation of the four components from the trial shown in Figure 1.7. Multiframe pseudocolor displays of the signals are overlaid on the image of the olfactory bulb. The red color and the black contour lines label the areas where the signals are larger than 80 and 20% of the peak. The DC component in this animal covers almost the entire bulb. The other three panels show the position and propagation of one cycle (indicated by the red and green lines) of the three different oscillations. The black horizontal bars indicate the time of the odor command pulse. The data are filtered by a high-pass digital RC (5 Hz) and low-pass Gaussian (30 Hz) filters. (Modified from Lam, Y-W., et al., *J. Neurosci.*, 20: 749, 1999).

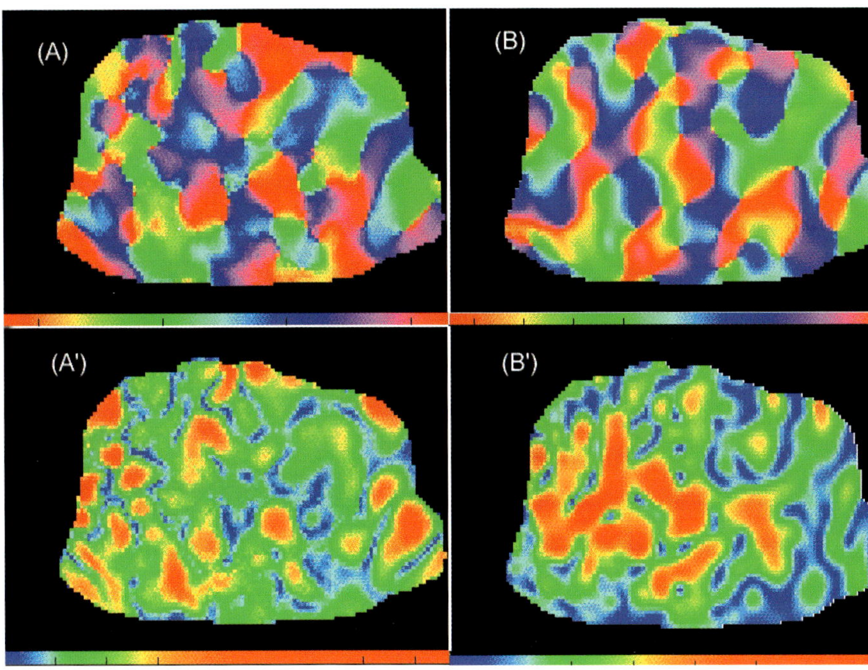

COLOR FIGURE 3.5 Orientation and direction from areas 17/18 in the cat cortex. Tissue size is 5 × 7 mm. Direction preference (A) and response strength (A') at each pixel. The upper color bar indicates a variation between $-\pi$ and π. The lower color bar refers to an order of magnitude change in amplitude. Orientation preference (B) and response amplitude (B'). The upper color bar indicates a variation between $-\pi/2$ and $\pi/2$. The lower color bar refers to an order of magnitude change in relative response amplitude.

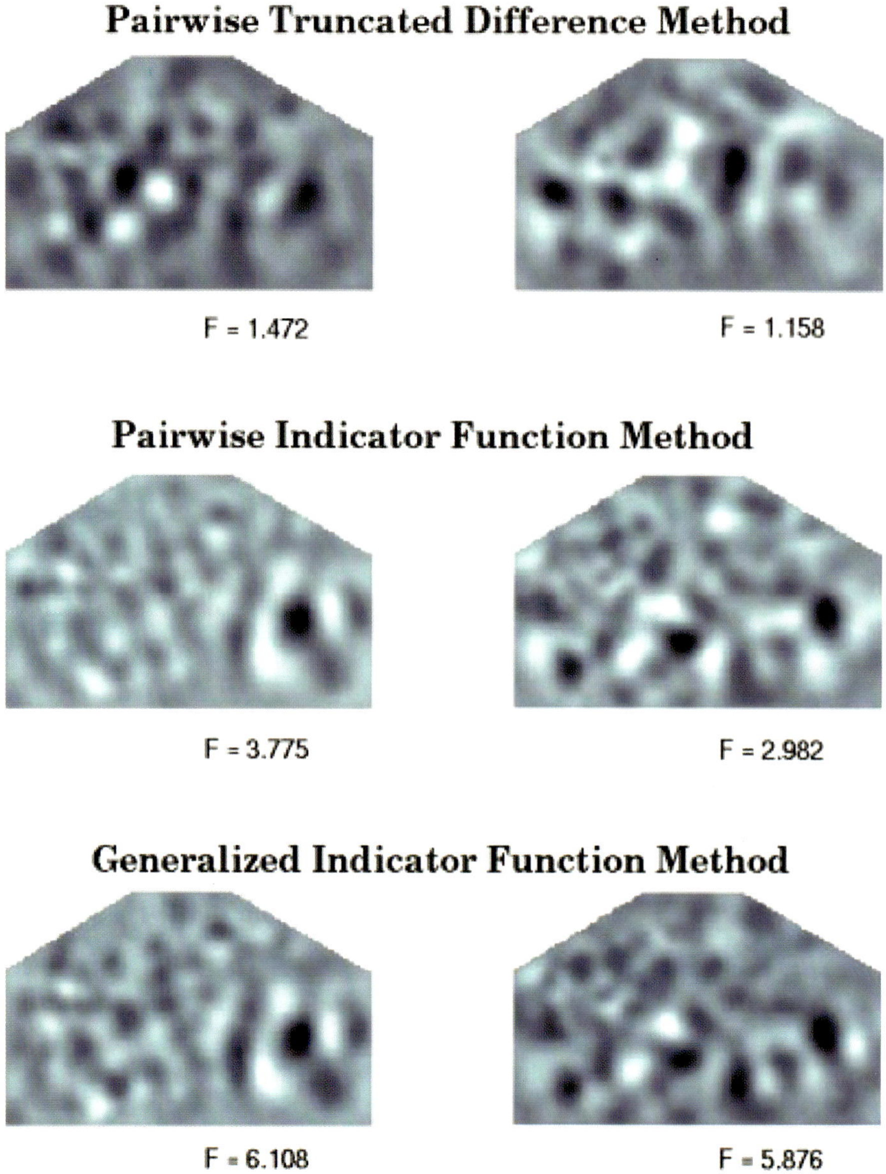

COLOR FIGURE 3.9 Orientation eigenfunction pairs, each shown with its signal-to-noise ratio, F, statistic as calculated by truncated differences (upper row); indicator functions (middle row); generalized indicator functions (lower row). These figures are based on data from the cat cortex reported in Everson et al., *PNAS*, 95, 8334–8338, 1998.

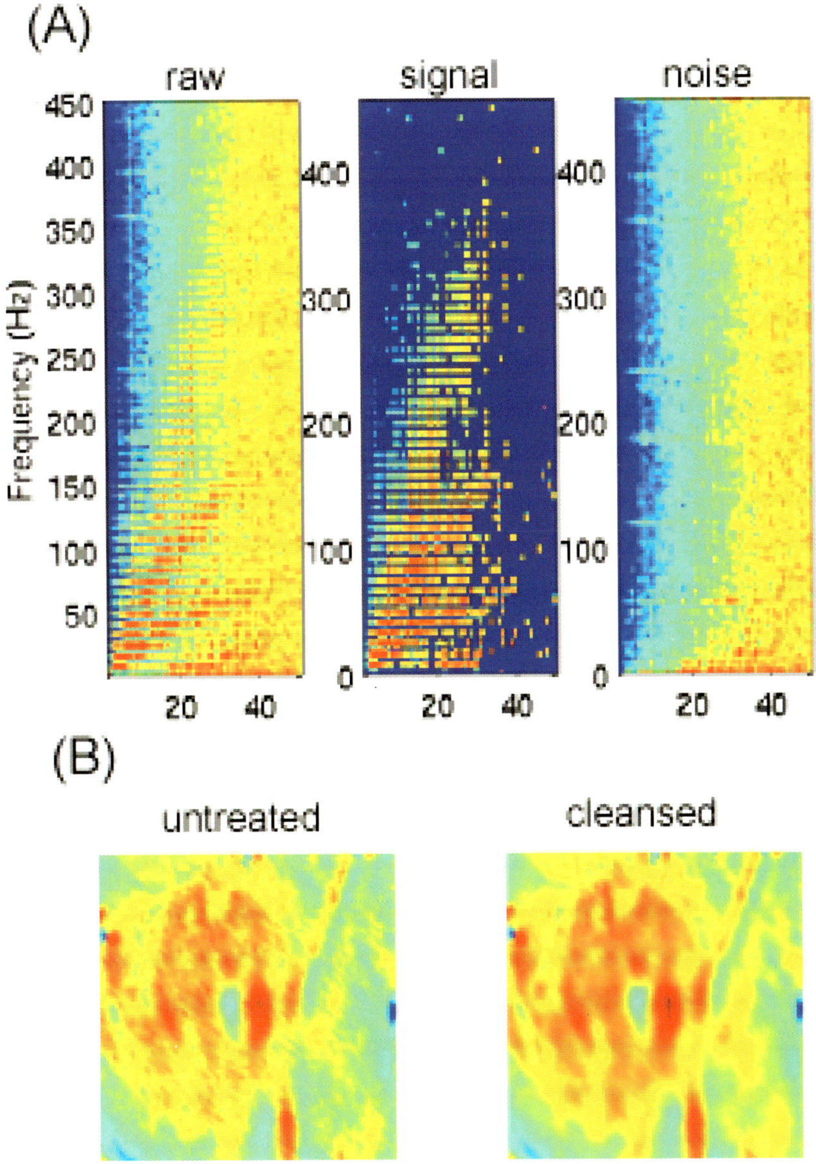

COLOR FIGURE 3.10 Analysis of fluorescence images of a voltage-sensitive dye from the *in vitro* mouse heart, as a function of time. The raw data analyzed were collected by Gregory Morley, Mount Sinai School of Medicine; image rate was 914 frames/sec. A: (raw) spectral concentration of data vs. PCA mode number; (signal) spectrum with noise removed; (noise) residual noise. B: mouse heart activity: (untreated) record; (cleansed) record.

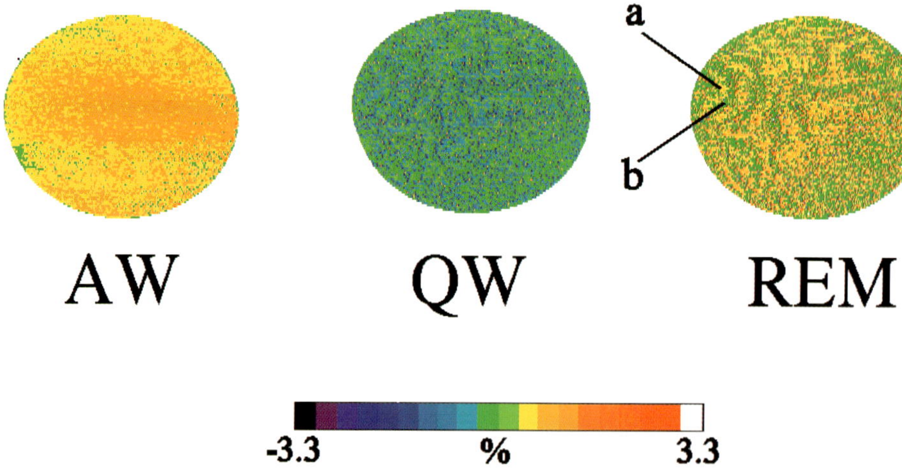

COLOR FIGURE 4.7 Representative averaged (n = 40) images of reflectance, collected from the dorsal hippocampus of a cat at 3-sec intervals, during waking (AW), quiet waking (QW), and rapid eye movement (REM) sleep. The hippocampal image is 1.6 mm in diameter. Note the curved columns of enhanced (a) or diminished (b) activity occurring preferentially during REM sleep, and the relative absence of such organization in AW. The color bar in the lower portion of the figure is the pseudocolor scale used to illustrate changes in activity. Yellow to red pixels indicate areas of increased activity (decreased reflectance), with white representing an activity increase of 3.3%. (Derived from Poe, G.R., et al., *J. Neurosci.*, 14(5), 2933, 1994.)

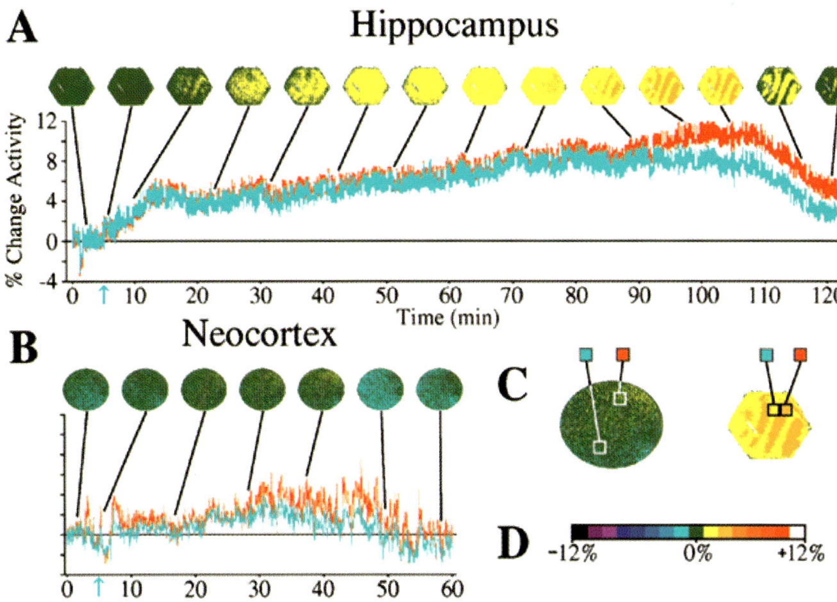

COLOR FIGURE 4.8 Representative images of activity changes from the hippocampus (A) of a cat and from the neocortex (B) of another cat before and after (arrows) administration of 3.5 mg/kg cocaine; traces of enhanced (red) and diminished (blue) activity are shown below the images. Each image is an average of 40 frames collected at 3-sec intervals and subtracted from a baseline pre-cocaine image. The images are pseudocolored, using the scale in D. In the hippocampal placement, banding of activity occurs (C, right), while the cortical site (C, left) showed no such organization and only modest activation. Images are derived from a 1 x 0.5 mm section from within the probe area. (From Rector, D.M., et al., *Neuroscience*, 54(3), 633, 1993).

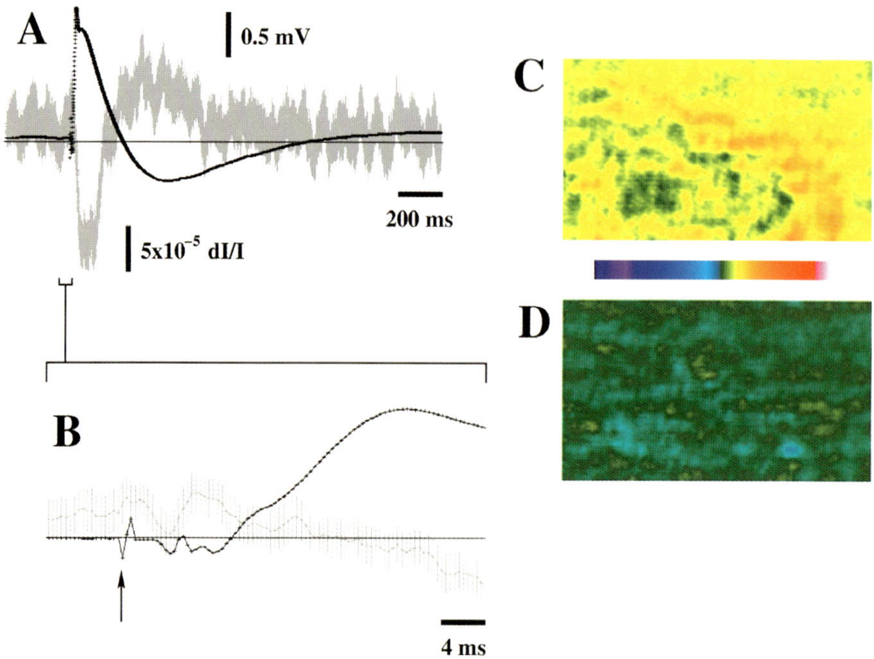

COLOR FIGURE 5.5 A fiber optic probe with perimeter illumination was placed on the surface of the dorsal hippocampus of a cat. Back-scattered light was collected with a photodiode at 2.4 kHz or a CCD camera at 100 Hz during stimulation of the contralateral hippocampus. Stimulation elicited evoked electrical responses (dark lines, Panels A and B) and concurrent optical responses. Panel B shows an expanded view of the electrical and optical response around the stimulus artifact (arrow). Images of peak optical changes from the dorsal hippocampus surface reveal significant spatial patterns during stimulation of the contralateral structure that projects directly with a laminar organization to the area under the image probe (Panel C). Stimulation 1 mm away from the optimal site shows little optical response (Panel D). Images are pseudocolored such that decreases in light intensity (increased activation) are colored with warm colors (yellow to red), increases in light intensity (decreased activation) are colored with cool colors (blue to purple), and no significant change from baseline is colored green.

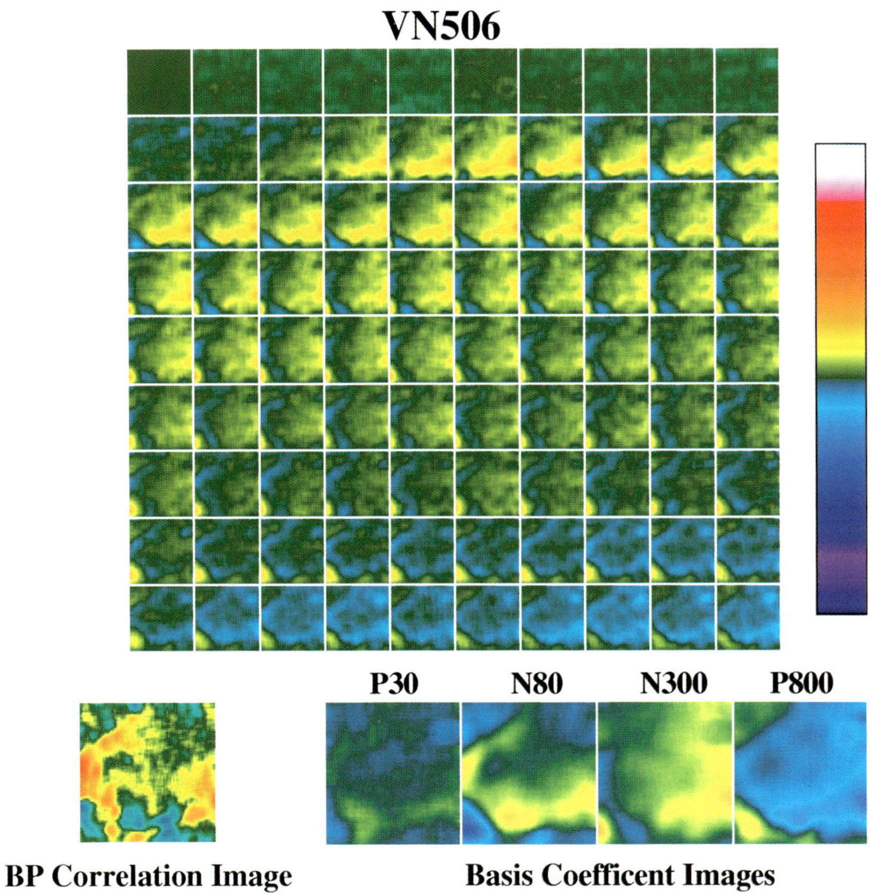

COLOR FIGURE 5.7 Average image sequences synchronized to electrical stimulation of the vagus nerve show the spatial representation of evoked activity in the dorsal medulla (1000 trials). The top panel shows 90 frames at 10-msec intervals representing 100 msec before and 800 msec after a shock is delivered to the ipsilateral vagus nerve on frame 10. Images are pseudocolored so that warm colors represent percent reflectance decrease (to -5×10^{-4}) and cool colors represent percent reflectance increase (to -5×10^{-4}), while green represents no significant change from baseline. The lower left panel shows a correlation image of the blood pressure signal compared an average image sequence triggered by the peak of the cardiac R-wave. Different regions of the stimulus-evoked sequence show unique response time courses that can be separated into four major components which model four physical processes associated with optical changes. Using four component temporal basis functions, four basis coefficient images were created on a pixel-by-pixel basis representing the fraction of the pixel intensity profile accounted for by each basis function (lower right panel). Since the P30 and P800 basis functions are opposite from the N80 and N300 curves, the corresponding basis function images were inverted to parallel the reflectance change. The correlation and basis coefficient images are pseudocolored so that warm colors represent positive correlation or positive fractional coefficients, cool colors represent negative correlation or negative fractional coefficients, and green represents no correlation or a fractional coefficient of zero.

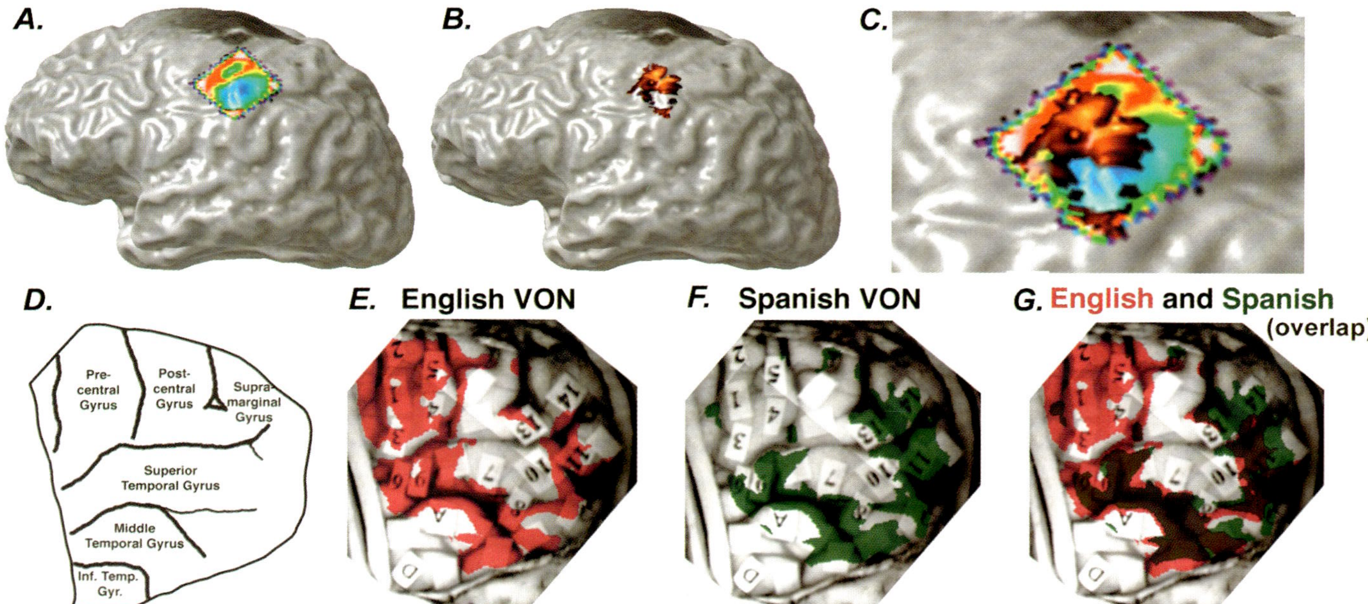
A.
B.
C.
D.
Pre-central Gyrus
Post-central Gyrus
Supra-marginal Gyrus
Superior Temporal Gyrus
Middle Temporal Gyrus
Inf. Temp. Gyr.
E. English VON
F. Spanish VON
G. English and Spanish (overlap)

COLOR FIGURE 7.5 Colocalization of optical intrinsic signals (OIS) and electrophysiological maps. (A, B, C) Colocalization of OIS and somatosensory evoked potentials (SSEPs). (A) Pseudocolor representation of SSEPs over left sensorimotor cortex is illustrated. The dark blue color indicates maximum amplitude negative peaking evoked potentials, indicating the location of the somatosensory cortex. The green areas, on the other hand, highlight cortical areas which demonstrate maximal phase reversal of the evoked potential response, corresponding to the motor cortex. Stimulus used was a transcutaneous right median and ulnar nerve stimulation (2-sec, 15-mA pulsed current). (B) Optical response to identical stimulation paradigm. Dark brown areas indicate areas of low intensity optical changes, while whiter areas in the middle of the activation region indicate higher intensity reflectance changes. (C) Superposition of optical response and SSEP, demonstrating a high degree of correlation. Regions of highest measured electrical change correspond to those areas where optical signals were largest in size and intensity. Both sensory and motor strips were activated. The stimulus intensity was sufficient to cause involuntary hand and finger movements. (D, E, F, G) Colocalization of optical signals and electrocortical stimulation maps. Tags indicate critical motor and language areas (numbers) and areas in which after-discharges were observed with stimulation (letters). Blank tags demarcate areas in which no effect of cortical stimulation was observed (ESM). (D) Schematic with labels identifying gyri observed in subsequent images. (E) Optical response to an English visual object-naming task (in which the patient is asked to name line drawings). Note that optical activity was observed over the same gyri in which critical ESM language sites were identified and no optical activity was observed over blank tags (which indicate areas not deemed essential for language processing by ESM). (F) Optical response to a Spanish visual object-naming (VON) task. (G) Comparison of English and Spanish VON. Optical maps from images (E) and (F) are thresholded (at one standard deviation above the mean optical response intensity) and displayed. Brown color indicates areas of activation overlap between the two languages tested. iOIS maps during English and Spanish VON in a single subject revealed cortical areas active during both tasks and areas specifically activated by a single language. (Adapted from Pouratian, N., et al. *J. Neurosurg.*, 93, 686, 2000.)

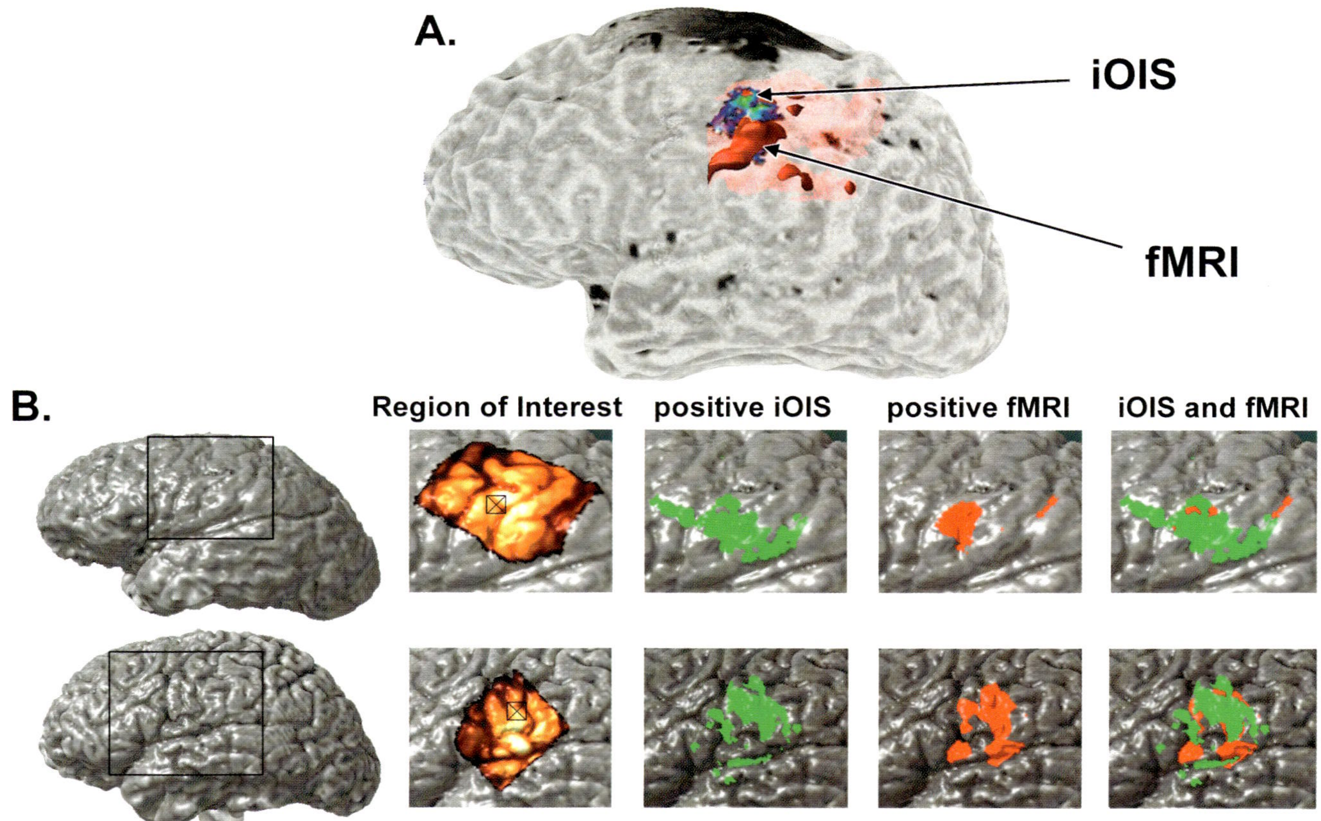
A.
iOIS
fMRI
B.
Region of Interest
positive iOIS
positive fMRI
iOIS and fMRI

COLOR FIGURE 7.7 Comparison of negative 610 nm OIS and BOLD fMRI signals in response to 2-sec somatosensory stimulus. Sensorimotor cortex was studied using identical stimulation parameters for fMRI and OIS: 2-sec 110-Hz index finger vibration in eight human subjects. The signals occurred adjacent to one another: the center of mass of the signals was separated by ~17 mm. While OIS signals were gyral, fMRI signals centered more superiorly and within sulci, consistent with a venous origin of BOLD fMRI signals. The gyral localization of the iOIS signal is consistent with optical signals emanating from capillary plexuses within the cortex. (The greener areas of the optical signal represent the most intense signal, –.3% reflectance change, while the purple areas represent the lowest-intensity optical reflectance changes observed.) The BOLD fMRI signal is displayed in red as a three-dimensional activation both at the surface of the brain and deep within the brain. The fMRI signal has been thresholded for display: all voxels with $p < .01$ for the correlation of the signal with a predetermined hemodynamic function are displayed. (B) Comparison of positive 610-nm response with BOLD fMRI using a block paradigm tongue motor task (20-sec blocks). Using both modalities, signals were identified near the inferior-lateral aspect of the central sulcus. The positive 610-nm response and the BOLD fMRI signal demonstrate high correlation (last column). These data support the notion that the two signals share similar etiologies, associated with the washout of deoxyhemoglobin by functional recruitment of cerebral blood flow and excess oxyhemoglobin. (Pouratian, N., et al., *Magn. Reson. Med.*, in press 2002.)

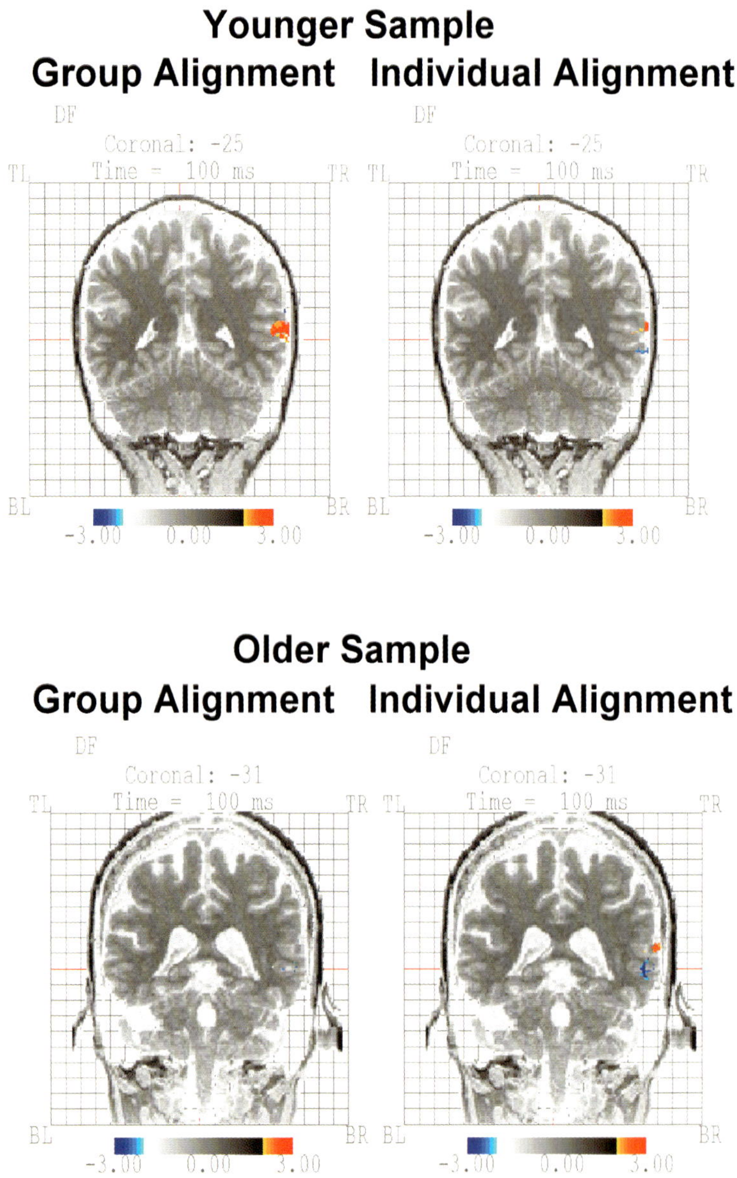

COLOR FIGURE 9.3 Effect of using a procedure to align optical data on individual brain anatomy in young (top row) and old (bottom row) subjects. The left panels refer to 3-D reconstructions based on group coordinates; the right panels refer to 3-D reconstructions based on individual subjects' coordinates. The data were collected during an auditory stimulation paradigm, at a latency of 100 msec from stimulation. Note the increased similarity of the effects in the two groups of subjects after alignment based on individual subject coordinates.

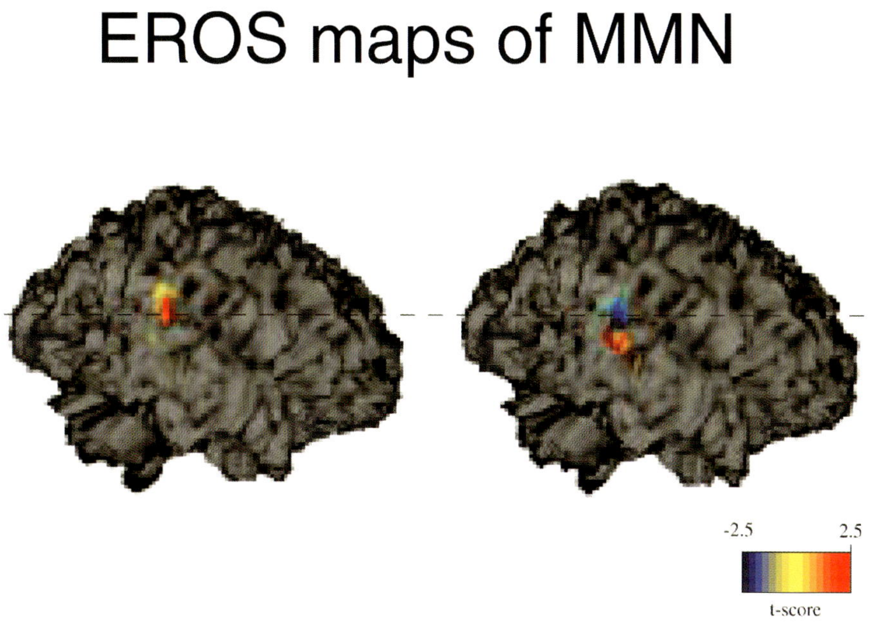

COLOR FIGURE 9.5 Maps of the right hemisphere optical response to stimuli in a passive auditory oddball paradigm. The response to standard stimuli (latency 100 msec) is presented on the left panel; the response to the deviant stimuli (latency 160 msec) is presented in the right panel. (From Rinne, T. et al., *NeuroImage*, 10, 620, 1999, with permission.)

6 Principles, Design, and Construction of a Two-Photon Laser-Scanning Microscope for *In Vitro* and *In Vivo* Brain Imaging

Philbert S. Tsai, Nozomi Nishimura, Elizabeth J. Yoder, Earl M. Dolnick, G. Allen White, and David Kleinfeld

CONTENTS

0-8493-2389-4/02/$0.00+$1.50
© 2002 by CRC Press LLC

6.1 INTRODUCTION

Many biological processes of current interest occur below the surface layers of accessible tissue. Often the surface layers cannot easily be removed without adversely affecting the physiology and function of the deeper layers. A variety of imaging techniques have been developed to perform sectioning deep to the surface using optical, electrical, and magnetic contrast agents and recording methods. Two-photon laser-scanning microscopy (TPLSM) provides optically sectioned images from depths of 500 μm or more below the surface in highly scattering brain tissue.[1-3] This method is unique in that it can provide images with submicron lateral resolution and micron axial resolution on the millisecond time scale,[4] as is required for the study of many dynamic biological processes.

This chapter provides an overview of the technical considerations relevant to the design and assembly of a system for TPLSM suitable for *in vivo* as well as *in vitro* imaging. Examples of data taken with the microscope and auxiliary components are presented. Comprehensive reviews of TPLSM have previously been given by Denk and Webb and colleagues.[5-8] Previous descriptions of the *de novo* construction of a system for TPLSM have been given,[9,10] along with descriptions of the conversion of commercially available confocal microscopes.[11-14]

6.2 OVERVIEW

6.2.1 Single-Photon Absorption for Imaging through Weakly Scattering Preparations

In traditional fluorescence microscopy, including confocal microscopy, fluorescent molecules that are either intrinsic to the tissue or artificially introduced are excited by a wavelength of light falling within the linear absorption spectrum of the molecule. A single photon is absorbed by the molecule and promotes it to an excited electronic state. Upon thermal equilibration and relaxation back to its ground state, the molecule will emit a photon that is shifted toward longer wavelengths with respect to the excitation light.

The probability of exciting fluorescence by the absorption of a single photon is linearly proportional to the spatial density of photons, i.e., the light intensity. The probability is therefore constant from any given axial plane, i.e., a plane taken normal to the propagation direction, regardless of the focusing of the incident beam. As one moves toward the focus of a beam, its intensity in a given axial plane increases as the power becomes distributed over a smaller cross-sectional area. At the same time, fewer fluorescent molecules in that plane are exposed to the illumination so that the lateral extent of the excitation is reduced. Thus, the total number of molecules excited in any plane is essentially a constant so that single photon absorption results in fluorescence throughout the entire axial extent of the incident beam.

6.2.1.1 Direct Imaging

Biological preparations that do not appreciably scatter or absorb the incident light, such as *in vitro* slice or cell culture, can be observed by a variety of direct imaging techniques,[15] typically with the use of a charge-coupled device (CCD) array detector. In whole-field fluorescent microscopy, the entire region of interest in the preparation is concurrently illuminated and all of the collected fluorescent light is recorded. This method has the advantage of a high rate of data collection — e.g., hundreds of frames per second with current array detectors — but suffers from the lack of optical sectioning. The fluorescence generated from planes outside the focal plane contributes in a nonuniform manner to the background level. The result is a lack of axial resolution and a lower signal-to-noise ratio than that for an image in which fluorescence from outside the focal plane is eliminated.

Axial resolution on the order of the diffraction limit may, in principle, be achieved with direct imaging methods in which light from regions outside the focal plane is systematically eliminated (Table 6.1). One method is to acquire data at multiple depths by shifting the focus above and below the object plane and then to use the images from out-of-focus planes, in conjunction with a deconvolution procedure, to model and subtract the out-of focus light.[16-18] This method and its extensions have the advantage of experimental simplicity in that only a motorized focus control needs to be added to a standard microscope. On the other hand, diffraction by the open aperture of the objective leads to a set of dark rings in the transfer function of the objective, so that this method is insensitive to the spatial frequencies of these rings. This is the so-called "missing-cone" problem in which the cone represents spatial frequencies in radial and axial momentum space (i.e., k_r and k_z).

TABLE 6.1
Nominal Properties of Techniques for Functional Imaging with Fluorescent Indicators

<table>
<tr><th>Mode</th><th>Technique</th><th>Axial Resolution</th><th>Lateral Resolution</th><th>Frame Period (100 by 100 pixels)</th><th>Depth Penetration</th></tr>
<tr><td rowspan="4">Direct Image Formation (CCD Acquisition)</td><td>Full Field</td><td>$\gg\lambda$</td><td rowspan="6">$\cong\lambda$</td><td>~100 Hz</td><td rowspan="3">$\sim\Lambda_{scattering}$</td></tr>
<tr><td>Vibrating Grating</td><td rowspan="2">$\sim\lambda$</td><td rowspan="3"><30 Hz</td></tr>
<tr><td>Deconvolution</td></tr>
<tr><td>Interferometric</td><td>$\ll\lambda$</td><td>$\ll\Lambda_{scattering}$</td></tr>
<tr><td rowspan="2">Point Scanning (PMT Acquisition)</td><td>Confocal</td><td rowspan="2">$\sim\lambda$</td><td>~10 Hz</td><td>$\sim\Lambda_{scattering}$</td></tr>
<tr><td>Multiphoton Excitation</td><td>>100 Hz (resonant or AOM line scan)</td><td>$\gg\Lambda_{scattering}$</td></tr>
</table>

An alternate procedure is to project a line grating into the object plane so that the projected image has a spacing on the order of the lateral diffraction limit. When this grating is modulated in time, only the part of the signal that emanates from the object plane is concurrently modulated. Thus, information about the intensity distribution in the object plane only may be extracted.[19,20] A final set of methods is applicable only to physically thin objects and relies on an interferometric cavity to set up an optical field that varies in the axial direction.[21-23] This may be used to achieve axial resolution on length scales of less than one wavelength.

6.2.1.2 Confocal Laser-Scanning Microscopy

Laser-scanning microscopy provides an improved means to achieve axial resolution compared with direct imaging techniques. An incoming laser beam is deflected by a pair of rotating mirrors, mounted at orthogonal angles, and subsequently focused by a microscope objective onto the preparation. The angular deflection of the collimated laser beam causes the focused spot to move across the preparation in a specified raster pattern. The fluorescent light generated at the focal volume is collected by a detector as a function of time; this time series is transformed back into a position in the preparation along the trajectory of the spot. The drawback to laser-scanning methods of microscopy is a reduced speed of acquisition compared with direct imaging techniques. The maximum acquisition rate is limited by the speed at which the scan mirrors can raster across the region of interest.

In confocal microscopy, axial resolution is obtained by preferentially collecting fluorescent light generated near the focal plane while rejecting light generated outside the focal plane. Specifically, the fluorescent light is refocused along the same beam path as the excitation light. After passing back through the scan mirrors, a process referred to as "descanning," the light is spatially filtered by a small aperture, or pinhole, whose radius is chosen so that its image in the preparation matches or exceeds the diffraction-limited radius of the illuminated region. The pinhole preferentially passes fluorescent light generated near the focal plane. Although confocal microscopy works well for optically thin preparations, it is inefficient for preparations in which substantial scattering occurs. As the fluorescent light becomes more heavily scattered, a greater portion of the photons from the focal plane are deflected from their original trajectories and thus blocked by the pinhole and remain undetected. Conversely, an increasing portion of the light from outside the focal plane will be deflected to pass through the pinhole. The result is a drop in the signal-to-noise ratio.

6.2.2 Two-Photon Laser-Scanning Microscopy for Strongly Scattering Preparations

Two-photon laser-scanning microscopy utilizes laser light at approximately twice the wavelength of that used in traditional, one-photon microscopy. These photons carry half of the energy required to promote a molecule to its excited state, and the excitation process requires simultaneous absorption of two photons from the

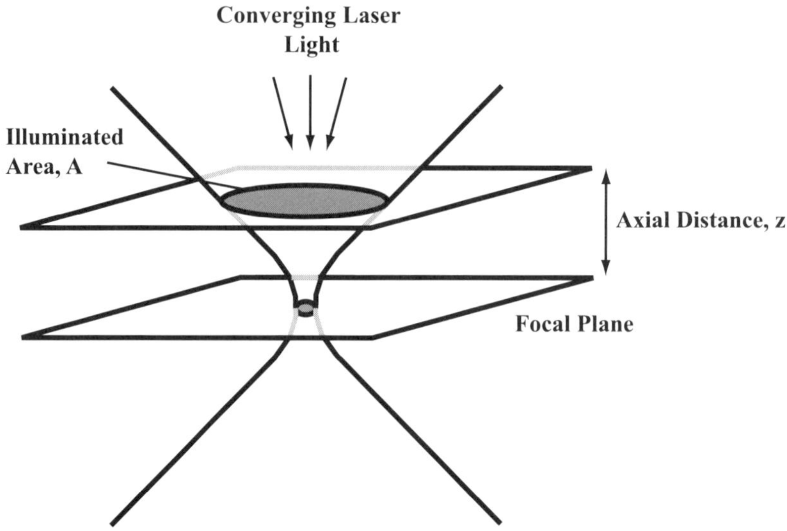

FIGURE 6.1 Cartoon of laser light near the focus of an objective. The shaded regions indicate the cross-sectional area in a given plane, normal to the direction of propagation, that is illuminated by the converging laser light. The illuminated cross-sectional area is seen to decrease as the distance to the focal plane decreases.

excitation beam. The probability of two-photon absorption thus scales with the square of the incident laser intensity. This nonlinear absorption process provides an intrinsic mechanism for optical sectioning solely by the incident beam.

As illustrated in Figure 6.1, the total fluorescence generated from a given cross-sectional plane, F_{total}, is proportional to the square of the incident intensity I and the illuminated area A. The intensity is defined as the incident laser power P divided by the illuminated area; i.e., $I = P/A$. The illuminated area in a given plane is proportional to the square of the axial distance z between that plane and the focal plane; i.e., $A \propto z^2$, at least for z larger than the confocal length, $z_{confocal}$, where

$$z_{confocal} \equiv \frac{1}{2\pi}\frac{\lambda_o}{n}\left(\frac{n}{NA}\right)^2 \tag{6.1}$$

and n is the optical index of the preparation at the excitation wavelength λ_o, NA is the numerical aperture of the objective, and the ratio (n/NA) is a purely geometric factor.[24] Thus,

$$F_{total} \propto I^2 A \propto \left(\frac{P}{z}\right)^2 \tag{6.2}$$

The amount of fluorescence generated from a given axial plane is seen to decrement as a quadratic function of the distance from the focal plane (Figure 6.2). The integrated fluorescence for $z \gtrsim z_{confocal}$ is thus finite; i.e.,

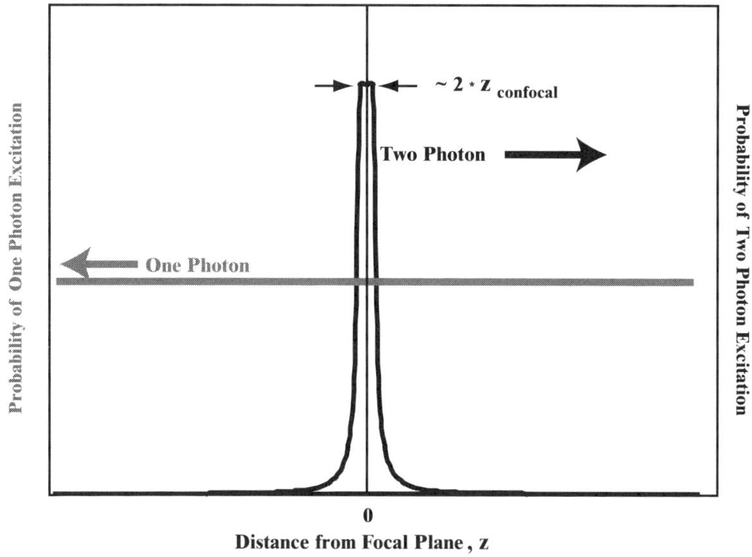

FIGURE 6.2 The integrated excitation of fluorophores within an axial plane as a function of the axial distance, z, from the focus. Single-photon excitation results in an excitation profile that is constant with axial distance (light line). Two-photon excitation results in an excitation profile that falls off quadratically with axial distance from the focal plane (dark line).

$$2\int_{z_{confocal}}^{+\infty} dz\, F_{total} \propto 2\int_{z_{confocal}}^{+\infty} dz \left(\frac{P}{z}\right)^2 \rightarrow \text{Constant} \tag{6.3}$$

and only light within a focal depth, $z \lesssim z_{confocal}$, contributes significantly to the signal. Thus, with TPLSM, all of the fluorescent light can be collected as signal, regardless of scattering. In contrast, the same integral (Equation 6.3) diverges for single-photon absorption.

The probability of simultaneously absorbing two photons is small compared to the probability of single-photon absorption at half the wavelength. Therefore, a pulsed laser source is necessary to provide the large incident intensities required to generate significant two-photon signals while concurrently maintaining the average power at levels that avoid heating of the preparation. The use of ultrashort laser pulses, i.e., 10 to 100 fsec, allows for the greatest efficiency and depth penetration while reducing the potential for thermal damage to the preparation (but note the conclusions of Konig et al.[25]). In particular, the extent of thermal damage depends on the average power delivered by the incident beam, while the depth penetration is a function of the instantaneous peak power.

Two-photon laser-scanning microscopy is uniquely suited for optically sectioning through thick preparations.[5,6,26-29] In brief:

1. The fluorescent signal is not descanned; i.e., the fluorescent light does not require spatial filtering with a pinhole, as is the case for confocal micros-

copy. This yields a higher collection efficiency for scattering preparations. It further requires a much simpler alignment of collection optics compared to confocal microscopy.

2. The use of near-infrared light for two-photon excitation, as opposed to blue/UV light for single-photon excitation, is preferable with biological preparations. Wavelengths in the near-infrared are less heavily scattered as well as less likely to be absorbed by biological molecules. In particular, flavins and other molecules involved in bioenergetics are far less excitable by near-infrared light than by UV/blue light.
3. The two-photon spectra of fluorophores are generally broader than their one-photon counterparts and are spectrally separated further from the fluorescence emission peak. Both of these characteristics simplify the technical task of separating the excitation and emission light and allow for the simultaneous excitation of multiple dyes with a single laser source.

6.3 BASICS

6.3.1 Resolution

The fundamental axial length scale is the confocal length (Equation 6.1), $z_{confocal}$. The axial resolution or depth of focus, denoted δz, is proportional to $z_{confocal}$. A worst-case estimate for δz is when the intensity goes through a first zero along the optical axis, which occurs for[24]

$$\delta z = 4 \cdot \pi \cdot z_{confocal} = 2 \frac{\lambda_o}{n} \left(\frac{n}{NA} \right)^2 \tag{6.4}$$

where λ_o/n is the wavelength in the preparation. An analogous expression holds for the lateral resolution, denoted δr. A worst-case estimate for δr is when the intensity goes through a first zero along a radius in the focal plane, which occurs for[24]

$$\delta r = 1.2 \cdot \pi \frac{NA}{n} z_{confocal} = 0.6 \frac{\lambda_o}{n} \left(\frac{n}{NA} \right) \tag{6.5}$$

For the practical case of a 40× magnification water immersion lens with NA = 0.8, the confocal length for 800-nm light is $z_{confocal} \cong 0.25$ µm, for which $\delta z = 3$ µm and $\delta r = 0.6$ µm.

6.3.2 Scattering by Brain Tissue

The beam will maintain a focus as long as the elastic scattering length, denoted $\Lambda_{scatter}$, is much greater than the depth of focus, i.e., for

$$\Lambda_{scatter} >> z_{confocal}$$

The scattering length corresponds to the length over which the incident light remains essentially undeflected. For neocortex in living rat, $\Lambda_{scatter} \sim 200\ \mu m$,[27,29,30] so that the depth penetration of TPLSM is limited only by the available power. However, for some preparations, such as the granule cell layer in mouse cerebellum, the interpretation of preliminary data (Denk and Kleinfeld, unpublished) suggests that $\Lambda_{scatter} \sim z_{confocal}$ and TPLSM is no longer effective.

6.3.3 Photobleaching and Photodamage

The nonlinear excitation profile of TPLSM results in localization of any photodamage, related to bleaching of the fluorophore or otherwise, that may occur. In contrast, single-photon excitation results in photodamage throughout the entire axial extent of the beam. Further, for both single- and two-photon imaging techniques, the incident beam can induce heating of the nervous tissue, although this effect may be substantial only when the average incident power exceeds ~0.1 W for the near-infrared wavelengths involved in TPLSM.[31]

Increased photobleaching at the focus, possibly driven by third-order or higher nonlinear processes, is a possibility with two- vs. single-photon excitation. Some experimenters have observed photobleaching rates that always scale linearly with two-photon absorption[25,32] or scale supralinearly with absorption only at relatively high intensities,[33] while other experimenters have observed the supralinear relation at all intensities.[34,35] At extremely high peak powers unlikely to be achieved under normal imaging conditions, the intensity of the laser pulse at the focus can cause ionization and result in permanent damage to the preparation.

For imaging through optically thick preparations, TPLSM offers fundamental advantages over single-photon excitation microscopy. For thin preparations, the possibility of increased but localized photodamage leaves the choice between the single- and two-photon methods unresolved.

6.3.4 Light Source

Pulsed laser light is not monochromatic. A temporally short pulse comprises a correspondingly broad range of wavelengths. While the exact width will depend on the precise shape of the pulse, the spectral width, $\Delta\lambda$, must scale with the temporal width, τ_P, as

$$\Delta\lambda \sim \frac{\lambda_o^2}{c\tau_P} \tag{6.6}$$

where λ_o is the center wavelength. For a sufficiently short pulse, the spectral width of the pulse can extend beyond that of the two-photon excitation spectrum for a given fluorescent molecule. In this limit, the excitation efficiency for an excessively short pulse can be lower than that for a longer pulse.

A laser with an appropriate pulse width as well as center wavelength, typically between 730 and 1000 nm, must be used in a TPLSM system (see Siegman[36] for a general review of lasers). Ti:Sapphire-based laser systems have been reported to produce pulses with widths shorter than 10 fsec, while commercially available Ti:Sapphire systems capable of generating roughly 100-fsec-duration pulses are readily available and are used for our system (Section 6.4). The pulse is broadened to 200 fsec or more at the exit of the objective. The corresponding full spectral bandwidth is ~10 nm. This width is less than the width of the two-photon absorption band of many relevant dyes, e.g., 80 nm for fluorescein.

The fluorescent light generated by TPLSM is sensitive to fluctuations in the laser power, the center wavelength of the pulse, and the spectral width of the pulse. Pulses with different spectral densities may be characterized by the same width yet provide different peak powers and thus different excitation strengths,[37] as demonstrated by the examples of Figure 6.3. On the one hand, the efficiency of two-photon excited fluorescence increases with increasing spectral width, i.e., as $\Delta\lambda$, or equivalently increases inversely with decreasing pulse width, i.e., as τ_P^{-1} (Equation 6.6). Thus, efficient pulses are associated with a large spectral bandwidth. On the other hand, a high level of spectral dispersion in the objective may lead to a drop in spatial resolution. This occurs when the optical index of the glasses in the objective significantly varies as a function of wavelength, although it is typically a small effect for the near-infrared wavelengths and 100- to 200-fsec pulse widths used in TPLSM.[38]

The spectral components of the pulse can be nonuniformly distributed through the pulse — a condition referred to as chirp — so that there is a systematic shift in frequency between the start and end of a pulse (Figure 6.3c). Under ideal conditions the chirp of the laser pulse should be zero at the focus of the objective so that the spectral composition of the pulse is uniform across the pulse. This results in a minimum pulse width and is usually achieved with external compensation for the group dispersion of the pulse.[39]

The average output power of the laser can be monitored with a simple photodiode. The center wavelength and the spectral width of the pulse of the laser can be determined with a basic spectrometer. While a fast oscilloscope can be used to display the repetition rate of the laser, even the fastest available oscilloscopes are currently unable to measure the laser pulse width directly. However, an interferometric autocorrelator can be built to monitor the shape of the pulse (Appendix B). The temporal distribution of wavelengths can be directly detected only by phase measurements, using a frequency-resolved optical gating or related technique.[40]

6.3.5 Delivery of Laser Light

The output parameters of a commercial laser system are unlikely to match the requirements of the microscope perfectly. The output beam may not be optimally collimated or may be collimated with a beam diameter too small or too large for use in the microscope. To compensate, the beam profile is calculated along the direction of propagation and can be reshaped using either lenses or curved mirrors.

6.3.5.1 Laser Beam Profile

The laser beam diameter and divergence can be calculated using the paraxial approximation, i.e., under the assumption that the intensity along the direction of propa-

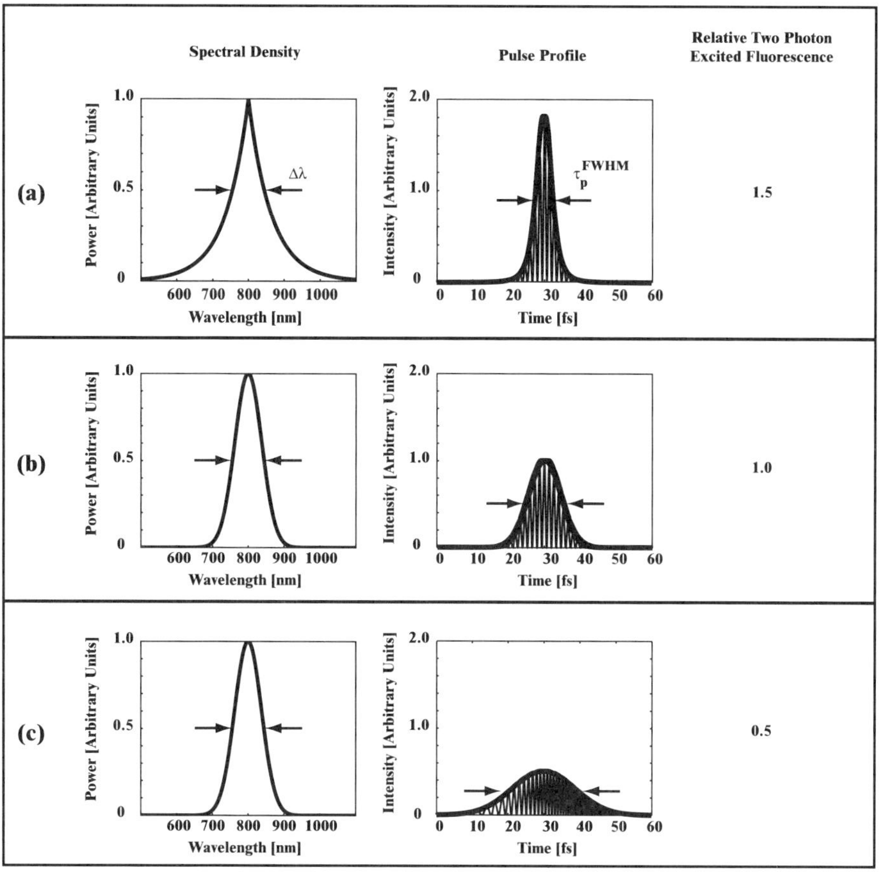

FIGURE 6.3 Pulses with different characteristics can have the same full-width half-maximal spectral width, indicated by $\Delta\lambda$. (a) A pulse with an exponential spectral envelope and the corresponding Lorentian pulse shape with a full-width half-maximal pulse width of τ_p^{FWHM}. (b) A pulse with a Gaussian spectral envelope and the corresponding Gaussian pulse shape. (c) A pulse in which the constituent frequency components are nonuniformly distributed, i.e., chirped. The extent of the chirp has been grossly exaggerated from typical conditions. Note that, for illustrative purposes, the pulse widths are about 10× shorter than those used in typical TPLSM experiments.

gation varies slowly on the scale of the wavelength, and the further assumption that the profile of the intensity follows a Gaussian distribution along the direction transverse to that of propagation.[41] This is depicted in Figure 6.4 for a divergent laser output that is refocused with an f = +500-mm lens; the beam diameter is denoted by D_{beam} and equals the diameter across the points where the intensity falls by a factor of e^{-1}. Superimposed on the Gaussian beam calculation is the calculation for the same system in the geometric optics limit. The result demonstrates that, with the exception of regions close to a focus, the Gaussian beam calculation can be approximated by a geometric optics calculation with modest accuracy.

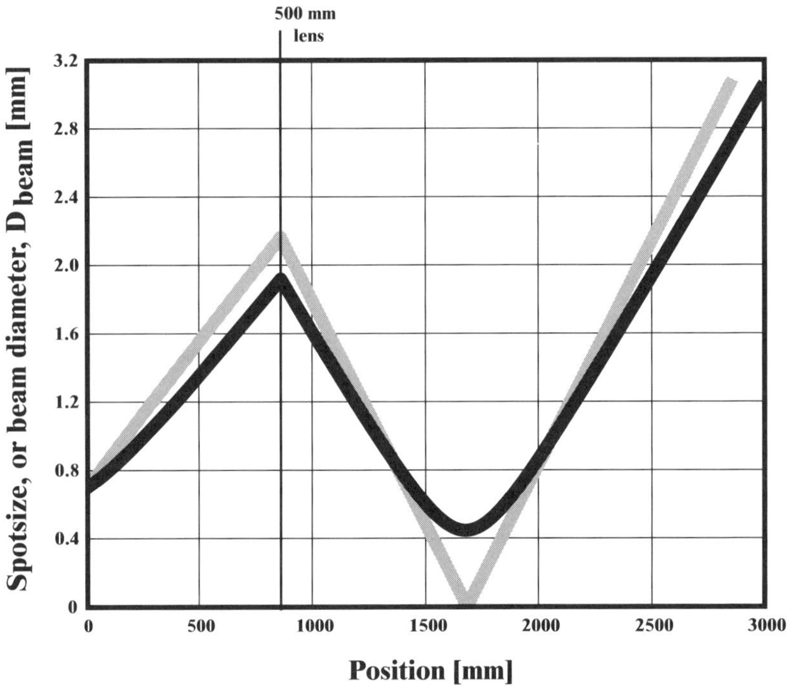

FIGURE 6.4 Calculations of the laser beam diameter, D_{beam}, as a function of distance for a beam with a starting diameter of D_{beam} = 0.70 mm and a divergence of 1.7 mrad. These values are typical for the output of a Ti:Sapphire laser. An f = +500-mm lens is placed 860 mm from the laser output aperture. The calculation for a Gaussian beam is shown as the dark line, while the geometric optics limit is depicted as the light line.

6.3.5.2 Reshaping the Laser Beam

Several different methods are available to change the divergence and diameter of a laser beam, as depicted in Figure 6.5. Lenses are simple to align, yet, as a general consideration, the glass will introduce an additional source of pulse dispersion (Figures 6.5a and 6.5b). Their effect on a 100-fsec pulse is small; thus, lenses may be used with commercial Ti:Sapphire systems, but lenses may introduce substantial dispersion for shorter, i.e., ~10-fsec, pulses. Curved metal mirrors do not introduce dispersion, yet they must be used off axis and thus introduce astigmatism into the laser beam. Specifically, for a two-curved-mirror telescope, typically sized mirrors must used substantially off axis to allow the beam to reflect from their centers yet pass the edge of the companion mirror (Figure 6.5c). The resulting astigmatism may degrade the sharpness of the laser focus at the preparation. A single-mirror scheme is a simple yet viable means to change the size of the beam (Figure 6.5d).

The technical characteristics involved with several telescope designs are:

1. A pair of achromatic doublets can be used as a Keplerian telescope (Figure 6.5b). The lens separation should be approximately the sum of the two

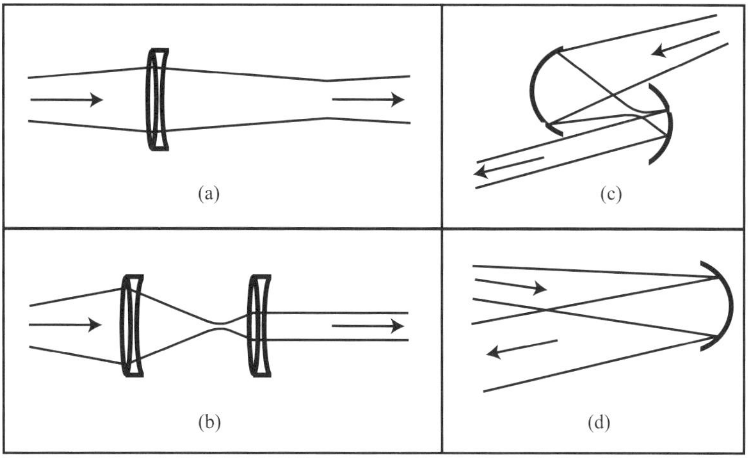

FIGURE 6.5 Four possible methods for reshaping a laser beam: (a) a single, long focal length lens, as used for the calculation of Figure 6.4; (b) a Keplerian telescope; (c) a telescope constructed with two curved mirrors; (d) a single, long focal length curved mirror.

focal lengths, but the actual location of the telescope along the beam path is flexible. The telescope can easily be adapted to provide different magnifications by choosing different combinations of lenses at the appropriate separation. The disadvantages of this configuration include the need for rails or precision translation stages for precise alignment and the dispersion produced by the lenses.

2. A curved-mirror telescope can be used with the mirror separation approximately equal to the sum of the two focal lengths (Figure 6.5c). In this Keplerian telescope, the actual location of the curved-mirror telescope is flexible, and the telescope can be adapted to provide different magnifications with the choice of different combinations of curved mirrors. This configuration also requires rails or precision translation stages for precise alignment. No dispersion is added to the system by the mirrors, but astigmatism is introduced into the laser pulse because the mirrors are used at substantially off-axis angles.
3. A single-focal length achromatic doublet can be used to reshape the beam (Figure 6.5a). An achromatic doublet is placed approximately one focal length from the original focus of the laser, which is typically inside the laser cavity. Although the alignment of a single lens is simplified compared to multielement configurations, the collimated beam diameter and placement of the lens are completely determined by the focal length of the lens. As with all lenses, some dispersion is added to the system by this configuration.
4. A single, long focal length curved mirror can be used to reshape the beam (Figure 6.5d). The lens is placed approximately one focal length from the original focus of the laser, which typically is inside the laser cavity. Alignment of the mirror is simplified compared to multielement

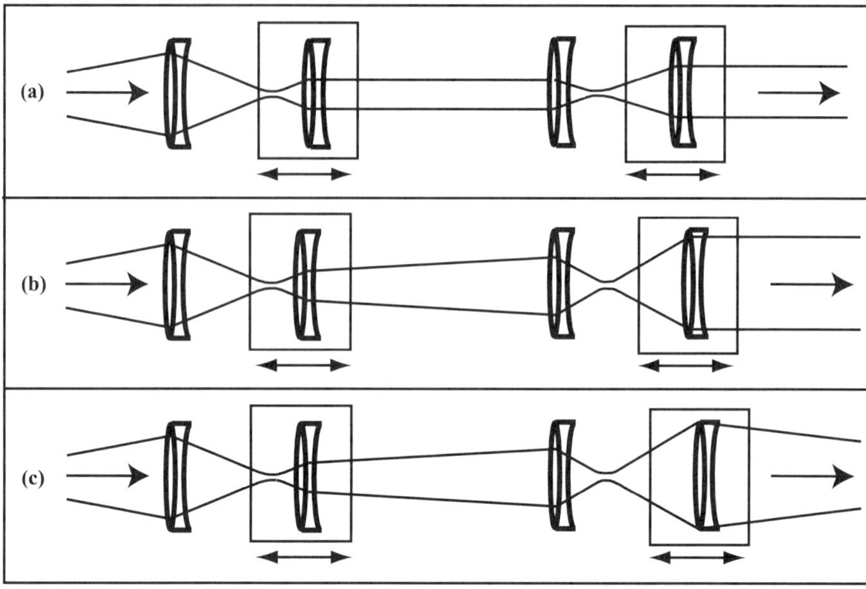

FIGURE 6.6 Ray diagrams for a dual telescope system with four lenses. The output beam diameter and divergence can be completely controlled by translating two of the four lenses. For illustration, we take the input beam to be divergent. (a) A collimated, smaller diameter output; (b) a collimated, larger diameter output; (c) a converging, larger diameter output.

configurations, but the collimated beam diameter and mirror placement are completely determined by the focal length of the mirror. Although no dispersion is added to the system by the mirror, a slight astigmatism is introduced into the laser pulse because the incoming and outgoing beams are off axis only in one plane.

5. For complete control of the beam profile with a fixed set of achromatic doublets or mirrors, a four-element, double-telescope system should be implemented (Figure 6.6). The laser beam diameter and the divergence can be altered by adjusting the location of two of the four elements. The deficit of this system is additional dispersion. Note that, in principle, a three-element system provides sufficient degrees of freedom but is somewhat more difficult to adjust.

In general, the laser beam should be collimated to minimize the spherical aberrations of the objective. As a practical matter, it may be necessary to introduce a slight convergence or divergence to the beam, as most lenses are designed to minimize spherical aberrations for visible (400 to 600 nm) light rather than near-infrared (700 to 1000 nm) light. The final size of the beam represents a compromise between resolution, which increases as the back aperture of the objective is increasingly filled, and throughput, which is maximal when the back aperture is underfilled. This tradeoff is illustrated in Figure 6.7 in terms of the normalized beam diameter, D_{beam}/D_{BA}, where D_{BA} is the diameter of the back aperture of the objective.

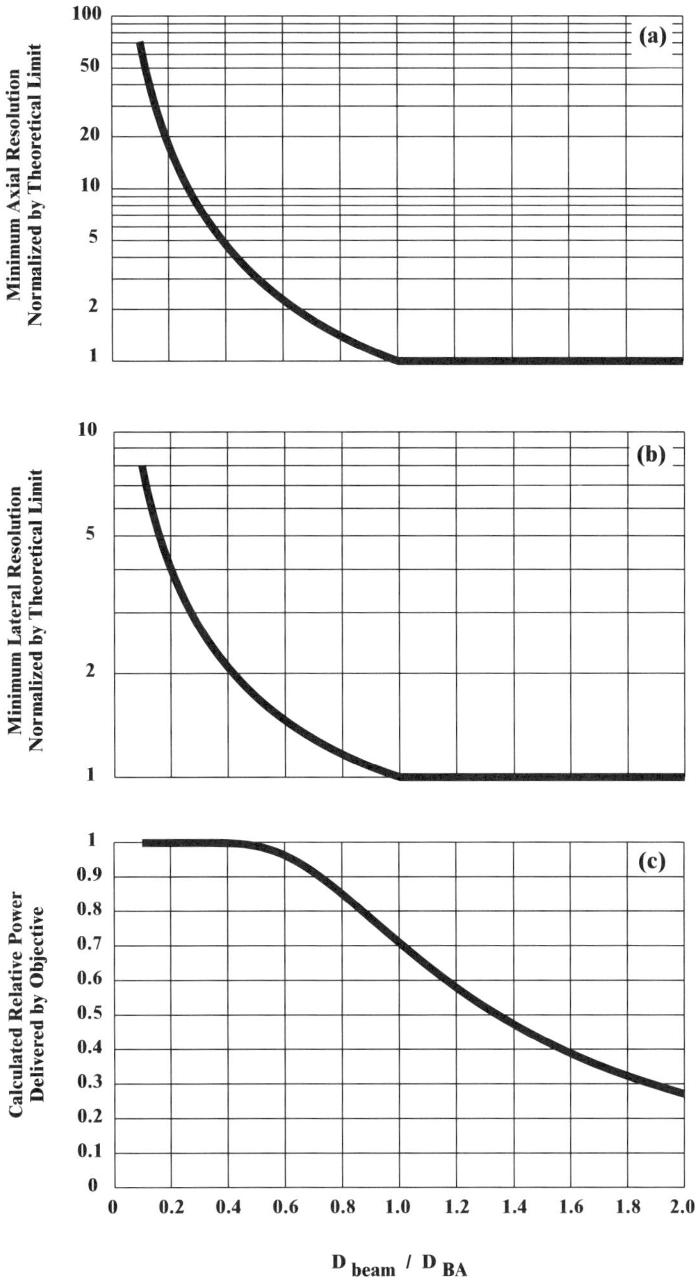

FIGURE 6.7 The tradeoff between power throughput and optical resolution as a function of the size of an incident Gaussian beam, normalized in terms of the filling factor D_{beam}/D_{BA}. The resolution was estimated in the plane wave limit, while the power was calculated in the Gaussian limit. (a) Lateral resolution vs. normalized radius; (b) axial resolution vs. normalized radius; (c) relative power vs. normalized radius.

6.3.6 Design of a Scan System

We consider in detail the manipulation of the path of the laser beam as a means to scan a focused spot across the preparation. The essential idea is that the objective converts the incident angle of a collimated beam into the position of a diffraction limited spot on the focal plane (Figure 6.8a). The focused laser spot can be moved across a stationary preparation by changing the angle of the incident beam. In order to change the angle without physically moving the source, one uses a system of mirrors and auxiliary lenses according to one of two schemes.[42] The first method involves a standard 160 mm tube length objective, i.e., a lens optimized for an image plane at 160 mm from the rear focal plane of the objective, in conjunction with a rotating mirror and a scan lens to deflect the rotated light back into the beam path (Figure 6.8b). A disadvantage to this configuration for TPLSM is that the incoming light to the objective is convergent rather than parallel. This can lead to astigmatism in the focused beam if a dichroic mirror, used for the collection of emitted light, is placed directly behind the objective.[5]

An alternative scheme for the scan system is to use an infinity-corrected objective, together with a rotating mirror and a tube lens as well as a scan lens (Figure 6.8c). In this configuration, the two auxiliary lenses serve to translate the angular deflection of the mirror into an incident angle at the objective while allowing the input to the mirror and the input to the objective to be collimated. This is the configuration used in our system, which will now be described in greater detail (Figure 6.9). The relevant variables are

f_1 ≡ Focal length of scan lens
f_2 ≡ Focal length of tube lens
d_1 ≡ Distance between the scan mirror pivot point and the scan lens
d_2 ≡ Distance between the scan lens and the tube lens
d_3 ≡ Distance between the tube lens and the back aperture of the objective
D_{SM} ≡ Effective diameter of the scan mirror
D_{BA} ≡ Diameter of the back aperture of the objective
D_{beam} ≡ Diameter of the beam at the back aperture
d_{SM} ≡ Distance between a pair of scan mirrors

The issue of one- vs. two-dimensional scanning can be separated from the issue of constraining the optics in the beam path. Thus, for clarity, we focus first on a one-dimensional scan system.

6.3.6.1 Constraints on Axial Distances and Optical Apertures

We consider the optimal alignment and layout of optics, with reference to Figure 6.9. The constraints are:

1. Infinity-corrected objectives require the incident light to be collimated at the back aperture of the objective. If the input to the scan system is a collimated laser beam, then the scan lens and tube lens must be configured as a telescope. Therefore, their separation is given by

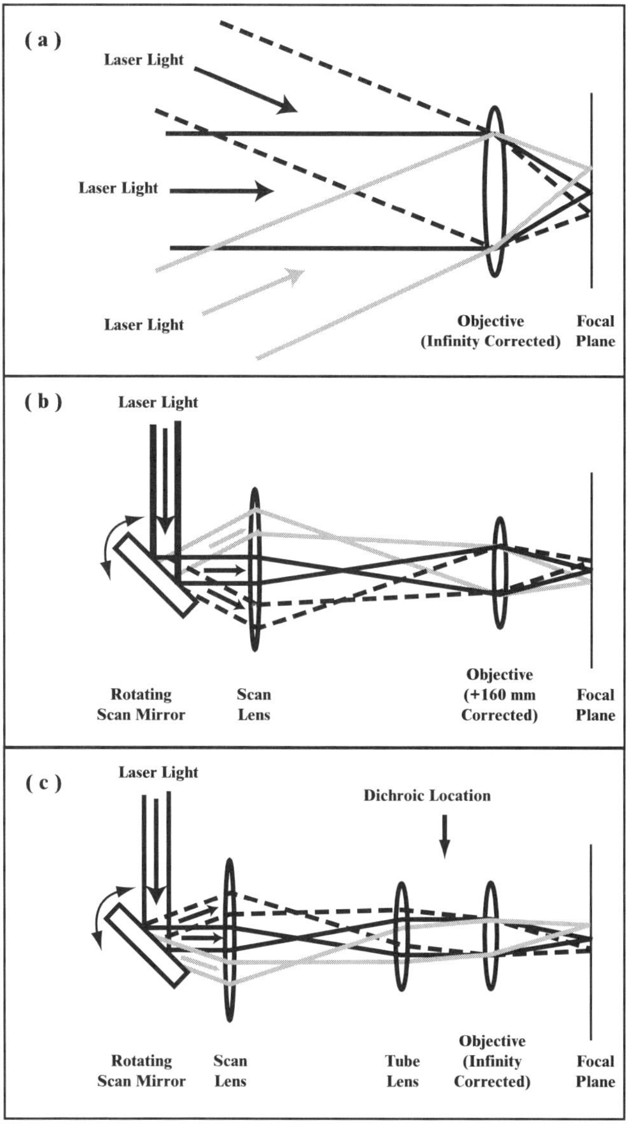

FIGURE 6.8 Illustration of different configurations for laser scanning. (a) A scanning system with only an objective. The collimated laser source must be physically moved and rotated behind an infinity-corrected objective. (b) A scanning system with a single intermediate lens, i.e., the scan lens, in addition to the objective. The objective is taken as optimized for a 160 mm image distance. The rotating mirror is imaged onto the back aperture of the objective. Note that the light is convergent behind the objective. (c) A scanning system with two intermediate lenses, i.e., the scan lens and the tube lens, in addition to the objective. The objective is taken as optimized for an infinite image distance. A scan lens and tube lens are placed between a rotating scan mirror and an infinity-corrected objective. The mirror surface is imaged onto the back aperture of the objective, and the light between the objective and the tube lens is collimated and thus suitable for the placement of a dichroic mirror.

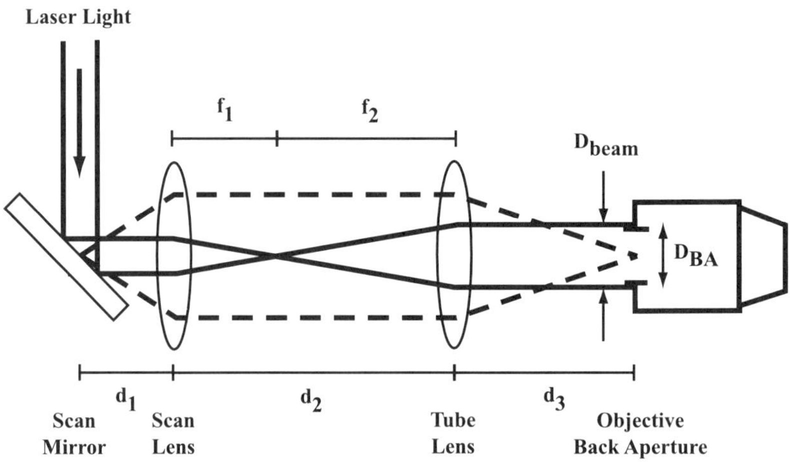

FIGURE 6.9 The alignment of optics in a scanning system consisting of an objective, tube lens, and scan lens. The actual laser pathway is indicated by the solid lines. The dashed lines depict the imaging condition between the center of the scan mirror and the center of the back aperture of the objective. The laser path is magnified to just overfill the back aperture, and the imaging condition (see text) prevents motion of the laser spot at the back aperture as the laser beam is scanned.

$$d_2 = f_1 + f_2 \tag{6.7}$$

2. The pivot point of the scan mirror should be imaged to the center of the back aperture of the objective to minimize motion (spatial deviations) of the laser path at the back aperture plane. This imaging condition is satisfied by

$$d_1 = \frac{(f_1)^2}{f_2} + f_1 - d_3\left(\frac{f_1}{f_2}\right)^2 \tag{6.8}$$

   This condition serves to maintain constant laser power through the back aperture of the objective as the scan mirror is rotated. A graph of the required separation d_1 as a function of the separation d_2 for six different choices of scan lens focal lengths and a tube lens with a focal length of $f_2 = +160$ mm is shown in Figure 6.10.
3. The incident laser beam should slightly overfill the back aperture of the objective in order to utilize its full NA; i.e.,

$$D_{beam} > D_{BA} \tag{6.9}$$

   This generates the highest resolution for a given objective.

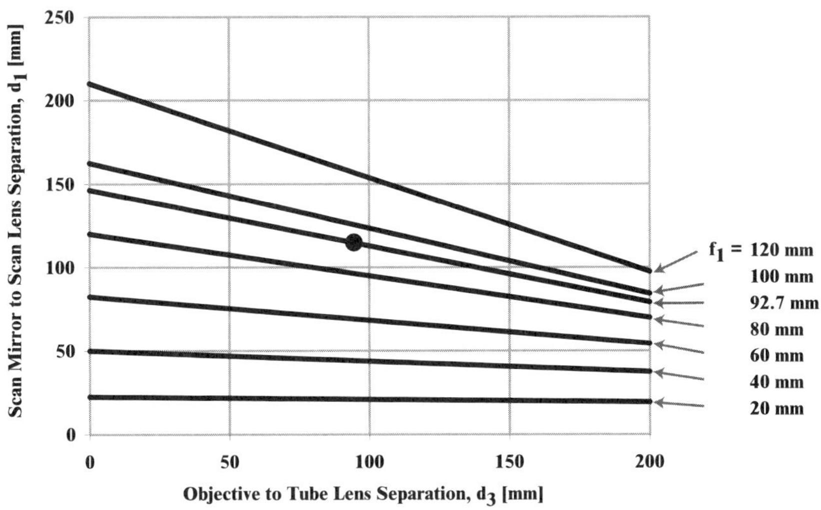

FIGURE 6.10 The calculated scan mirror to scan lens separation distance, d_1, as a function of the objective to tube lens separation distance, d_3. The calculation is shown for several choices of scan lens focal lengths, f_1. The focal length of the tube lens is assumed to be $f_2 = +160$ mm. The dot marks the actual values, $d_3 = 95$ mm and $f_1 = +92.7$ mm, used in our system.

4. Typical scan mirrors are only a few millimeters in dimension. The simultaneous requirements that the beam should underfill the scan mirrors as a means to avoid losses due to diffraction, yet slightly overfill the back aperture of the objective as a means to maximize the spatial resolution at the focus, constitutes a magnification constraint on the scan system. The required magnification, denoted m, is given by

$$|m| \equiv \frac{D_{BA}}{D_{SM}} \cong 2 \text{ to } 4 \tag{6.10}$$

where the numerical values are appropriate for typical dimensions of objectives' back apertures and scan mirrors. For the case in which a standard $f_2 = +160$-mm tube lens is used in the scan system, the required focal length of the scan lens is determined by a second constraint on the magnification; i.e.,

$$f_1 \equiv \frac{f_2}{|m|} \cong 40 \text{ to } 80 \text{ mm} \tag{6.11}$$

5. The above constraints (Equations 6.7 to 6.11) reduce the scan system to a single degree of freedom. For a given separation distance between the tube lens and the objective d_3, the imaging constraint determines the scan

lens–to–scan mirror separation d_1 or vice versa (Figure 6.10). In practice, many objectives are designed to be corrected for chromatic aberration at a fixed value of d_3. By convention, this is taken to be

$$d_3 \cong 95 \text{ mm} \tag{6.12}$$

This constraint will further increase the collection efficiency of the tube lens for visual inspection of the preparation. In practice, d_3 is also constrained by the physical size of the detector assembly.

The scan lens should be a lens with a small f-number, i.e., ratio of focal length to clear aperture, that is well corrected for spherical, chromatic, and off-axis aberrations. A minimal scan lens could be an achromatic doublet with an f-number of 2. More precise scan lenses are available from specialized manufacturers and are particularly valuable for applications involving large scan angles. In practice, stereoscope objectives are well corrected for off-axis aberrations and work well as scan lenses, as discussed in Section 6.4.

6.3.6.2 Two-Dimensional Scanning

A true, distortion-free, two-dimensional scan can be obtained with two scan mirrors and the use of additional optics between the scan mirrors to insure that the centers of both mirrors and the objective back aperture are all conjugate planes of one another. The addition of intermediary optics, while conceptually straightforward and employed in commercial instruments, e.g., the Biorad MRC series, makes proper alignment more difficult and can result in additional dispersion or astigmatism.

A simpler two-dimensional scan configuration can be used that utilizes no intermediary optics if the separation between the scan mirrors can be made small compared to focal length of the scan lens; i.e.,

$$f_1 >> d_{SM} \tag{6.13}$$

The conjugate plane of the back aperture is placed halfway between the two scan mirrors. The small deviation from the exact imaging conditions results principally in two effects. The first is a slightly increased sensitivity of the focused beam profile to defects on the scan mirrors. The second is a slight lateral motion of the beam at the back aperture of the objective while the beam is scanned. This motion results in increased clipping, i.e., vignetting, of the beam at larger scan angles. The excitation intensity will therefore be lower at the edges of the scan than at its center (Figure 6.11).

The focal length of the scan lens is constrained from above by the requirement of sufficient magnification (Equation 6.11). For a tube lens with a focal length of $f_2 = +160$ mm, we have $f_1 = |m|^{-1} f_2 \approx 80$ mm. As a compromise, our system (Section 6.4) uses a scan lens of focal length $f_1 = +92.7$ mm. This distance is large compared to the spacing between the two scan mirrors in our system, i.e., $d_{SM} = 7.6$ mm, but

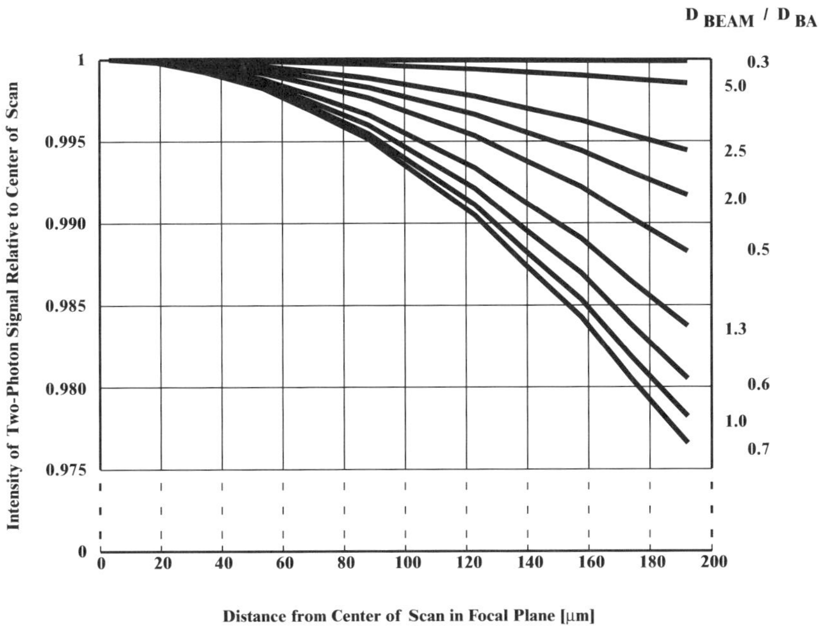

FIGURE 6.11 Theoretical calculation of the spatial inhomogeneity of the squared power, which corresponds to the two-photon excitation rate, due to the lateral motion of the scanned laser spot at the back aperture of the objective. The calculation was carried out in the geometric optics limit with the assumption of a Gaussian transverse beam profile. The different curves correspond to filling factors D_{beam}/D_{BA}. All parameters as specified in Section 6.4.

slightly larger than the constraint in Equation 6.11, so that we slightly underfill the back aperture. A typical imaging experiment utilizes a 40× magnification objective with a back aperture of $D_{BA} = 8$ mm. The theoretical two-photon intensity profile for different filling rations D_{beam}/D_{BA} is estimated by modeling this system with geometric optics (Figure 6.11). The calculated intensity deviations due to motion of the laser path at the objective back aperture plane are seen to be small for typical scanning conditions.

6.3.7 Fluorescent Light Detection

The fluorescent signal generated in the illuminated region of the preparation is incoherent and emitted over the entire 4π solid angle. A portion of the signal is collected in the backward direction by the objective and deflected by a dichroic beamsplitting mirror and focused onto a photomultiplier tube (PMT) (Figure 6.12). Filters and additional dichroic mirrors and detectors can be used to further isolate the fluorescent signal and, if required, separate different fluorescent signals according to wavelength. For preparations that are not too optically thick, e.g., slice or cell culture, an additional detector can be placed beneath the preparation to collect the signal emitted in the forward direction[43] (Figure 6.12).

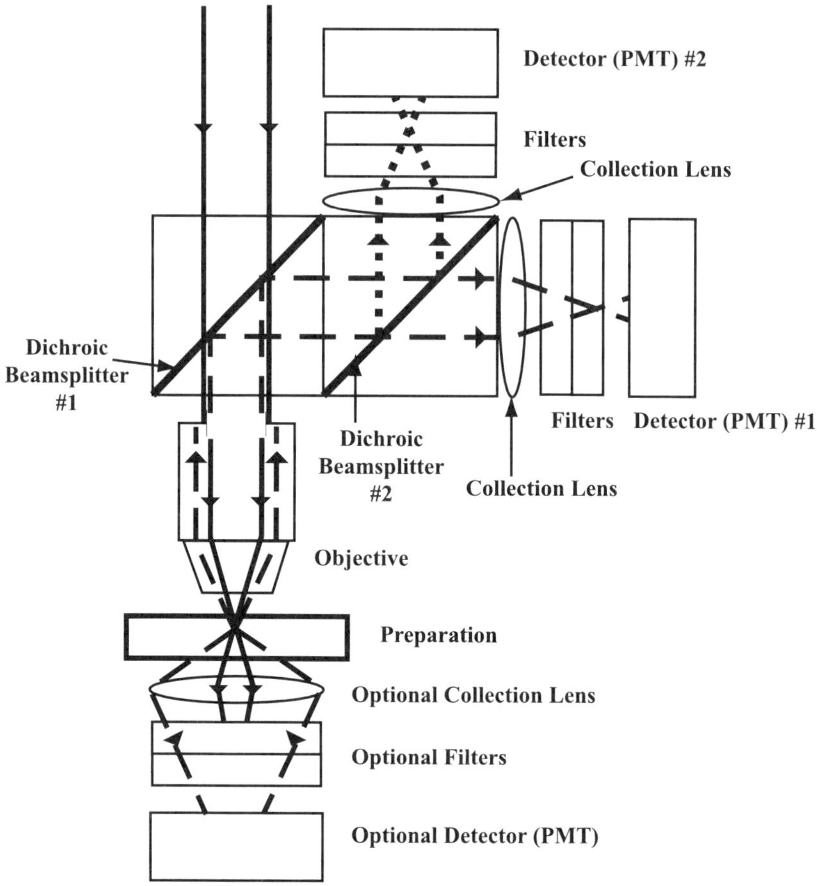

FIGURE 6.12 Schematic of the detection configuration for two-photon excited fluorescence. The excitation laser light is depicted as a solid line, while the fluorescent signals are shown as dashed and dotted lines. The collection system shown below the sample stage can be added for preparations, such as brain slice, that are not too optically thick.

6.3.7.1 Placement of Detector Assembly

The detector assembly should be placed directly above the objective. The two main considerations in this choice of detector position are:

1. Collection efficiency is increased by placing the collection assembly directly above the objective. Although light emitted from the focal plane of the infinity-corrected objective should ideally be collimated as it returns through the back aperture, the actual fluorescence will diverge due to scattering and diffraction. In particular, it is mainly the scattered light that will miss the clear aperture of the tube lens. The unscattered light collected by the objective will almost exactly retrace the beam path of the excitation laser. Therefore, placing the detector assembly as close to

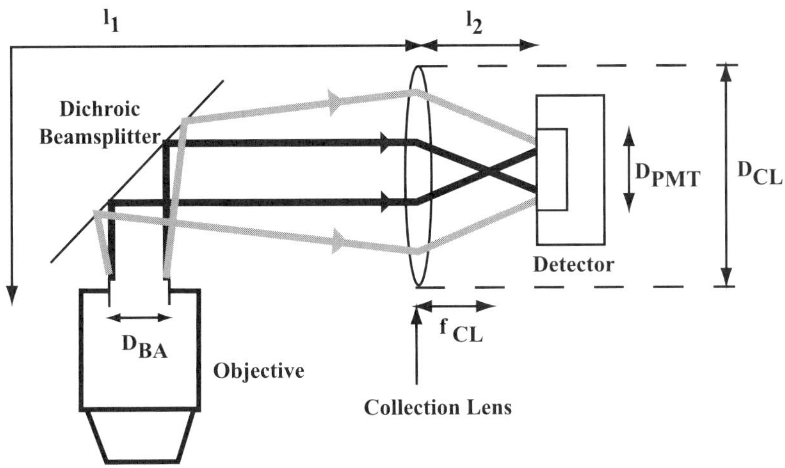

FIGURE 6.13 Placement and separations of detector elements for two-photon excited fluorescence. Unscattered fluorescent light is depicted as a dark line, while scattered fluorescent light is depicted as a gray line.

the objective as possible should result in improvements for more strongly scattering preparations.

2. Fluorescent light collected from an infinity-corrected objective will be nearly collimated just above the objective. Placing the dichroic mirror in the path at this point, rather than at a location where the beam is diverging, reduces the sensitivity of the signal to local defects on the dichroic mirror.

6.3.7.2 Constraints for Detection Elements

The backward collection pathway is shown in Figure 6.13 with the second dichroic mirror and PMT pathway eliminated for clarity. We note the following definitions:

l_1 ≡ Distance between collection lens and back aperture of the objective
l_2 ≡ Distance between active surface of the detector and collection lens
D_{PMT} ≡ Effective diameter of the detector active surface
f_{CL} ≡ Focal length of the collection lens
D_{CL} ≡ Clear aperture of the collection lens

The following conditions should be maintained for the choice and placement of collection optics:

1. The back aperture of the objective should be imaged onto the active surface of the PMT to minimize motion of the fluorescent signal on the detection surface. This imaging condition is given by

$$l_2 = \frac{l_1 f_{CL}}{l_1 - f_{CL}} \tag{6.14}$$

and is graphed in Figure 6.14 for five choices of collection lens focal lengths, f_{CL}.

2. The collection lens should have a large clear aperture to maximize collection of scattered light, some of which diverges as it exits the back aperture of the objective. In practice, physical constraints of the detector assembly (described in detail later) limit the clear aperture to

$$D_{CL} \approx 30 \text{ mm} \tag{6.15}$$

3. The separation between the collection lens and the objective should be minimized to within physical constraints. This also serves to increase the collection efficiency of the scattered light. As a maximal limit, we arbitrarily choose a distance at which a clear aperture is at least half of the empirically determined maximal divergent spot size from the back of a typical objective. For example, this corresponds to a half-angle spread of 11.3° from the D_{BA} = 8-mm back aperture of a 40× magnification objective. For a collection lens with a clear aperture of D_{CL} = 30 mm, this implies a spacing of

$$l_1 \cong 130 \text{ mm} \tag{6.16}$$

4. The focal length of the collection lens should be as short as possible. This, in combination with the imaging constraint of Equation 6.14, assures that

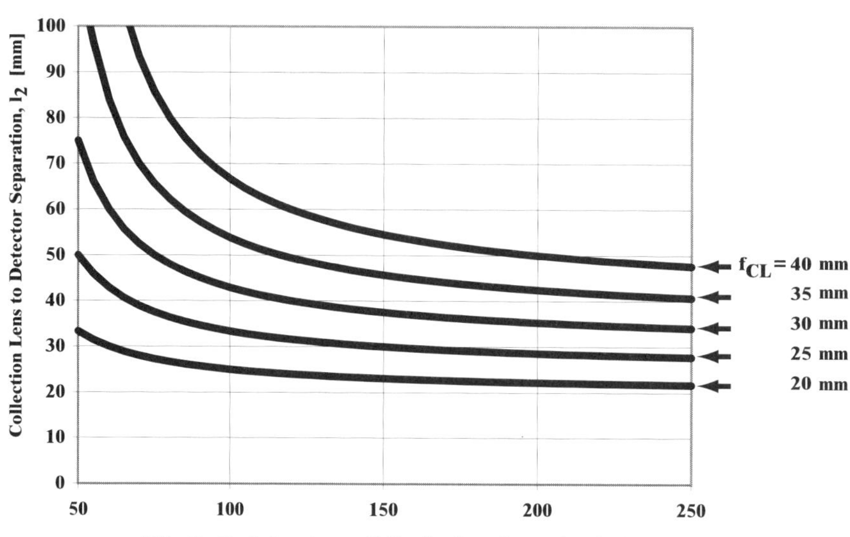

FIGURE 6.14 Graph of the imaging constraint for the distance between the collection lens and the detector, l_2, vs. the distance between the detector and the back aperture of the objective, l_1, for five choices of focal length for the collection lens.

the image of the back aperture is minified to fit onto the smaller active area of the PMT. The magnification is given by

$$|m| = \frac{f_{CL}}{l_1 - f_{CL}} \tag{6.17}$$

The active area of the PMT described in this chapter is 4 mm. The objective back apertures range up to 12 mm in diameter. Therefore, the magnification requirement for the described system is $|m| = 1/3$. Together with $l_1 = 130$ mm (Equation 6.17), we have

$$f_{CL} = 32.5 \text{ mm} \tag{6.18}$$

In many imaging applications, aberrations and misalignment of optical elements distort the spatial distribution of the collected light. This directly affects the resolution of the final image. In laser-scanning microscopy, the final image is produced as a time series of collected light intensities. All of the light collected during a given time interval contributes equally to the intensity value of the corresponding spatial pixel, at least insofar as the efficiency of the detector is spatially uniform. Thus, it is more important to collect a consistently large portion of the fluorescent light at the detector than to minimize aberrations in the collection optics.

For completeness, we note that an alternate detection system could be used in which the collected light must pass through two lenses on its way to the detector. The collection optics would then be placed in a configuration similar to that described for the scanning system. The main advantage to this configuration is that the unscattered light, which constitutes the predominant part of the signal in weakly scattering preparations, arrives at the detector as a collimated beam rather than as a focused spot. This would reduce the sensitivity of the signal to any spatial inhomogeneities in the responsivity of the active surface of the detector. The two-lens collection system suffers from the disadvantage of increased complexity and increased loss, as each added optical element produces additional reflection and diffraction.

6.4 REALIZATION

6.4.1 Optical and Mechanical

Our system is laid out on a 4 × 8 × 1 ft air table, which provides room for the lasers, ancillary optics, an interferometric autocorrelator, a two-photon spectrometer, and two experimental setups (Figure 6.15). Room for dispersion compensation optics, should they be required, is also available. Wherever possible, standard, commercially available optics and electronics are used to maximize the reproducibility of the system; these are summarized in Appendix A. Various adapters are fabricated to combine parts of noncompatible design. Machine drawings of these adapters are available by contacting the senior author. Lastly, ancillary electronics are supported by a hanging shelf above the air table.

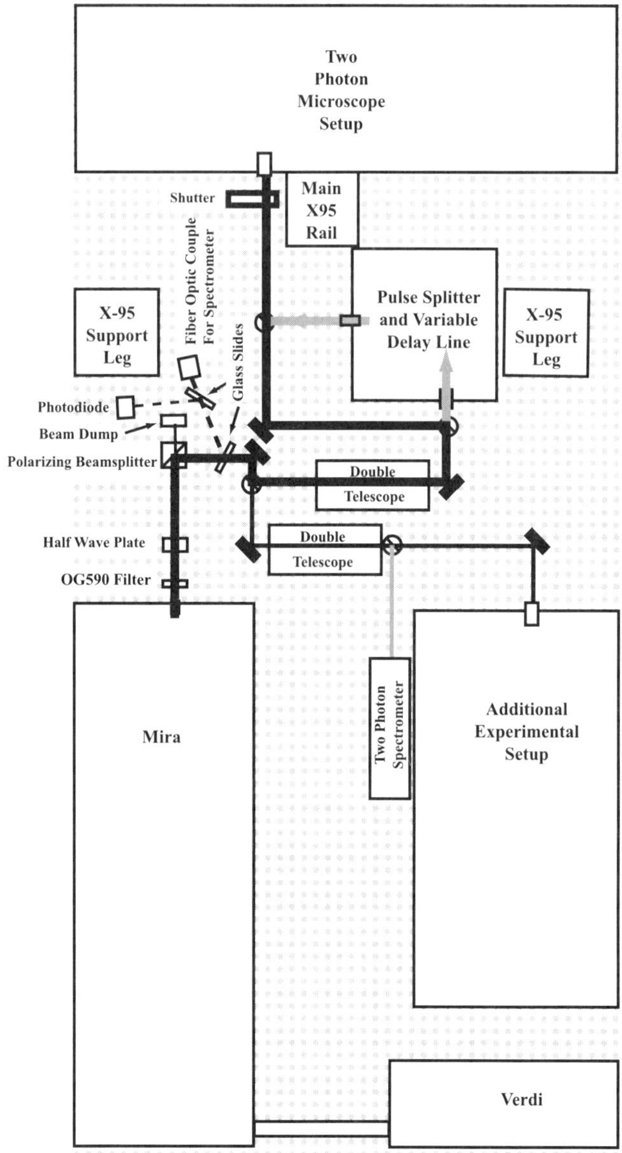

FIGURE 6.15 Schematic diagram, drawn approximately to scale, of the layout on an optical table for delivery of laser light to our two-photon laser-scanning microscope. The main beam path is depicted by the thick dark line. The beam path through the pulse splitter and delay line, used for the optical autocorrelator, is depicted by the thick gray lines. The beam path to the additional experimental setup is depicted by the thin dark line. The beam path to the two-photon spectrometer is depicted by the thin gray line. The dashed lines represent a small fraction (~4%) of the beam that is picked off with glass slides and used to monitor the output power and wavelength of the laser. Circles with diagonal lines represent protected silver mirrors in flipper mounts. Solid rectangles at 45° represent protected silver mirrors in standard mirror mounts. The dimensions of the table are 4 by 8 feet.

6.4.2 Laser System

Our laser system consists of an Nd:Vanadate diode pump laser (Coherent Verdi V10) and a Ti:Sapphire laser (Coherent Mira 900) equipped with a broadband mirror set. Some of the important characteristics of this laser system are:

1. The diode pump laser delivers up to 10 W of pump power at a wavelength of $\lambda = 532$ nm, close to the approximately 15 W maximum thermal limit of the Ti:Sapphire crystal.
2. The Ti:Sapphire laser delivers 100- to 300-fsec pulses, at the preparation, that are tunable over the wavelength range of $\lambda_o = 700$ to 990 nm.
3. These pulses are delivered at a repetition rate of 76 MHz. Therefore, the time between the pulses is ~13 ns, which is long compared with typical fluorescent lifetimes of 1 to 10 ns.
4. The maximum average output power of the laser is approximately 1.5 W at $\lambda_o = 800$ nm, or upwards of 100× the saturation power for typical dyes.[5] This allows for more than two orders of magnitude of loss due to scattering as one probes deep into brain tissue.

Additional equipment required for laser operation include two chillers, typically supplied with the lasers, and a dry nitrogen gas purge to facilitate laser operation for wavelengths longer than 900 nm. In addition, various diagnostic equipment is essential, including an optical spectrometer, an ~500-MHz oscilloscope, an infrared view scope, and infrared view cards.

6.4.3 Path from Laser to Microscope

The delivery pathway of the laser light from the source to the scanners is depicted in Figure 6.15. A colored glass filter, type OG590, is placed immediately after the laser output port as a means to attenuate any residual $\lambda = 532$-nm pump light along the beam path. The output of the Ti:Sapphire laser is horizontally polarized; thus, the intensity of the laser output may be controlled by manipulating the polarization state. The polarization angle is rotated by a zero-order half-wave plate and a polarizing beamsplitter is used to separate the polarized light into its horizontal, or P, and vertical, or S, components. The vertical component, which has a slightly higher reflectance by metal mirror surfaces mounted on the table top and positioned at an angle of 45° relative to the beam path, is deflected toward the microscope. Rotation of the output polarization from horizontal to vertical, via rotation of the half-wave plate, increases the intensity of the light deflected to the microscope according to the cosine of the angle. Lastly, we note for completeness that intensity control on the millisecond time scale can be achieved using a scheme in which a birefringent plate is mounted to a fast optical scanner (W. Denk, patent pending).

A fraction of the laser light is used for beam diagnostics. A first glass slide positioned at 45° relative to the beam is used to pick off roughly 4% of the beam power and deflects it to a PC-based spectrometer via a fiber optic cable. A second glass slide is used to pick off an additional roughly 4% of the beam power and deflects it to a photodiode as a relative measure of the intensity of the laser output.

Here, as in general, the pick-off should be located downstream of the intensity control mechanism but upstream of the shutter.

Either a two-lens or four-lens telescope is used to reshape the laser beam for input to the microscope as required. The output of the Ti:Sapphire laser has a 1.7-milliradian (full-angle) divergence and an initial beam diameter of $D_{beam} = 0.7$ mm. This is reshaped to a beam diameter equal to that of the scan mirrors (i.e., $D_{SM} \approx$ 4 mm), with an f = +25-mm and +100-mm pair as well as a dual f = +50-mm pair for the case of a four-lens telescope.

A mechanical shutter is placed in the beam path just before the beam is deflected upward by a periscope toward the scan mirrors (Figure 6.15). The shutter position is computer controlled and, for safety reasons, its default position in the absence of electrical power blocks the beam from the scanners.

As a practical issue, for purposes of alignment we use iris diaphragms as variable apertures to define the height and position of the beam. Unless otherwise stated, all mirrors have a flat, protected silver surface and are held in standard kinematic mounts. Certain mirrors are placed on translation stages for transverse alignment. Last, safety precautions must be maintained in the use of laser light for TPLSM. Thus, black anodized metal shields are used to surround the scanners and some optics located near normal eye level.

6.4.4 Path along Microscope: Excitation

The microscope is built as a tripod with X-95 optical rail (Figure 6.16). One leg serves as an optical bench, and the other two legs serve as supports. A customized adapter atop the tripod serves to connect the legs. A pair of mirrors mounted at 45° is used as a periscope to steer the beam to the height of the scan mirrors. The top periscope mirror mount is connected horizontally to the main X-95 rail and angled at 45°. The laser light deflected from the top periscope mirror travels horizontally to the alignment mirror and then to the two scan mirrors. The laser light then travels downward through the scan lens, the tube lens, and the dichroic beamsplitter before being focused by the objective onto the preparation (Figure 6.17).

There are five main attachments to the main X-95 rail (Figure 6.17). At the top, two carriages are attached, one in front and one in back. The front carriage holds the scanners and an alignment mirror, and the back carriage holds the top half of the periscope. The next attachment is the scan lens mount. A 100%/0% switchable trinocular headstage, which provides the eyepieces for visual inspection of the preparation and holds the tube lens, is mounted on a separate X-95 carriage below the scan lens. The bottom attachment is a modular focusing unit, to which the detector assembly and the objective holder are mounted. Last, the sample stage allows for "x" and "y" translation of the preparation and sits directly on the air table.

6.4.4.1 Optical Scanner

The back X-95 carriage holds the top mirror of the periscope. An L bracket attaches to the mirror mount and holds it in the appropriate orientation (Figure 6.17). The front X-95 carriage is connected to a rod assembly that holds both the scanners and

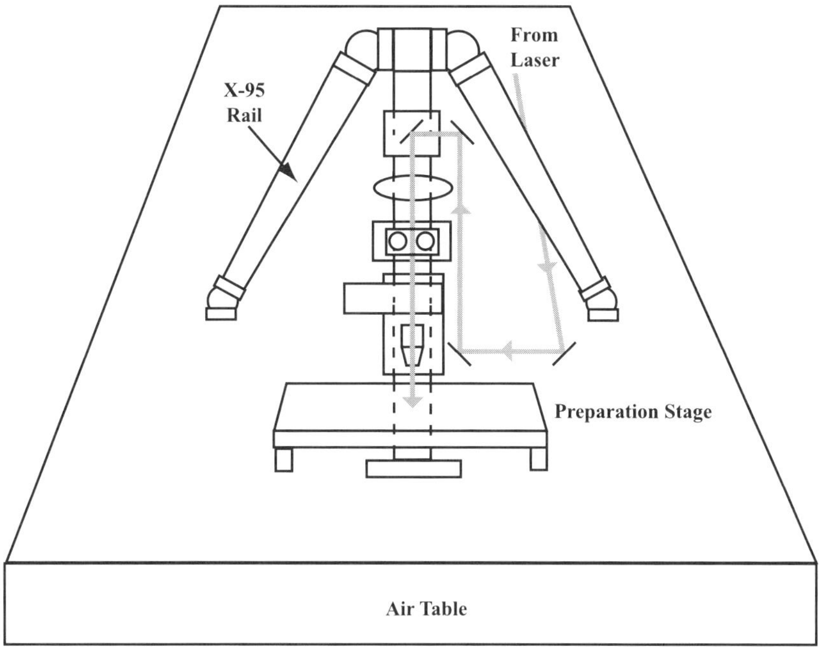

FIGURE 6.16 Illustration of the physical superstructure of the two-photon laser-scanning microscope. The tripod is constructed of X-95 rail; the front leg forms the optical rail of the microscope and the side legs act as supports. The incident laser path is shown in gray.

an alignment mirror (Figure 6.18). The alignment mirror is located at the orthogonal intersection point of the planes defined by the periscope mirrors and the first scan mirror. The scanners are mounted in a small, orthogonally positioned mirror assembly that is an accessory to the scanners. The mirror assembly is attached to a custom-designed, water-cooled base that, in turn, is attached to the rod assembly; water cooling is necessary only for large scan angles at the highest speeds, an atypical situation. The rod assembly itself consists of seven posts, i.e., 2 of 8 in., 4 of 6 in., and 1 of 4 in., and seven cross-post adapters.

6.4.4.2 Scan Lens

The scan lens is a 1.0× magnification Plan-Apo dissection microscope lens (Leica). It is held and centered on a custom-designed mount and attached to the main X-95 carriage. A major consideration for the design of the scan lens holder is mechanical stability, and thus our mount is reinforced with diagonal bracing.

6.4.4.3 Headstage

The headstage is an upright trinocular (Zeiss) in which the phototube pathway is used as the laser beam pathway. The headstage contains a 100%/0% beam switcher that is used to change between visual inspection with the binocular and laser-

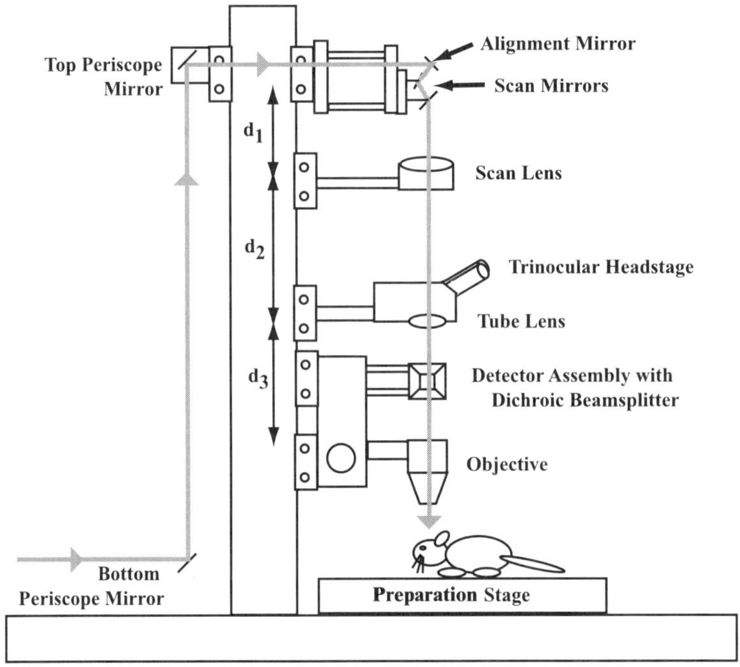

FIGURE 6.17 The main attachments to the main X-95 rail in the two-photon laser-scanning microscope. The laser pathway from the periscope to the objective along the optical rail that forms the microscope includes the alignment mirror, the scan mirrors, the scan lens, the tube lens, and the detector assembly. The collection optics for the fluorescent light are not shown. The values of the separation distances for our system are d_1 = 114.5 mm, d_2 = 252.7 mm, and d_3 = 95 mm.

scanning microscopy. The trinocular contains an f_2 = +160-mm tube lens that is used as the second lens of the scan system telescope. The trinocular is directly mounted to the custom-designed headstage adapter, reinforced with diagonal bracing, that is in turn mounted on an X-95 carriage.

6.4.4.4 Focusing Stage

Axial alignment of the objective is achieved with a commercial modular focusing unit (Nikon). This unit allows 30 mm of vertical motion, which facilitates changing preparations and objectives. A custom-made adapter plate is attached to the focusing unit and the objective holder is mounted to the adapter plate. The objective holder consists of a custom-made groove plate and five Microbench™ (Linos Photonics) parts in the configuration shown in Figure 6.19.

6.4.5 Path along Microscope: Detection

6.4.5.1 Detector Assembly

The detector assembly is attached to the focusing unit via the adapter plate. The entire detector assembly is placed on a slider plate so that the dichroic beamsplitter

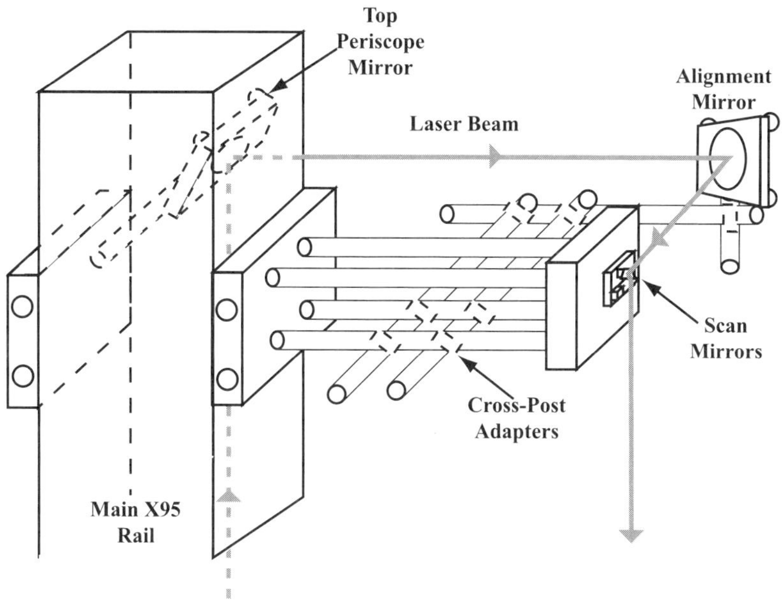

FIGURE 6.18 The uppermost attachments to the X-95 rail that forms the microscope. The attachment at the rear holds the top periscope mirror. The attachment at the front holds the rod assembly and is constructed of stainless steel posts, right-angle cross-post adapters, and an X-95 carriage that connects to the scan mirrors and the alignment mirror.

can be removed from the visual path and then replaced without changing its alignment relative to the other collection optics. The back panel of a slider plate is mounted to the adapter plate, and a Microbench™ mounting plate is attached to the front panel of the slider plate. Four rods inserted into the mounting plate are used to attach the detector assembly by its rear left face.

The detector assembly is constructed mainly with Microbench™ parts in a double-cube configuration (Figure 6.20a). Nine square plates are formed into a rectangle using four long rails and two short rails. Two f_{CL} = +27-mm lenses serve as the collectors and are loaded into the far right plate and the right rear plate. Eight 2-in. rods are inserted into the same plates to provide a mounting area for the glass filters and detectors. Infrared blocking filters and appropriate bandpass filters are mounted in additional square plates that slide directly onto the rods. The PMT detectors are mounted in a customized holder (Figure 6.20b), which also slides directly onto the rods.

The two dichroic beamsplitter mirrors, one of which separates the incident laser light from the emitted fluorescent light (beamsplitter #1) and one of which resolves the fluorescent light into separate spectral bands (beamsplitter #2), are held on prism mounts inserted through the front left plate and bottom right plate. The laser enters through the top left plate and exits through the bottom left plate of the double cube. The far left plate, front right plate, and the open top right region are covered with black tape to block stray light from reaching the detectors.

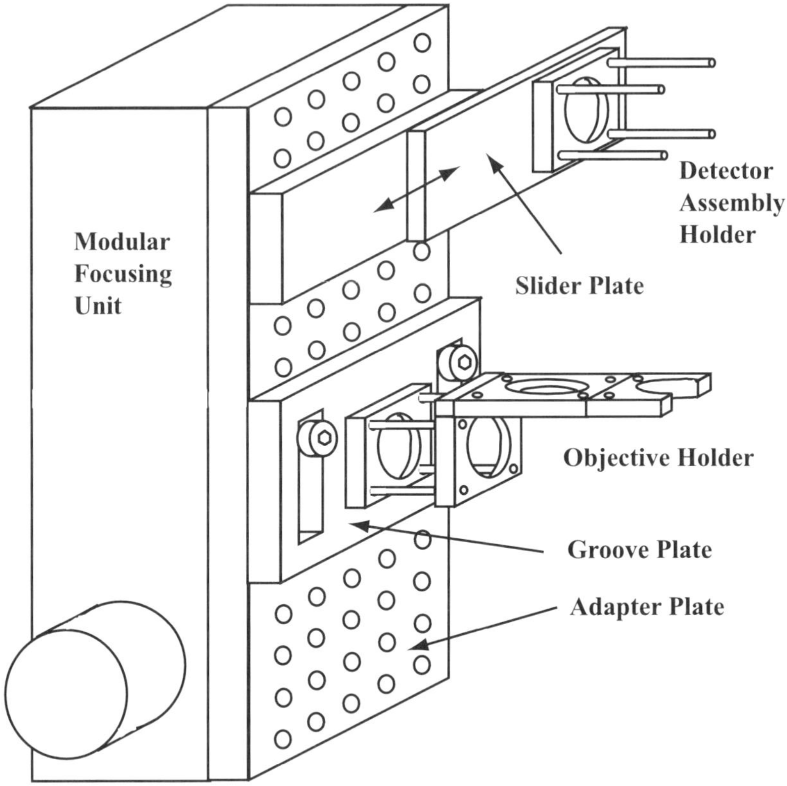

FIGURE 6.19 The focusing unit that supports the objective and detector assembly on the main X-95 rail. The objective holder consists of Microbench™ parts (Linos Photonics) connected to a modular focusing unit (Nikon) by two custom-made plates; an adapter plate with 1/4-20 holes spaced on ½-in. centers that is driven by the modular focusing unit, and a groove plate that supports the objective holder. The detector assembly holder consists of a slider plate and Microbench™ parts and is mounted on the adapter plate.

6.4.5.2 Collection Optics

The first element in the collection pathway (Figure 6.21) is the long-pass dichroic beamsplitter mirror, i.e., $\lambda_{cut} = 600$-nm DCLP, that transmits light at the laser excitation wavelength while deflecting fluorescent light toward the detectors. A second dichroic may be used, if necessary, to separate the fluorescent signals from two different fluorescent indicators. For example, we use a $\lambda_{cut} = 505$-nm DRLP to resolve light from yellow fluorescent protein (YFP) vs. cyan fluorescent protein (CFP). The signal in either path is then focused onto the active area of a PMT by a short-focal-length lens. A laser-blocking filter, comprising 2 mm of Corning no. BG40 glass, is used to further reduce the background level and associated noise caused by scattered incident laser light. Additional filters are placed in front of the detector as is appropriate for the given experiment, e.g., $\lambda_{pass} = 525 \pm 5$ nm to isolate YFP and $\lambda_{pass} = 485 \pm 11$ nm to isolate CFP.

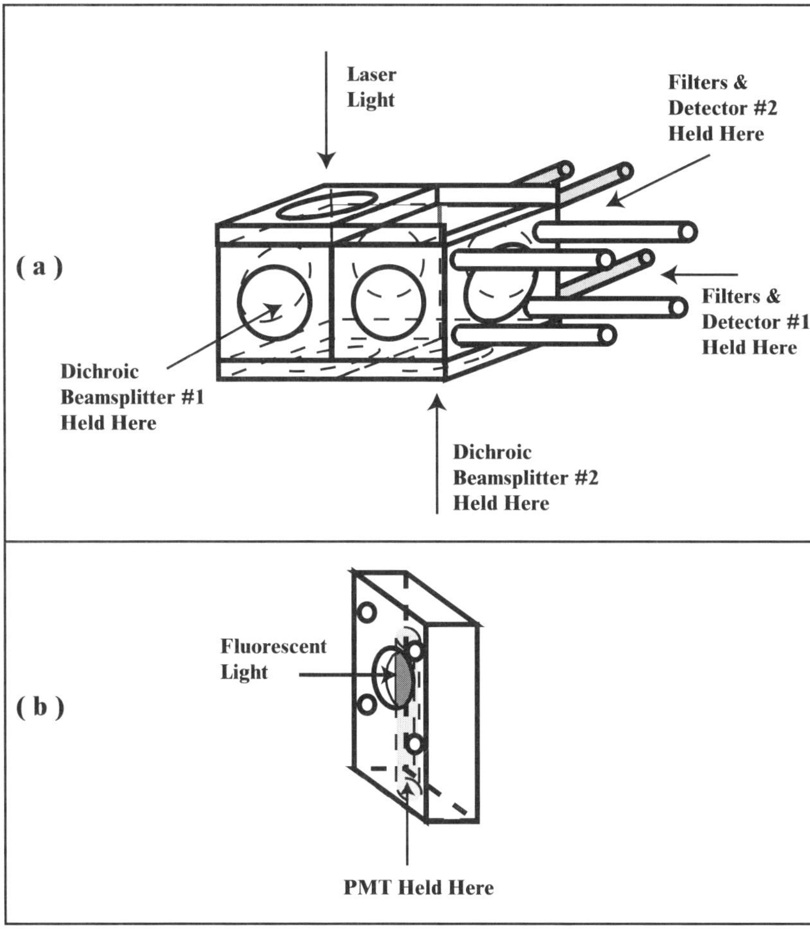

FIGURE 6.20 Illustration of the detector assembly. (a) The main assembly is fabricated from Microbench™ parts. Dichroic beamsplitters are mounted on adjustable prism mounts and placed in the labeled openings. Filter blocks and the photomultiplier tubes (PMTs) are held by the four rods protruding along each path. (b) The PMT case is a custom-made aluminum cage designed to be compatible with Microbench™ parts and the compact Hamamatsu PMTs.

In the collection assembly, an f_{CL} = +27-mm lens of D_{CL} = 30-mm diameter is placed at l_1 = 120 mm from the back aperture of the objective and at l_2 = 35 mm from the detector. In this configuration, the back aperture is imaged onto the detector surface with a magnification of 0.29×. Therefore, a D_{BA} = 8-mm back aperture is imaged to a 2.3-mm diameter circle at the PMT active surface.

6.4.5.3 Photodetectors

The detectors in our system are Hamamatsu R6357 photomultiplier tubes. We chose these PMTs because of their spectral properties, high sensitivity, and relatively small

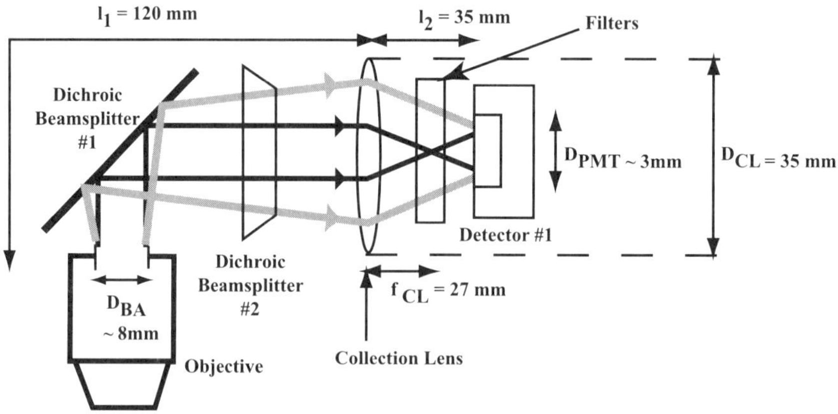

FIGURE 6.21 Detailed schematic of the detector assembly. The positions of both dichroic beamsplitters are shown; however, for clarity we omit the optics and detector for the second, otherwise equivalent, pathway. Unscattered fluorescent light is depicted as dark lines, while scattered fluorescent light is depicted as gray lines.

size. They are powered by adjustable high-voltage supplies. The output of each PMT is converted to a voltage with a current-to-voltage converter of local design (no. OPA111 operational amplifier, Burr Brown) with a measured time constant of $\tau = 0.8$ μsec and adjustable conversion gains of 1×10^4 to 8×10^4 V/A. An alternative and possibly superior solution is to use a charge-to-voltage switched integrator gated by the pixel clock of the acquisition program.

6.4.6 Scan and Acquisition Electronics

The scanners and associated electronics and computer boards must be chosen and configured to provide horizontal and vertical scans that are synchronized to each other, so that they define a raster, as well as to the data acquisition. Additionally, in order to make the scan and acquisition run independently of the computer's CPU, both the scan functions and the acquisition function should be synchronized to a pixel clock on one of the computer boards, as depicted in Figure 6.22.

6.4.6.1 Scan Mirrors and Drivers

We use state-of-the-art galvanometer optical scanners from Cambridge Technology Inc. (model 6210). These scanners are capable of rotations up to ± 20° and rates of up to 1.7 kHz at reduced deflection angles. We typically use angles of ± 5° or less; for a 40× magnification, deflections of ± 1° correspond to distances of ± 40 μm in the focal plane. The scanners hold $D_{SM} = \simeq 4$ mm aperture mirrors (6 min at 45°) provided by the manufacturer. Each scanner is driven by a MicroMax™ Series 670 Single Axis Board Level Mirror Positioning System that takes a standard analog ±10-V signal as the input and transforms the voltage into an angle.

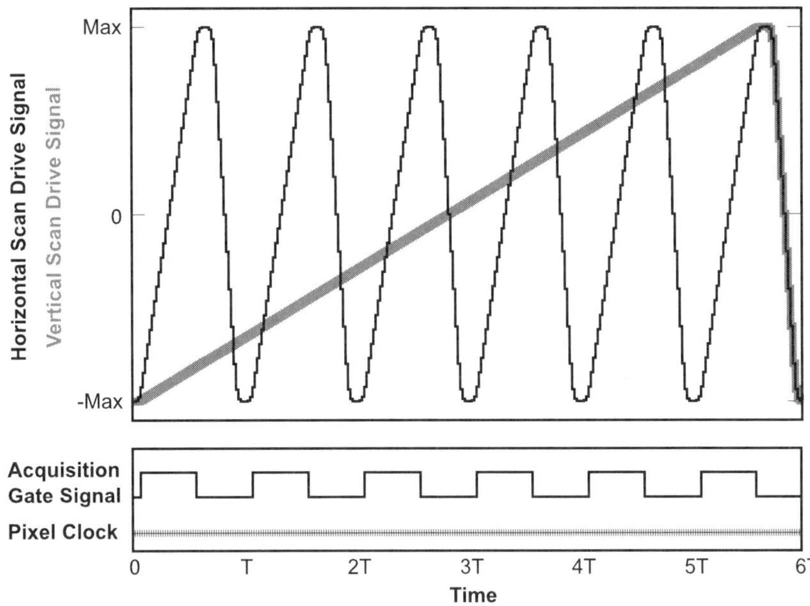

FIGURE 6.22 Cartoon illustration of the drive signals for the scan mirrors and data acquisition over the period of a full frame of acquisition. The time is in units of horizontal scan periods, denoted T. All functions are synchronized to the continuously running pixel clock. The data acquisition gate is active for a portion of the fast horizontal scan. During this time window, data are acquired at the pixel clock rate. The horizontal (fast) scan signal is updated with each pixel clock tick, while the vertical (slow) scan signal is updated once per fast scan cycle. The number of steps and the number of horizontal scans per vertical scan have been greatly reduced for illustrative purposes.

6.4.6.2 Computer Cards

The horizontal- and vertical-scan system is composed of a scan generator consisting of a pair of National Instruments model PCI-6711 analog output boards. The horizontal- and vertical-scan boards have first-in-first-out (FIFO) memories so that one-dimensional waveforms may be downloaded to the boards. The boards can then be commanded to transmit those waveforms repetitively. Digital counters, also located on the boards, can be programmed to set up the sync functions, so that the horizontal and vertical scans will run in a synchronized manner. Although each board has four analog output channels, the PCI-6711 has only a single common clock and a single common FIFO memory. Therefore, two boards are needed for very fast scan operation. Eight digital output lines are also available on each board, which may be used for various control output functions such as a shutter control for the laser.

Data acquisition is performed by a National Instruments model PCI-6110E four-channel analog input board. Two channels are used for the two photomultipliers and two are reserved for electrophysiological data or additonal detectors. This board has four independent analog-to-digital converters. Therefore, it can acquire data at very fast rates and directly stream the data into memory. The board is synchronized with the two scan boards (Figure 6.22).

The three interface boards communicate with each other via the real-time system integration (RTSI) bus accessed via a connector on the top surface of the board (Table 6.2). A 34-pin cable connects all of the RTSI bus lines. The pixel clock, line clock, and frame clock signals from the raster generator are passed to the PCI-6110E, so that data collection is automatically synchronized to the scan (Figure 6.22). The host computer needs only to manage the data flow.

The scan and acquisition software sets up the control counters and downloads the scan waveforms to the respective boards. The start signal is then issued from the host machine, and the horizontal generation begins and repeats continuously without any further intervention by the software. This method of semiautonomous operation allows instrument operation and data collection to occur in a manner independent of the processor speed of the computer. The computer needs only to have enough memory to buffer the data. Alternately, the data can stream directly to disk, depending on the data rate required for a particular experiment.

6.4.6.3 LabVIEW™-Based Software

The scan and data collection system is controlled by a master program written in LabVIEW™ 5.1. At the start of the program, the ramp is generated by means of a LabVIEW™ subroutine. The ramp waveform is divided into six separate sections: a rising cubic acceleration region, a rising linear region, a rising cubic deceleration region, a falling cubic acceleration region, a linear falling region, and a falling cubic deceleration region. The ends of the separate regions match in value and slope so that a continuous, smooth waveform is formed (Figure 6.22). The resulting waveform reduces the slew rate of the signals that will be applied to the mechanical scanners and thus facilitates the fastest possible scan rates. The horizontal waveform is calculated by means of a FOR LOOP with six separate paths, selected in turn as the FOR LOOP steps from zero to the final value, which is the number of pixels desired for one line. Each path is constructed out of standard LabVIEW™ mathematical components. The vertical waveform utilizes the same waveform calculation method; the period of the vertical waveform must be a multiple of the period of the horizontal waveform. For each board, the AO Config.vi routine supplied by LabVIEW™ is used to assign the on-board FIFO memory that stores the calculated waveform values.

Each of the scan boards has a system timing controller circuit, denoted as the DAQ STC, that makes a common set of control signals available on each board (Table 6.2). Thus, the control signals and the counter signals from one board are available to the other boards via common connections provided by the RTSI bus. In addition to the control signals, each DAQ STC circuit provides two counters and eight general-purpose digital I/O lines. The RTSI bus carries various control and clock signals as well as eight lines, designated RTSI 1 through 8, that are software assigned by the user. Most of the I/O and control signals for each of the counters can be assigned to any of the RTSI bus lines.

For the horizontal-scan ramp, the route signal.vi routine supplied by LabVIEW™ is used to assign the AO update signal to one of the RTSI lines, i.e., RTSI 4. This

TABLE 6.2
Assignments for National Instruments Boards

Real-Time System Integration (RTSI) Bus Assignments

RTSI #	Signal	Source	Purpose
4	Pixel clock	Board #1	Master clock for counters in the scan generators
3	Pixel clock gate	Board #3	Asserts the start of an acquisition period during a horizontal scan
2	Frame clock	Board #1	Asserts the beginning of each frame
1	Line clock	Board #2	Asserts the beginning of each line scan

Connection Assignments

Board #	Name	Board Connection	Function
1	Horizontal scan generator (PCI-6711)	Counter 0 input	Counts down line clock (Input from RTSI #1)
		Counter 0 output	Generates frame clock (Output to RTSI #2)
		Analog output DAC0	Horizontal scan signal
2	Vertical scan generator (PCI-6711)	Counter 0 input	Counts down pixel clock (Input from RTSI #4)
		Counter 0 output	Generates line clock (Output to RTSI #1)
		Analog output DAC0	Vertical scan signal
		Digital I/O output 0	Shutter control
3	Analog input (PCI-6110E)	Counter 0 input	Counts down pixel clock (Input from RTSI #4)
		Counter 0 output	Generates pixel clock gate (Output to RTSI #3)
		Counter 0 gate	Gated by line clock (Input from RTSI #1)
		Trigger 1	Acquisition gate (Input from RTSI #3)
		Startscan	Acquisition clock (Input from RTSI #4)
		Digital I/O lines 0–4	Output to focus stepper motor controller
		Analog input 0	Detector, PMT #1
		Analog input 1	Detector, PMT #2
		Analog input 2	Reserved for electrophysiological data or optional detector
		Analog input 3	Reserved for electrophysiological data

signal serves as the system master pixel clock. Counter 0 on the vertical-scan board is initialized with the index value of the horizontal-scan generator and then counts down the pixel clock, which reaches zero at the end of the horizontal scan. This way, the counter generates an output pulse when the horizontal scan completes. This signal is called the line clock and is assigned to RTSI 1 by means of the route signal.vi routine. By means of the down counter.vi routine, the line clock signal is used as the update clock for the vertical-scan ramp generator. Note that, as distributed by National Instruments, the down counter.vi cannot be assigned to a signal on the RTSI bus. To modify the routine, access the diagram of the down counter.vi, remove the case structure associated with the time-base assignments, and configure the cluster containing the time-base source and time-base signal so that these two parameters are selected by inputs to the routine. This way, these parameters can be selected in the main vi routine. The time-base source is set to 6, which selects the RTSI, and the time-base signal is set to the desired RTSI pin.

With the scan boards configured as indicated, the horizontal-scan generator is the master and the vertical-scan generator is synchronized to the horizontal-scan pixel clock and steps up every time the horizontal-scan ramp completes one cycle. Thus, the horizontal-scan and vertical-scan ramps are always synchronized (Figure 6.22). In a similar manner, counter 0 is initialized with the vertical-scan ramp index value and counts down the line clock on RTSI 1, so the resultant output is triggered when the vertical-scan ramp finishes. This signal is called the frame clock and, by means of a route signal.vi, is attached to RTSI 2. As configured, pixel clock, line clock, and frame clock are all available on the RTSI bus.

Analog data collection is best understood by assuming that the horizontal-scan and vertical-scan ramps are both active and that the pixel clock, line clock, and frame clock are being generated. Using a modified delayed pulse config.vi routine supplied by LabVIEW™, counter 0 is arranged so that it is gated when line clock is asserted and then begins counting the pixel clock on RTSI 4 until the horizontal ramp reaches the desired value. Then the counter 0 output, denoted by the pixel clock gate and assigned to RTSI 3, opens the analog gate. Data collection is clocked by the pixel clock and is initiated by gating the pixel clock on RTSI 4 via the Startscan input. Then counter 0 is reset by the pixel clock gate and begins counting pixels until the requested number of data points has been reached. Then the output of counter 0 turns off the analog data collection using the delayed pulse config.vi routine. The horizontal scan generator completes the scan and then resets, and the process begins again. The delayed pulse config.vi routine is modified in the same manner as the down counter.vi routine. The diagram is modified by removing the structure in front of the time-base source and time-base signal cluster and adding direct assignments, i.e., for RTSI 4.

The analog data are read by using the services of the AI read.vi routine and displayed by using the National Instruments IMAQ™ imaging software. AI read.vi generates a two-dimensional array of data, but that array is not in a form suitable for generating a display since it contains no horizontal-scan indexing information. One solution is to place the AI read.vi routine in a FOR LOOP indexed by the number of lines in each frame. The AI read.vi is commanded to read the pixel data for one scan line; a complete frame of data from the input buffer is acquired by calling the routine at each cycle of the FOR LOOP. For display purposes, one channel is selected from the AI read output array by means of the index array.vi routine. This effectively converts the two-dimensional array to a one-dimensional array, formed back into a two-dimensional array as the index values are automatically incremented by the FOR LOOP as the array passes through the loop boundary. The array is presented to the array-to-image.vi routine, along with an image definition from the create image.vi routine. The image data then pass to the windraw.vi routine that displays the image on the monitor screen.

When the main program is started, the sub vi routine calculates the horizontal-scan and vertical-scan values, the boards are programmed, and the various counters are initialized. Then the system stops and waits for the start signal. All functions receive a common start so that all operations start together and remain synchronized. Once started, the horizontal-scan and vertical-scan ramps operate without using any control signals. The computer resources can then be devoted to managing data collection and storage and the display monitor. The system also waits for the stop button to initiate a controlled shutdown procedure. In shutdown, all the counters are cleared along with the RTSI bus lines, which allows a fresh start.

6.4.7 Focus Control

It is often desirable to step along the focus axis either as a means to automate the choice of focal plane or to acquire a set of successive images systematically. The modular focus unit is easily coupled to a stepping motor by capturing the fine focus knob in a friction-held cap. The motor is a two-phase stepping motor with two windings driven by an SN754410 (Texas Instruments) motor driver integrated circuit. The motor signals for the driver are generated in a LabVIEW™ sub program that allows both manual and automatic stepping. To increase the angular resolution, the stepping motor is operated in half-step mode, so the stepping pattern is 4 bits wide and 8 bits long. The pattern is installed in an eight-element array by means of the build array.vi routine. The eight steps described in binary are:

$$\begin{aligned}&\text{Step 0: 1000}\\&\text{Step 1: 1001}\\&\text{Step 2: 0001}\\&\text{Step 3: 0101}\\&\text{Step 4: 0100}\\&\text{Step 5: 0110}\\&\text{Step 6: 0010}\\&\text{Step 7: 1010}\end{aligned} \tag{6.19}$$

In automatic mode, one of the counters in one of the DAC STC chips is used to count the frame clock and the counter value is then converted to modulo 8 by means of a LabVIEW™ math function. The resultant number varies between zero and seven and is used to index the array by means of the index array.vi. If the reverse motion is desired, a formula node is used to subtract the value from seven before indexing. For manual mode, the stepping is controlled by switches on the

LabVIEW™ front panel that enable a FOR LOOP that steps from zero to seven once every 0.1 sec. The stepping pattern selected is presented to the DIO write.vi and the DIO start.vi routines. The 4 bits then appear on four lines of the digital port selected by the port config.vi routine; a fifth line is used as an on/off control (Table 6.2). Each binary bit corresponds to a single pin of the port.

In this scheme, reading the counter is done after the image is displayed and stored. In the unfortunate event that the data rate is high and the computer processor speed is slow, the counter may increment more than once before being read. In an unusual case such as this, it might be necessary to implement control of the stepping motor with hardware.

6.4.8 Image Rotator

For many experiments it is desirable to rotate the image. For example, in line-scan operation, in which the vertical-scan axis is degenerated into a single line, image rotation allows the region of interest to be aligned along the horizontal-scan axis. The image is rotated by reorienting the raster scan. The rotation is performed in hardware by four analog multipliers arranged in a rotation matrix, wired to conform to

$$\begin{aligned} X' &\leftarrow X\cos\Theta - Y\sin\Theta \\ Y' &\leftarrow X\sin\Theta + Y\cos\Theta \end{aligned} \tag{6.20}$$

where X and Y are the original horizontal- and vertical-scan waveforms, X′ and Y′ are the rotated horizontal- and vertical-scan waveforms, and Θ is the desired rotation angle. The sinΘ and cosΘ terms are calculated by a microprocessor, converted to analog by digital-to-analog converters, and presented to the analog multipliers. Two more multipliers are used, one in each output channel, to act as gain controls. The system is capable of rotation by any angle without limitation. The microprocessor is controlled by means of a keypad and a rotation control knob; the operation status is displayed on a liquid crystal display panel. This system is similar to all-analog designs implemented in the past.[44] The virtue of the present system is the use of a single knob and keypad for ease of control and the use of an external serial RS232 input to the microprocessor to give the main LabVIEW™ program control of the rotator as well.

6.5 PERFORMANCE AND APPLICATIONS

6.5.1 System Validation

After our system was constructed and tested, various calibrations were performed to determine the true lateral resolution, axial resolution, exact values of scale factors, and uniformity of the image.

6.5.1.1 Spatial Scales

The horizontal and vertical scale factors are determined by imaging a circular opening of known diameter, as shown for a 50-μm metal pinhole illuminated with

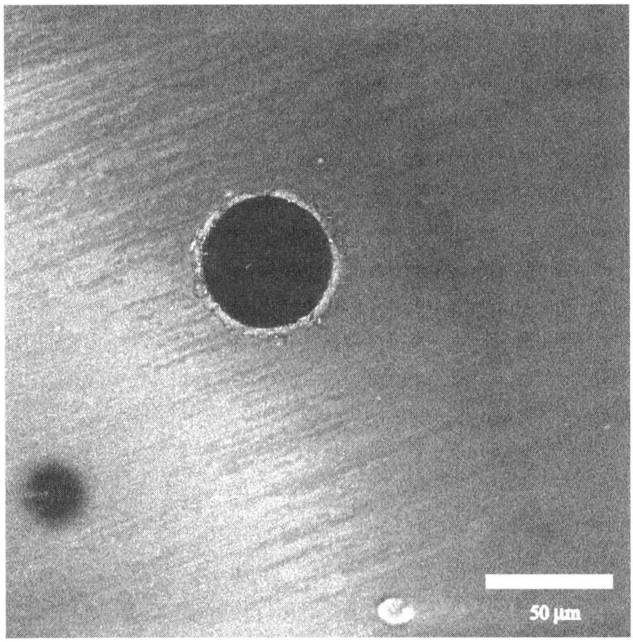

FIGURE 6.23 Image of an aperture used to calibrate the lateral dimension of our two-photon laser-scanning microscope. The aperture corresponds to a 50-μm pinhole and the contrast is generated by reflected light from the polished metal surface.

800-nm light (Figure 6.23). No fluorescent dye is used; the reflected light from the polished surface is sufficiently strong so that enough 800-nm light leaks through the dichroic beamsplitter and blocking filters to provide a signal. The diameter along the horizontal and vertical axes, as measured in pixels, yields the scale factor at a particular zoom setting. As the image is best viewed without distortion, the gain for the horizontal and vertical directions should be identical and can be adjusted via the digital multipliers in the "rotator box." This calibration is repeated at several magnifications to ensure the linearity of the scale factor as a function of magnification.

6.5.1.2 Resolution

The resolving power of the microscope can be determined by imaging fluorescent beads whose diameter is below the optical resolution limit. The beads must be dilute, so that an individual bead can be identified, and they should be suspended in agarose to reduce motion during imaging. Figure 6.24 contains an image of 0.2-μm yellow-green fluorescent microspheres (Polysciences Inc., PA), suspended in 2% agarose, that was obtained with our microscope using a 40× water immersion objective (Zeiss). The full width at half-maximal intensity from individual beads in the lateral dimension was calculated from the acquired images (Matlab™; Mathworks, MA). The axial resolution is obtained by compiling a series of images taken at different axial positions and calculating the intensity profile of a single bead as a function of axial position. We determined that the spatial resolution (FWHM) for our system was

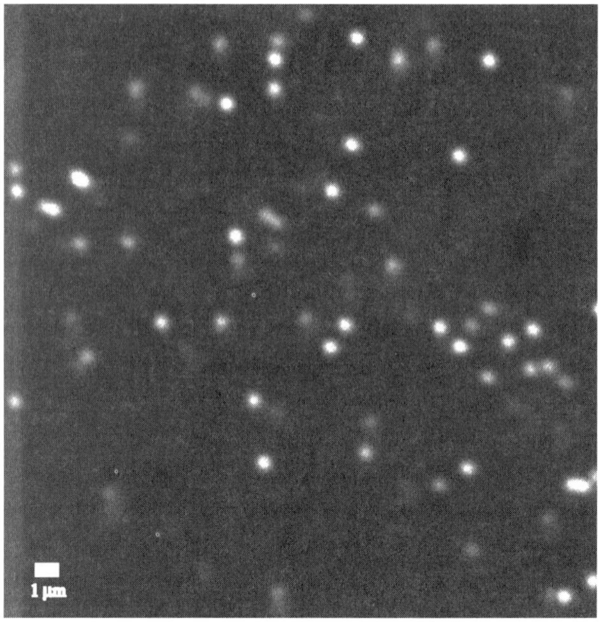

FIGURE 6.24 Image of fluorescent microspheres used to establish the resolution of our two-photon laser-scanning microscope. The microspheres are 0.2-µm latex balls (no. 17151; Polysciences, Inc., Warrington, PA) coated with fluorescein.

$$d_{lateral} \cong 0.5\ \mu m \quad \text{and} \quad d_{axial} \cong 2\ \mu m$$

These are consistent with the calculated, worst-case values (Equations 6.4 and 6.5) and a previous characterization.[45]

6.5.1.3 Spatial Uniformity

The homogeneity of the scan was determined by imaging a spatially uniform field of fluorescence, i.e., a Petri dish filled with a 100-µM solution of fluorescein. This method detects inhomogeneities in the detection pathway as well as in the scan pathway. The profile obtained in this manner, shown in Figure 6.25, is observed to be asymmetric and to decay more extensively, by nearly 20% over 100 µm along some directions, than the prediction for the scan pathway alone (Figure 6.11). Nonetheless, the profile is stable for a given configuration of the microscope, the intensity profile in the central 100 µm of the field is uniform to within 10%, and the measured profile can be used as a calibrated filter for images taken with this system. Postprocessing with this filter yields a fluorescent image free of spatial intensity distortions.

6.5.2 Pulse Width

The shape and duration of the incident laser pulse critically affects the efficiency of nonlinear optical processes (Figure 6.3). The ultrashort pulses used for multiphoton

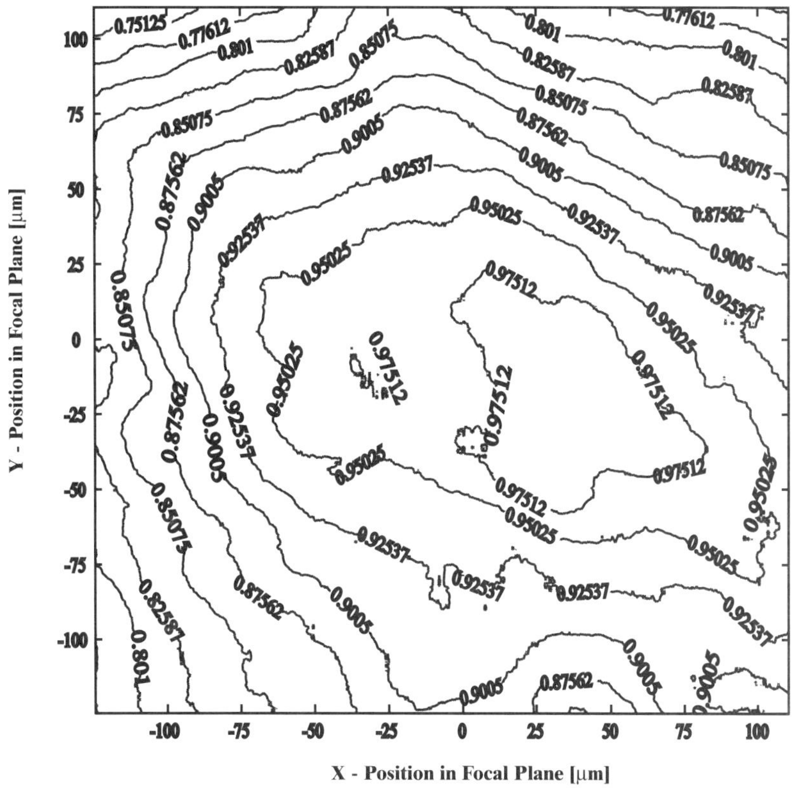

FIGURE 6.25 Contour plot of the spatial uniformity of the measured two-photon fluorescence measured with a 40× magnification lens. Note that the decrement across the entire field is due to both nonideal scanning and detection conditions and exceeds that calculated for the scanning alone (Figure 6.11). The decrement across the central 100 μm of the field is less than 10%.

microscopy are typically one hundred to a few hundred femtoseconds in duration. Since no electronic photodetector has the necessary time resolution to measure such time scales, optical pulse measurement techniques must be used. An autocorrelation measurement is an effective and relatively simple method of monitoring the pulse length.

6.5.2.1 Autocorrelation

In an autocorrelation measurement, a laser pulse is split into two identical copies. One pulse is delayed relative to the other by a time τ, which results in two pulses with electric fields $E(t)$ and $E(t + \tau)$. The two pulses are recombined and used to excite a nonlinear process such as two-photon fluorescence. The interferometric autocorrelation function, denoted $I_{AC}(\tau)$, is given by the sum of $E(t)$ and $E(t + \tau)$, raised to the fourth power and averaged over time. This appears as an oscillatory signal with an envelope that depends on the delay time τ (Appendix B). A convenient

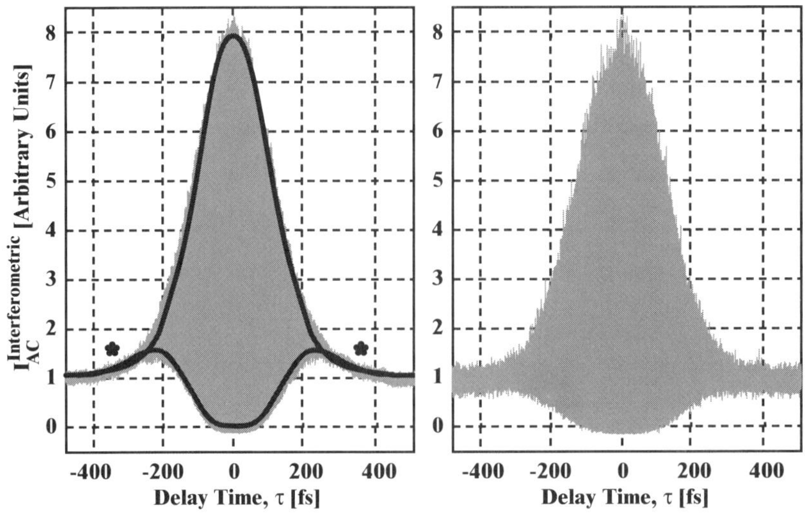

FIGURE 6.26 The interferometric autocorrelation function measured for laser pulses at the focus of the objective with our two-photon laser-scanning microscope. (a) The autocorrelation under typical levels of performance. The asterisks indicate additional curvature as a result of chirp in the pulse. The envelope of the autocorrelation is fit with the calculated autocorrelation for a chirped pulse (thick envelope), from which we estimate the FWHM of the laser pulse to be 250 fsec. The chirp parameter is $b = 5 \times 10^{-5}$ fsec^{-2}; this corresponds to a fractional shift in frequency of $2 \times b \times \tau_P^{FWHM} \times \omega = 0.01$, with $\omega = \lambda_0/2 \times \pi \times c$, where $\lambda_0 = 800$ nm and c is the speed of light. (b) The autocorrelation for a pulse with essentially no chirp, i.e., $b \cong 0$. For this case we estimate the FWHM of the laser pulse to be 180 fsec (Equation A5).

feature of this autocorrelation is an 8:1 ratio of the amplitude of the maximum from zero to the amplitude of the constant background produced by widely separated pulses. Achieving this ratio indicates that the apparatus is well aligned. In the case that the laser pulse has essentially no chirp, the FWHM of the autocorrelation is equal to 1.7-times the FWHM of the intensity of the original pulse.

Our laser system and microscope were set up without external pulse compression. We constructed an autocorrelator that could be switched into the beam path (Figure 6.15), as described.[46,47] We found that, although the commercially purchased Ti:Sapphire laser should output pulses around 120 fsec, pulses at the microscope objective are typically closer to

$$\tau_P^{FWHM} = 250 \text{ fsec}$$

as shown by a fit of a model autocorrelation (Appendix B) to the measured autocorrelation function (Figure 6.26a). This pulse is chirped, as evidenced by the curvature in the wings of the autocorrelation function (* in Figure 6.26a). The increase in pulse width represents the cumulative effect of optics throughout the entire TPLSM system (Figure 6.15). By judicious tuning of the laser, we could achieve pulses without chirp and with widths of $\tau_P^{FWHM} = 180$ fsec (Figure 6.26b and Equation A5).

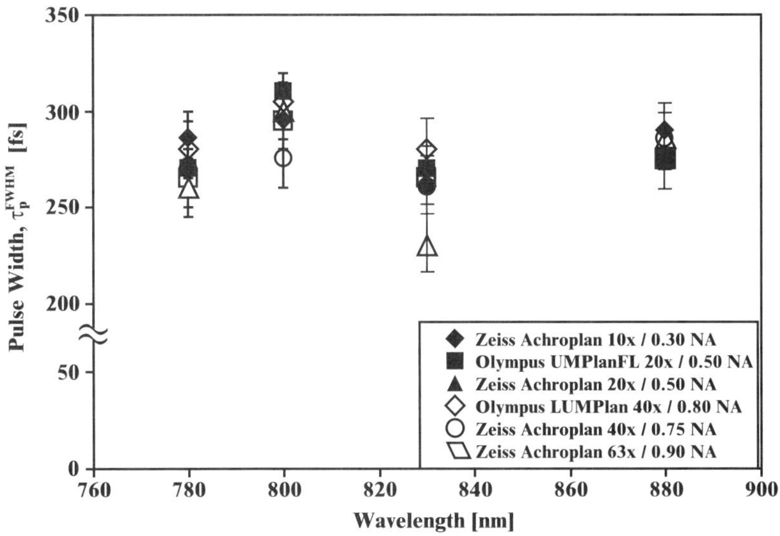

FIGURE 6.27 Plot of the FWHM of the laser pulse measured for six different common objectives as a function of excitation wavelength. There is no significant difference in pulse width between lenses or as a function of frequency.

6.5.2.2 Comparison of Objectives

To compare the performance of various brands and types of objectives, each with different glasses and detailed construction, identical laser parameters were measured using a number of objectives on the microscope. We found that the variations of the measured pulse width from the objectives were not significant at a given wavelength and no significant change in pulse width over the 780- to 880-nm range of wavelengths for the incident light (Figure 6.27) occurred.

6.5.2.3 Effect of the Scan Lens

One concern in the construction of the microscope is dispersion in the large amount of glass of the scan lens that is required to achieve superior off-axis optical characteristics. Pulse widths were compared for measurements with the normal scan lens (Leica, Appendix A) in place as opposed to a thinner achromat used as a scan lens. With the initial pulse width near $\tau_P^{FWHM} = 250$ fsec, and for wavelengths near 790 nm, we found that the use of an achromat shortens the pulse width to $\tau_P^{FWHM} = 230$ fsec. This is less than a 10% improvement at the cost of decreased off-axis optical quality. For comparison, removal of the scan lens and the tube lens so that only the objective remains in the pathway results in a value of $\tau_P^{FWHM} = 180$ fsec.

6.5.3 Two-Photon Spectra

Single-photon excitation spectra have been published for most common dyes and can be experimentally obtained with a noncoherent source. In contrast, the two-photon excitation spectra of relatively few molecules have been published to date. Two-photon excitation spectra cannot be trivially derived from their corresponding single-photon excitation spectra, partially due to different quantum selection rules governing one- vs. two-photon absorption. Therefore, two-photon excitation spectra must be determined experimentally whenever one uses a new dye or indicator.

We constructed a spectrophotometer to measure the two-photon action spectra, i.e., the extent of integrated fluorescence, as a function of the excitation wavelength. The spectrum is uncompensated for the relative extent of radiative vs. nonradiative decay. It is obtained by measuring the two-photon excited fluorescence from an unknown sample relative to that of a dye with a known two-photon spectrum[48]; in our case we used fluorescein at pH 11 (the tabulated spectrum for pH 13[49] matches an earlier measurement taken at pH 11[48]). The measurements are performed and compared on a wavelength-by-wavelength basis; thus, the final action spectrum does not depend on detailed knowledge of the pulse characteristics of the laser (Figure 6.28a and Appendix C). For most fluorescent studies, in which the signal is based on the amount of generated fluorescence, the action spectra contain more relevant information than the excitation spectra. On the other hand, for certain photobleaching experiments, the actual excitation spectra may contain the more relevant information.

The two-photon action spectrum of the fluorescent dye, cascade blue, is shown in Figure 6.28b (dark lines and symbols). Superimposed on the graph (gray lines and symbols) is the published two-photon excitation spectrum of the dye taken by direct measurement. The two curves are well matched to within a scale factor, i.e., the two-photon fluorescent efficiency. This result demonstrates the validity of the comparative method. The two-photon spectra for yellow fluorescent protein (YFP) and cyan fluorescent protein (CFP) are shown in Figure 6.28c. These two genetically encodable variants of green fluorescent proteins are used in several physiological indicators, e.g., Ca^{2+}-sensitive probes,[50,51] that can be delivered by viral vectors.

6.5.4 *In Vitro* [Ca^{2+}] Measurements

We describe the use of our two-photon laser-scanning microscope to probe the change in internal calcium concentration of plated HeLa cells infected with the Ca^{2+}-sensitive cameleon protein via an adenovirus vector.[52] Cameleon is a chimeric protein that consists of two genetically encodable fluorophores, CFP and YFP, separated by a calcium-binding protein complex derived from calmodulin;[50] the spectra of these fluorophores are described in Figure 6.28c. At low calcium concentrations, the fluorophores are held distal from one another, and excitation of the cyan fluorophore results primarily in emission characteristic of CFP. At high calcium concentrations, the calcium-induced conformational change of calmodulin brings the two fluorophores in close proximity to each other. This allows fluorescence resonance energy transfer to occur from the cyan to the yellow fluorescent protein. In this conformation, excitation of the cyan fluorophores results in some emission characteristic of yellow fluorescent protein. The ratio of the two emission signals can therefore be used as an indicator of calcium concentration.

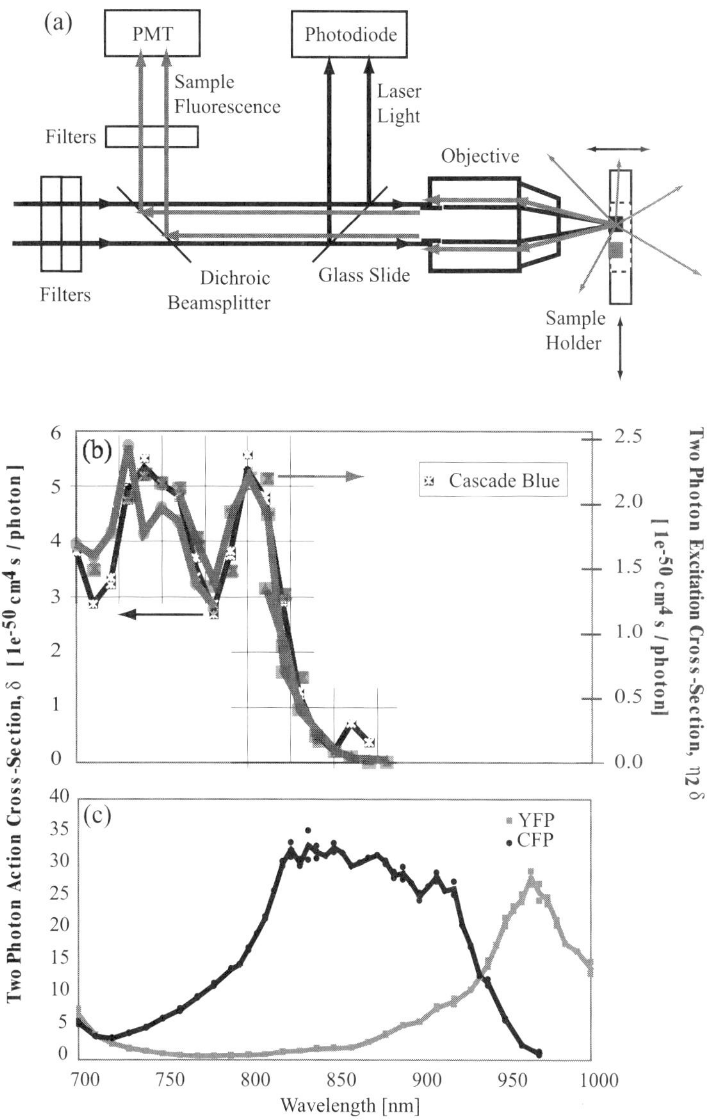

FIGURE 6.28 Two-photon action spectra. (a) Schematic of the two-photon spectrometer. The laser light path is represented by dark lines and the fluorescence detection pathway is represented by gray lines. The test sample is depicted as a gray box and the reference sample is depicted as a dark box. The samples can be translated in the axial direction and one lateral direction as a means to make sequential measurements of the two samples and thus compare their relative two-photon generated fluorescence. (b) Two-photon action spectrum of cascade blue (black symbols and connecting lines). The published two-photon excitation spectrum[48] is shown for comparison (gray lines). The relative scale between the two data sets was chosen to match the two curves. (c) The two-photon action spectra for the genetically encodable yellow fluorescent protein (YFP; gray squares) and cyan fluorescent protein (CFP; dark circles). The lines connect the average of the data values at each wavelength.

The images shown in Figures 6.29a and 6.29b correspond to the fluorescence from HeLa cells that express cameleon. The image shown in Figure 6.29a is formed from emission centered on 485 nm and represents a mixture of emission from the CFP and the YFP, with the dominant emission from the CFP. The image shown in Figure 6.29b corresponds to fluorescence centered on 525 nm and consists primarily of emission from yellow fluorescent protein. A change in calcium concentration may be induced in HeLa cells by the bath application of histamine, which triggers an intracellular release of calcium. This change may be quantified by measuring, for the 485-nm and 525-nm bands, the integrated fluorescence in a region of interest vs. time. These signals are shown in Figure 6.29c; note the counter-correlated signal change in the two wavelength bands. The ratio of the two fluorescent signals is clearly seen to increase in response to the histamine application.

6.5.5 *In Vivo* Cortical Blood Flow

A unique aspect of TPLSM is the ability to observe anatomical and functional signals in the brains of living animals deep to their pial surface.[3] As an example, we consider the flow of blood in individual capillaries in rat cerebral cortex[2] (Figure 6.30). A cranial window is prepared over whisker barrel sensory cortex, the dura is removed, and a metal frame and closed surgical window are attached to the skull for stable mounting of the preparation to the microscope table.[20] The blood stream is labeled with fluorescein isothiocyanate-dextran (77 kD). A single optical section obtained 100 μm below the pial surface is shown in Figure 6.30a; the maximal projection reconstructed from 41 individual images, ranging in depth from 80 to 120 μm below the pial surface of the brain, is shown in Figure 6.30b. Individual red blood cells appear in the capillaries as dark stripes that correspond to locations where the fluorescent dye was excluded from the vessel during the scan. Depending on the optical properties of the preparation, images from depths exceeding 600 μm have been achieved using TPLSM.[30] It may also be possible to image flow in individual capillaries through a thinned but intact skull.[53]

Line scans may be taken repeatedly along an individual capillary to produce a continuous measure of blood flow (Figure 6.31). The spatial coordinate is taken to be along the direction of the scan, while the perpendicular direction represents the temporal coordinate. As the blood cell moves along the capillary, its position along the line scan changes from scan to scan, thereby tracing out a dark line, as shown for three examples (Figures 6.31a to 6.31c). The data of Figure 6.31a show continuous flow; analysis of these data provides information regarding three parameters of capillary blood flow: blood cell speed, blood cell flux, and blood cell linear density, where flux = linear density × speed. The slope of the line is the inverse of the instantaneous speed; i.e., speed = $|dt/dx|^{-1}$. The spacing between the lines along the temporal coordinate direction is a measure of the cell flux, while the spacing between the lines along the spatial coordinate direction is a measure of the cell density. The line scan data of Figure 6.31b contain a short vertical region, indicative of a momentary stall of the red blood cells, while those of Figure 6.31c illustrate some of the high variability in flow that may be observed.

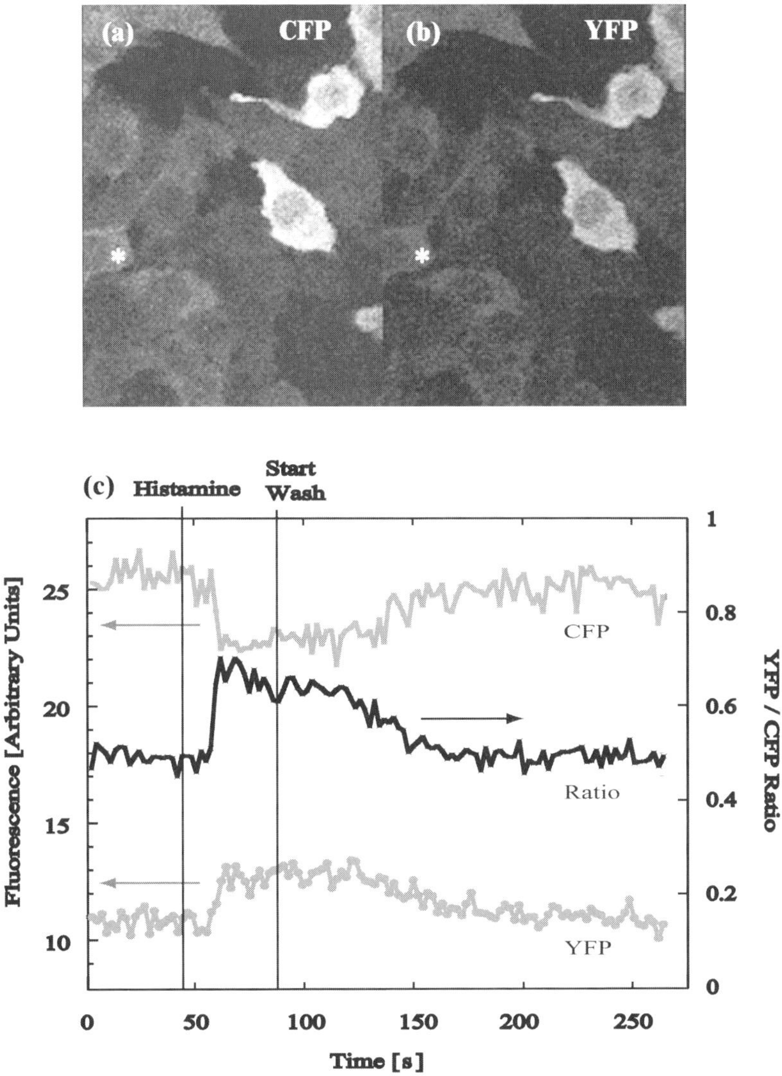

FIGURE 6.29 *In vitro* TPLSM used to measure Ca^{2+} dynamics in HeLa cells infected with the cameleon protein construct via an adenovirus vector. (a) Image of the HeLa cells at emission wavelengths centered on 485 nm. The majority of the emission that forms the image is from cyan fluorescent protein. (b) The same field simultaneously imaged at emission wavelengths centered on 525 nm. Almost all of the emission that forms the image is from yellow fluorescent protein. (c) Graph of intensity vs. time in channels centered on 485 and 525 nm for an isolated region of interest, designated by the * in parts a and b, as a function of time. Also shown (dark line) is the ratio of the two channels. Bath application of 100-μM histamine was used to trigger an intracellular release of free Ca^{2+}. Note the counter-correlated signal change in the two channels and the resultant increase in the fluorescent ratio in response to the histamine application. The initial emission values are seen to recover subsequent to a wash with saline.

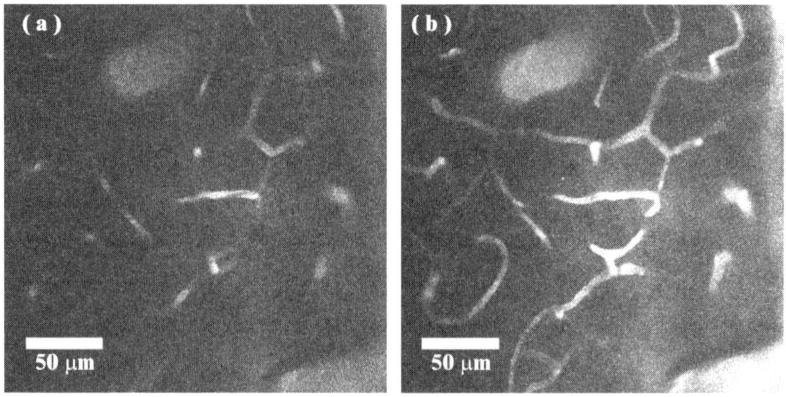

FIGURE 6.30 *In vivo* TPLSM used to image a capillary bed in rat vibrissa primary sensory cortex. An optical window was prepared over this region and the blood stream was labeled with 77-kD fluorescein conjugated-dextran. (a) A single planar image from a depth of 100 μm below the pial surface of cortex. Notice that individual red blood cells, which occlude fluorescent dye, appear in the capillaries as dark stripes. (b) A maximal projection reconstructed from 41 planar images, as in part a, taken between 80 and 120 μm below the pia.

APPENDIX A: PARTS AND VENDORS FOR THE TPLSM

Mechanical

The 8 × 4 × 1 ft anti-vibration air table was purchased from Newport (Irvine, CA). For mirror mounts, we used the Ultima® series from Newport and the Flipper Mounts from New Focus (Santa Clara, CA). All mirror mounts were attached to standard $^1/_2$-in. posts and post holders from Newport held to the air table with clamping plates and forks from New Focus. Microbench™ parts were obtained from Linos Photonics (Milford, MA). The large X-95 rails and assorted clamps and brackets were purchased from a variety of vendors. Crossed roller bearing translation stages were obtained from Newport, the slider for the detector assembly was from New England Affiliated Technologies (SD Series; Lawrence, MA), and the Modular Focusing Unit was obtained from Nikon (part no. 883773; Garden City, NY).

Optical

All mirrors were λ/10 flatness, protected silver mirrors obtained from Newport. Water-immersion objectives were purchased from Zeiss (Thornwood, NY) and Olympus. Blocking filters were obtained from Andover Corporation (Salem, NH), while dichroic beamsplitters and interference filters were from Omega Optical Inc. (Brattleboro, VT) and Oriel Instruments (Stratford, CT). The wave plate, polarizing beamsplitter, and intermediary optics were obtained from Newport and Linos Photonics. The scan lens, a Plan-Apo 1.0× magnification dissection stereoscope objective (focal length of 92.7 mm, physical length of 60 mm, and measured front and back working distances of 56 mm) was from Leica (McBain Instruments, Chatsworth, CA).

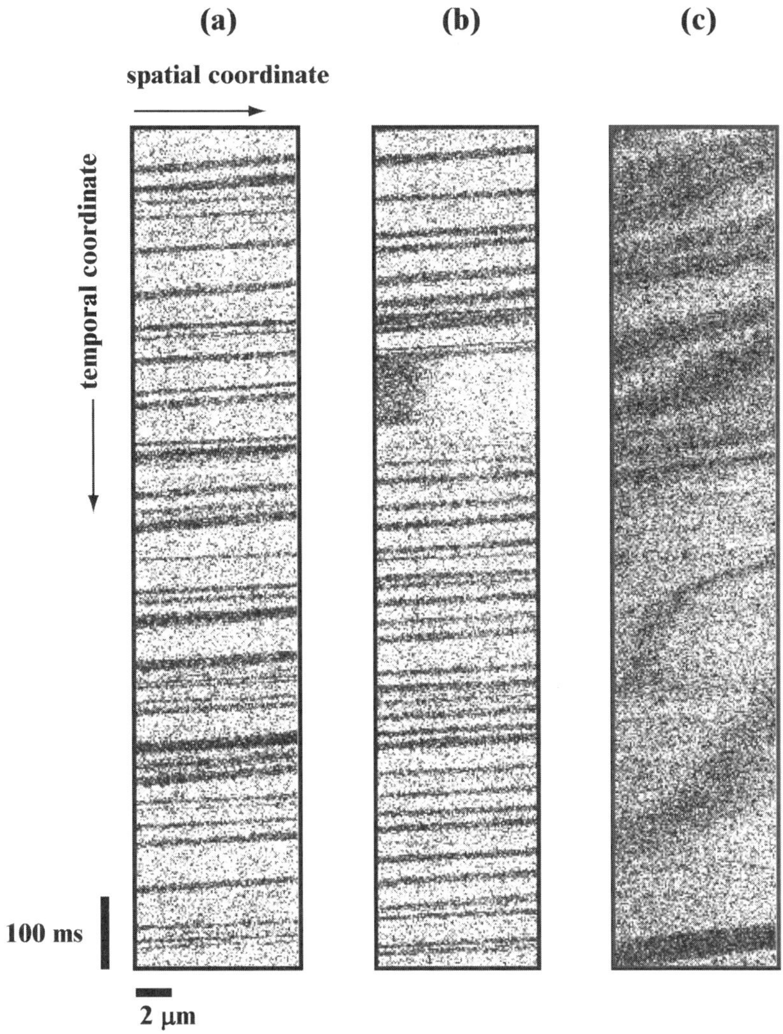

FIGURE 6.31 *In vivo* TPLSM used for functional imaging of capillary blood flow. Shown are line scan data taken along single capillaries in rat vibrissa primary sensory cortex through an optical window, as in Figure 6.30. The motion of red blood cells appears as dark stripes against a bright plasma background in these time-space plots of repeated line scans through the capillary; see text for details. (a) Relatively uniform flow. The thickness of the dark lines may reflect the orientation of the red blood cell. (b) Flow that is interrupted by a brief stall. (c) Flow that is highly irregular and that changes in speed within the field of view.

Scanners

The scanning galvanometers and their drivers were obtained from Cambridge Technologies (Boston, MA). The scanners are Model 6210 Galvanometer Optical Scanners capable of ±20° maximum deflection. They have a 300-μsec set time. We used

MicroMax™ 670 scan drivers, which take a ±10-V analog input. The power supplies for the scan drivers were obtained from Acopian (Easton, PA).

Photomultiplier Tubes

The detectors in our system were selected for high sensitivity with low dark current (model R6357 specified for dark currents below 2 nA; Hamamatsu Corporation, Bridgewater, NJ). Each photomultiplier tube is powered by a model E850-13 base (Hamamatsu Corporation) that is energized with a 0- to 2000-VDC adjustable power supply (Model PMT-20C-N; Bertan Associates, Hicksville, NY), although the tube is used only below the maximum rating of 1250 V required for this particular PMT.

Electronics

The waveform generator for the scanning galvanometers consists of two model PCI-6711 Arbitrary Waveform Boards from National Instruments (Austin, TX). The data acquisition board is a model PCI-6110E four-channel analog input board from National Instruments. The stepping motor that controls the axial translation of the focus is a Vexta model PH268M-M1.5 (Herbach and Rademan Company, Moorestown, NJ). It has 400 steps per revolution, which results in 0.25-μm axial steps in combination with the modular focusing unit.

Laser and Ancillary Equipment

The Ti:Sapphire laser is a Mira-900 (Coherent Laser Group, Santa Clara, CA) equipped with a broadband mirror set. The pump laser is a Verdi-V10 (Coherent Laser Group) 10-W pump laser. Two chillers (NESLAB Instruments, Portsmouth, NH) are required for stable temperature operation of the laser and are supplied by Coherent. The spectrometer consists of a PC2000-ISA computer card (Ocean Optics Inc., Dunedin, FL) and an associated fiber optic cable. The fast oscilloscope used to monitor the repetition rate of the Ti:Sapphire laser is a 400-MHz Tektronix model TDS430A oscilloscope (Newark Electronics). The infrared viewer is a Find-R-Scope from FJW Optical Systems, Inc. (Palatine, IL). The photodiode used to monitor the average laser power is a Si-based PIN diode obtained from Hamamatsu Corporation. A calibrated energy meter (model 818T-10, Newport) is used to calibrate the photodiode.

APPENDIX B: BASICS OF INTERFEROMETRIC AUTOCORRELATION

In the autocorrelation process the laser pulse is split into two identical copies and one is delayed relative to the other by a time τ. This results in two pulses with electric fields $E(t)$ and $E(t + \tau)$. These are subsequently multiplied together through a nonlinear process such as two-photon excitation. For such a process, the instantaneous two-photon excited fluorescence for a given delay time is

$$I(t,\tau) = \left| [E(t) + E(t+\tau)]^2 \right|^2 \tag{A1}$$

The time resolution of the photodetector is much slower than the fluorescence duration; thus, the output of the photodiode is the time average of $I(t, \tau)$ (Equation A1). This signal corresponds to the autocorrelation function, $I_{AC}(\tau)$; i.e.,

$$I_{AC}(\tau) = \int_{-\infty}^{\infty} dt \left| [E(t) + E(t+\tau)]^2 \right|^2 \tag{A2}$$

To deduce the shape of the original pulses from the autocorrelation signal $I_{AC}(\tau)$, we assume a known original pulse shape. The autocorrelation signal can be calculated in terms of parameters describing the original pulse. The measured autocorrelation signal shape can then be used to deduce the original pulse parameters. Ultrashort laser pulses usually have Gaussian or secant-squared time envelopes. Since the Gaussian is a good estimate of the actual pulse shape and facilitates calculations, we assume a Gaussian profile for our beam. The electric field of the laser pulses can be described as

$$E(t) = E_0 \exp\left\{ -\frac{t^2}{a^2} + i\omega t \right\} \tag{A3}$$

where the full-width-half-maximum (FWHM) of the envelope of the intensity of the pulse is $a\sqrt{2\ln 2}$, ω is the center wavelength of the pulse, and the origin of time is arbitrary, as $E(t)$ and $E(t + \tau)$ appear only in integrals over all of time, e.g., Equation A2.

The autocorrelation measurement can be set up in two ways. In an interferometric autocorrelation, fringes caused by the interference of the two pulses are observed. The two pulses are incident on the fluorescent material so that they are exactly colinear and are polarized in the same direction. In a noninterferometric autocorrelation, the fringes are smeared out either by introducing an angle between the beams or a difference in the polarization direction of the two pulses. While both autocorrelation methods give a measure of pulse length, the interferometric autocorrelation will also indicate the presence of chirp in the pulse.

The interferometric autocorrelation signal with $E(t)$ given by Equation A3 is

$$I_{AC}^{\text{interferometric}}(\tau) \propto 1 + 2e^{-\frac{\tau^2}{a^2}} + 4e^{-\frac{3\tau^2}{4a^2}}\cos(\omega\tau) + e^{-\frac{\tau^2}{a^2}}\cos(2\omega\tau) \tag{A4}$$

The upper envelope and lower envelope can be found by setting $\omega\tau = \pi$ or 2π. A convenient feature of interferometric autocorrelation is the 8:1 ratio of the amplitude of the maximum from zero to the amplitude of the constant background. Achieving this ratio indicates that the apparatus is well aligned. In the case with unchirped pulses, the FWHM of $I_{AC}^{interferometric}(\tau)$, denoted τ_{AC}^{FWHM}, is related to the FWHM of the original pulse by

$$\tau_{P}^{FWHM} \approx \sqrt{\frac{ln2}{2}} \cdot \tau_{AC}^{FWHM} \tag{A5}$$

In the presence of chirp, the interferometric autocorrelation function develops curvature in the wings of the signal (* in Figure 6.26). When this feature is present, the FWHM of the autocorrelation and the original pulse no longer have the simple relation given by Equation A5. The central peak of the autocorrelation signal will become narrower because the red-shifted, leading portion of the pulse will only incompletely interfere with the blue-shifted tail. The electric field of a linearly chirped pulse can be described as

$$E(t) = E_0 \exp\left\{-\frac{t^2}{a^2} + i\omega t - ibt^2\right\} \tag{A6}$$

where b is a measure of the chirp. The experimentally obtained autocorrelation must be fit to the autocorrelation that is calculated (Equations A2 and A6) with the two parameters, a and b, taken as variables in order to determine the pulse length. In practice, for two-photon microscopy an absolute measurement of the chirp is generally not necessary. One typically adjusts only the laser to minimize the curvature in the wings of the autocorrelation signal.

APPENDIX C: SPECTROMETER FOR TWO-PHOTON ACTION SPECTRA

The high optical intensities necessary for generation of significant two-photon excitation typically require a pulsed laser source. The Ti:Sapphire is one of few currently available pulsed lasers tunable over a wavelength range that is suitably large for spectroscopy. Unfortunately, the pulse characteristics of this laser do not remain constant as it is tuned from 700 to 1000 nm. Two-photon excitation is extremely sensitive to changes in the pulse width and pulse shape of the laser output. Thus, a direct measurement of a two-photon-induced excitation spectrum would require that the pulse characteristics be simultaneously monitored, such as with an autocorrelation technique (Appendix B). On the other hand, two-photon excitation spectra can be obtained without detailed knowledge of the pulse characteristics by comparing the two-photon excitation spectrum of an unknown sample with that of a dye of known two-photon spectrum. Direct measurements of the two-photon excitation spectra for some common dyes have been reported by Xu.[7] Based on this past work, our two-photon excitation spectrometer is designed to take measurements against dyes of known two-photon spectra.

The comparative method for obtaining two-photon excitation spectra makes the assumption that the pulse characteristics do not fluctuate significantly over the time scale of the measurement at any given wavelength and output power of the laser. Although characteristics of the laser will change as the laser is retuned, the measurement of the test sample is taken directly after measurement of the reference sample at each wavelength without retuning the laser between the two measurements. Thus, although the exact pulse characteristics are unknown and may differ between wavelengths, they are the same for both samples at any given wavelength.

The excitation cross section for the test sample at a given wavelength can be calculated by taking the ratio of its collected fluorescence to that collected from the reference sample and multiplying by the known excitation cross section of the reference sample. If the emission spectra of the test and reference samples are different and the efficiencies of the objective, dichroic beamsplitter, detector, or filters differ for the two emission spectra, then these efficiencies ratios must also be factored into the final calculation.

Design

The design of the comparative two-photon excitation spectrometer is shown in Figure 6.28A. The laser light passes through a long-pass dichroic beamsplitter, chosen to transmit the laser wavelengths and reflect the fluorescent signals. A portion of the laser beam is deflected to a photodiode to monitor the average laser power. The remainder of the beam passes to the objective and is focused into the sample, which is free to translate both laterally and in the axial direction. The transmission curve of the dichroic is typically not constant across the tunable laser wavelengths. It is, therefore, important that the photodiode pick-off is placed after the dichroic so that it monitors the actual laser power transmitted to the objective. Small inaccuracies in the average power reading do not affect the two-photon spectra, as the average power is used solely to check that the fluorescent signal varies with the square of the incident power and thus verify that neither the dye nor the PMT are saturated.

Construction

The frame for the two-photon spectrometer is built primarily from Microbench™ parts. The dichroic beamsplitter and glass slide are mounted on prism mounts. The objective is held horizontally by a threaded Microbench™ plate and the filters, photodiode, and PMT are mounted in Microbench™ parts and attached via protruding rods (Figure 6.20B). The sample holder is a single-axis Microbench™ translation stage situated to move in the transverse direction. The sample holder is part of a spring-loaded translation assembly that allows translation in the axial direction.

The samples are held in microslide cuvets (VitroCom Inc; Mountain Lakes, N.J.) with inner dimensions of 0.1×2.0 mm, chosen to minimize the sample volume. The microslides are held with bone wax to a 25-mm adapter plate (Microbench™) in which we machined parallel grooves for the microslides to seat. An OG590 colored glass filter (Oriel Instruments) was placed at the entrance of the spectrometer to prevent single-photon excitation by residual 532-nm light from the pump laser. We use a 600-DCLP dichroic beamsplitter (Omega Optical Inc.) to separate the incident and emitted beams and a BG40 glass filter (Andover Corp.) to block remaining laser light in the emitted pathway. The detector is a Hamamatsu R6353 compact PMT, chosen for its low dark current.

Measurement Procedures

The microslide cuvets are filled with the test and reference samples, respectively, sealed with bone wax, and placed into the cuvet holder. The cuvets are laterally translated until the beam strikes the center of the reference cuvet. The fluorescent signal from the reference cuvet is then maximized by translating the cuvet in the axial direction. All fluorescent measurements are taken at this axial position. The laser is tuned to a particular wavelength and the fluorescent signal from the center of the reference cuvet is recorded. The cuvets are then laterally translated a predetermined amount so that the beam strikes the center of the test cuvet and the fluorescent signal is recorded. Finally, the samples are translated laterally so that the beam falls between the cuvets and a background reading is taken. The average laser output power is monitored and recorded during each measurement. The laser is then tuned to a new wavelength and this process is repeated.

After the process has been repeated for all desired wavelengths, the laser is tuned to a wavelength at which the excitation of test and reference dyes is reasonably strong. The test cuvet is centered laterally on the beam and the fluorescent signal is recorded. The signal is then maximized in the axial direction. Ideally, the signal should already be maximized, but small variations in the cuvet holder or the cuvet thickness could result in a displacement of the optimal axial positions of the two cuvets. The ratio of the signals taken at the two axial positions is recorded and used to obtain a true test-to-reference fluorescence ratio.

ACKNOWLEDGMENTS

We thank Kerry R. Delaney, Winfried Denk, and Jeffrey A. Squier for useful discussions, Oliver Griesbeck and Roger Y. Tsien for supplying material for the *in vitro* cameleon studies, Samar Mehta for involvement with the spectroscopic measurements, and Kenneth Duff of the Scripps Institute of Oceanography Development Shop for his careful machining. Our work was supported by grants from the David and Lucile Packard Foundation, the NINDS, and, in part, by an NIMH predoctoral training award to Philbert Tsai, an NSF predoctoral fellowship to Nozomi Nishimura, and an NSRA postdoctoral fellowship to Elizabeth Yoder.

REFERENCES

1. Svoboda, K. et al., *In vivo* dendritic calcium dynamics in neocortical pyramidal neurons, *Nature*, 385, 161–165, 1997.
2. Kleinfeld, D. et al., Fluctuations and stimulus-induced changes in blood flow observed in individual capillaries in layers 2 through 4 of rat neocortex, *Proc. Natl. Acad. Sci. U.S.A.*, 95, 15741–15746, 1998.

3. Denk, W. et al., Anatomical and functional imaging of neurons and circuits using two photon laser scanning microscopy, *J. Neurosci. Methods*, 54, 151–162, 1994.
4. Denk, W., Strickler, J.H., and Webb, W.W., Two-photon laser scanning fluorescence microscopy, *Science*, 248, 73–76, 1990.
5. Denk, W., Piston, D.W., and Webb, W., Two-photon molecular excitation in laser-scanning microscopy, in *Handbook of Biological Confocal Microscopy*, Pawley, J.W., Ed., Plenum Press, New York, 1995, 445–458.
6. Denk, W. and Svoboda, K., Photon upmanship: why multiphoton imaging is more than a gimmick, *Neuron*, 18, 351–357, 1997.
7. Xu, C. et al., Multiphoton fluorescence excitation: new spectral windows for biological nonlinear microscopy, *Proc. Natl. Acad. Sci. U.S.A.*, 93, 10763–10768, 1996.
8. Williams, R.M., Piston, D.W., and Webb, W.W., Two-photon molecular excitation provides intrinsic 3-dimensional resolution for laser-based microscopy and microphotochemistry, *FASEB J.*, 8, 804–813, 1994.
9. Soeller, C. and Cannell, M.B., Construction of a two-photon microscope and optimisation of illumination pulse duration, *Pflugers Arch.*, 432, 555–561, 1996.
10. Wier, W.G. et al., A custom confocal and two-photon digital laser scanning microscope, *Am. J. Physiol. Heart Circ. Physiol.*, 278, H2150–H2156, 2000.
11. Denk, W., Two-photon scanning photochemical microscopy: mapping ligand-gated ion channel distributions, *Proc. Natl. Acad. Sci. U.S.A.*, 91, 6629–6633, 1994.
12. Potter, S.M., Two-photon microscopy for 4D imaging of living neurons, in *Imaging Neurons: A Laboratory Manual*, Yuste, R., Lanni, F., and Konnerth, A., Eds., Cold Spring Harbor Laboratory Press, Cold Spring Harbor, New York, 2000.
13. Majewska, A., Yiu, G., and Yuste, R., A custom-made two-photon microscope and deconvolution system, *Pflugers Arch.*, 441, 398–408, 2000.
14. Fan, G.Y. et al., Video-rate scanning two-photon excitation fluorescence microscopy and ratio imaging with cameleons, *Biophys. J.*, 76, 2412–2420, 1999.
15. Lanni, F. and Keller, H.E., Microscopy and microscopic optical systems, in *Imaging Neurons: A Laboratory Manual*, Yuste, R., Lanni, F., and Konnerth, A., Eds., Cold Spring Harbor Laboratory Press, Cold Spring Harbor, New York, 2000.
16. Hiraoka, Y., Sedat, J.W., and Agard, D.A., Determination of three-dimensional imaging properties of a light microscope system. Partial confocal behavior in epifluorescence microscopy, *Biophys. J.*, 57, 325–333, 1990.
17. Agard, D.A. et al., Fluorescence microscopy in three dimensions, *Methods Cell Biol.*, 30, 353–377, 1989.
18. Holmes, T.J. et al., Light microscopic images reconstructed by maximum likelihood deconvolution, in *Handbook of Confocal Microscopy*, 2nd ed., Pawley, J.B., Ed., Plenum Press, New York, 389–402.
19. Neil, M.A.A., Juskaitis, R., and Wilson, T., Method of obtaining optical sectioning by using structured light in a conventional microscope, *Opt. Lett.*, 22, 1905–1907, 1997.
20. Lanni, F. and Wilson, T., Grating image systems for optical sectioning fluorescence microscopy of cells, tissues, and small organisms, in *Imaging Neurons: A Laboratory Manual*, Yuste, R., Lanni, F., and Konnerth, A., Eds., Cold Spring Harbor Laboratory Press, Cold Spring Harbor, New York, 2000.
21. Bailey, B. et al., Enhancement of axial resolution in fluorescence microscopy by standing-wave excitation, *Nature*, 366, 44–48, 1993.
22. Gustafsson, M.G., Agard, D.A., and Sedat, J.W., I^5M: 3D widefield light microscopy with better than 100 nm axial resolution, *J. Microscopy*, 195, 10–16, 1999.

23. Bahlmann, K., Jakobs, S., and Hell, S.W., 1: 4Pi-confocal microscopy of live cells, *Ultramicroscopy,* 87, 155–164, 2001.
24. Born, M. and Wolf, E., *Principles of Optics: Electromagnetic Theory of Propagation Interference and Diffraction of Light,* 6th ed., Pergamon Press, Oxford, 1980.
25. Konig, K. et al., Pulse-length dependence of cellular response to intense near-infrared laser pulses in multiphoton microscopes, *Opt. Lett.*, 24, 113–115, 1999.
26. Centonze, V.E. and White, J.G., Multiphoton excitation provides optical sections from deeper within scattering specimens than confocal imaging, *Biophys. J.*, 75, 2015–2024, 1998.
27. Oheim, M. et al., Two-photon microscopy in brain tissue: parameters influencing the imaging depth, *J. Neurosci. Methods*, 111, 29–37, 2001.
28. Beaurepaire, E., Oheim, M., and Mertz, J., Ultra-deep two-photon fluorescence excitation in turbid media, *Opt. Commun.*, 188, 25–29, 2001.
29. Dunn, A.K. et al., Influence of optical properties on two-photon fluorescence imaging in turbid samples, *Appl. Opt.*, 39, 1194–1201, 2000.
30. Kleinfeld, D. and Denk, W., Two-photon imaging of neocortical microcirculation, in *Imaging Neurons: A Laboratory Manual*, Yuste, R., Lanni, F., and Konnerth, A., Eds., Cold Spring Harbor Laboratory Press, Cold Spring Harbor, New York, 2000, 23.1–23.15.
31. Schonle, A. and Hell, S.W., Heating by absorption in the focus of an objective lens, *Opt. Lett.*, 23, 325–327, 1998.
32. Aquirrell, J.M. et al., Long-term two-photon fluorescence imaging of mammalian embryos without compromising viability, *Nat. Biotechnol.*, 17, 763–767, 1999.
33. Koester, H.J. et al., Ca^{2+} fluorescence imaging with pico- and femtosecond two-photon excitation: signal and photodamage, *Biophys. J.*, 77, 2226–2236, 1999.
34. Piston, D.W. and Patterson, G.H., Photobleaching in two-photon excitation microscopy, *Biophys. J.*, 78, 2159–2162, 2000.
35. Neher, E. and Hopt, A., Highly nonlinear photodamage in the two-photon fluorescence microscopy, *Biophys. J.*, 80, 2029–2036, 2001.
36. Siegman, A.E., *Lasers*, University Science Books, Sausalito, CA, 1986.
37. Bardeen, C.J. et al., Effect of pulse shape on the efficiency of multi-photon processes: implication for biological microscopy, *J. Biomed. Opt.*, 4, 362–367, 1999.
38. Hecht, E., *Optics,* 2nd ed., Addison-Wesley Publishing, Reading, PA, 1987.
39. Fork, R.L., Martinez, O.E., and Gordon, J.P., Negative dispersion using pairs of prisms, *Opt. Lett.*, 9, 150–152, 1984.
40. Trebino, R. et al., Measuring ultrashort laser pulses in the time-frequency domain using frequency-resolved optical gating, *Rev. Sci. Instrum.*, 68, 3277–3295, 1997.
41. Yariv, A., *Optical Electronics*, Holt, Reinehart and Winston, New York, 1985.
42. Wilson, T. and Sheppard, C., *Theory and Practice of Scanning Optical Microscopy*, Academic Press, London, 1984.
43. Mainen, Z.F. et al., Two-photon imaging in living brain slices, *Methods*, 18, 231–239, 1999.
44. Yuste, R. and Denk, W., Dendritic spines as basic functional units of neuronal integration, *Nature*, 375, 682–684, 1995.
45. Soeller, C. and Cannell, M.B., Two-photon microscopy: imaging in scattering samples and three-dimensionally resolved flash photolysis, *Microscopy Res. Tech.*, 47, 182–195, 1999.
46. Muller, M., Squier, J., and Brankenhoff, G.J., Measurement of femtosecond pulses in the focal point of a high-numerical-aperture lens by two-photon absorption, *Opt. Lett.*, 20, 1038–1040, 1995.

47. Diels, J.-C.M. et al., Control and measurement of ultrashort pulse shapes (in amplitude and phase) with femtosecond accuracy, *Appl. Opt.*, 24, 1270–1282, 1985.
48. Xu, C. and Webb, W.W., Measurement of two-photon excitation cross sections of molecular fluorophores with data from 690 to 1050 nm, *J. Opt. Soc. Am.*, 13, 481–491, 1996.
49. Xu, C., Two-photon cross section of indicators, in *Imaging Neurons: A Laboratory Manual*, Yuste, R., Lanni, F., and Konnerth, A., Eds., Cold Spring Harbor Laboratory Press, Cold Spring Harbor, New York, 2000.
50. Miyawaki, A. et al., Fluorescent indicators for Ca^{2+} based on green fluorescent proteins and calmodulin, *Nature*, 388, 882–885, 1997.
51. Romoser, V.A., Hinkle, P.M., and Persechini, A., Detection in living cells of Ca^{2+}-dependent changes in the fluorescence emission of an indicator composed of two green fluorescent protein variants linked by a calmodulin-binding sequence, *J. Biol. Chem.*, 272, 13270–13274, 1997.
52. Miyawaki, A. et al., Dynamic and quantitative Ca^{2+} measurements using improved cameleons, *Proc. Natl. Acad. Sci. U.S.A.*, 96, 2135–2140, 1999.
53. Yoder, E.J. and Kleinfeld, D., Cortical imaging through the intact mouse skull using two-photon excitation scanning microscopy, *Microscopy Res. Tech.*, 56, 304–305, 2002.

7 Intraoperative Optical Imaging

Arthur W. Toga and Nader Pouratian

CONTENTS

7.1 INTRODUCTION

Intraoperative mapping provides an unparalleled opportunity to examine the basic physiology of the functioning human brain. At the same time, it poses unique challenges for acquisition, analysis, and interpretation over and above those issues that are part of any other intact *in vivo* brain mapping. These include the operating room environment, time constraints, spatial resolution, the status of the brain under anesthesia or performance during awake protocols, a dynamic cortical geometry, and other considerations.

Despite these challenges, intraoperative brain mapping provides extraordinary opportunities to examine brain function unencumbered by other tissues. Tremendous insights into language, sensory, and motor representations have been achieved as a result of the advantages of intraoperative investigation. With the advent of intraoperative optical imaging, the opportunity exists to study brain function and physiology with greater spatial resolution, temporal resolution, and sensitivity than was previously possible. In this chapter, we will discuss the methodology of intraoperative

0-8493-2389-4/02/$0.00+$1.50
© 2002 by CRC Press LLC

optical imaging of intrinsic signals and address how this imaging modality overcomes many of the challenges presented by the operating room environment.

Haglund et al. were the first to observe optical signals in humans, reporting activity-related changes in cortical light reflectance during seizure and cognitive tasks.[1] This was a landmark paper because it demonstrated the use in humans of optical methods that up until this point had been employed only in monkeys, cats, and rodents. Since then, there have been reports describing the evolution of optical signals in the human cortex,[2] the mapping of primary sensory and motor cortices,[3] and the delineation of language cortices within[4] and across languages.[5] Optical signals can be measured using intrinsic signals, dyes restricted to physiological compartments, such as intravascular dyes,[6,7] or dyes sensitive to physiological events, such as oxygen-dependent phosphorescence quenching dyes[8]; however, only intrinsic signal changes have been reported in humans. Optical signals can also be detected that are specific to pathological tissue, such as tumors, by using tracers specific to tissue properties or even metabolic rates.[9]

In terms of clinical utility, the most obvious benefits of intraoperative optical imaging are obtained when it is used to map "eloquent" cortices (i.e., language, sensory, motor, or visual areas) adjacent to or involved in pathology. This allows maximal surgical resection of pathological tissue while leaving eloquent cortices intact. These eloquent regions provide the opportunity to utilize simple activation protocols compatible with an operating room environment or, in the case of passive stimulation, the ability to create functional maps in anesthetized patients.

This chapter surveys the development and methodology of intraoperative optical imaging of intrinsic signals. We include comparisons between pre- and intraoperative maps along with methods to equate them in space. We also review how intraoperative optical intrinsic signal imaging (iOIS) has been applied to address questions of basic physiology and cognitive function. Finally, we comment on the potential use of iOIS as a clinical brain mapping tool for neurosurgical guidance.

7.2 SOURCES OF INTRINSIC SIGNALS

Optical intrinsic signals are activity-related changes in the reflectance of cortex (and other tissues) to visible light. They were first demonstrated *in vitro* by Hill and Keynes[10] and have been subsequently used *in vivo*[11,12] for functional imaging in rodent,[13,14] cat,[15] primate,[6] and human.[1,2,4,5,16,17] The etiology of the intrinsic signals[50] includes reflectance changes from several optically active processes indirectly coupled to neuronal firing,[18] including changes in blood volume,[6,7] cellular swelling,[19] hemoglobin concentrations,[20] and cytochrome activity,[21] depending on the wavelength measured.[20]

Imaging at wavelengths sensitive to oxygen extraction (i.e., 600 to 630 nm) produces maps more spatially correlated with underlying neuronal activity than at wavelengths influenced by blood volume changes (i.e., 850 and 550 nm).[6,8,22,23] This may be because fast changes in oxidative metabolism are more tightly coupled to electrical activity than the more delayed perfusion-related responses.[8] Imaging at 600 to 630 nm also offers the greatest signal-to-noise ratio and, when focused 2.0 mm beneath the cortical surface, least emphasizes blood vessel artifacts.[23] Due to the excellent spatial correlation of 610-nm signals with electrophysiological maps and its superior signal-to-noise ratio, our laboratory has focused on imaging optical signals in humans at 610 nm.[2-5] Optical reflectance changes in humans have also been observed using a 695-nm long pass filter.[1] The opportunity also exists to image at other wavelengths and use optical spectroscopy to explore questions of basic physiology in the human brain.

7.3 METHODS

Optical imaging in humans is performed following craniotomy and dural reflection according to clinical standards. The CCD camera is mounted either onto the operating microscope or directly to the skull via a bracket over the region of the craniotomy. In order to detect reflectance changes, the cortex is first epi-illuminated with white light. The light reflected off the cortex is then filtered at a particular wavelength before it is captured by the CCD camera (Figure 7.1). At the Laboratory of Neuro Imaging at the University of California, Los Angeles, we have successfully imaged 41 cases: 16 language mapping, 4 motor mapping, 24 somatosensory mapping (in three cases more than one type of mapping was possible), 9 AVM, 2 hemangiomas, and 30 tumors.

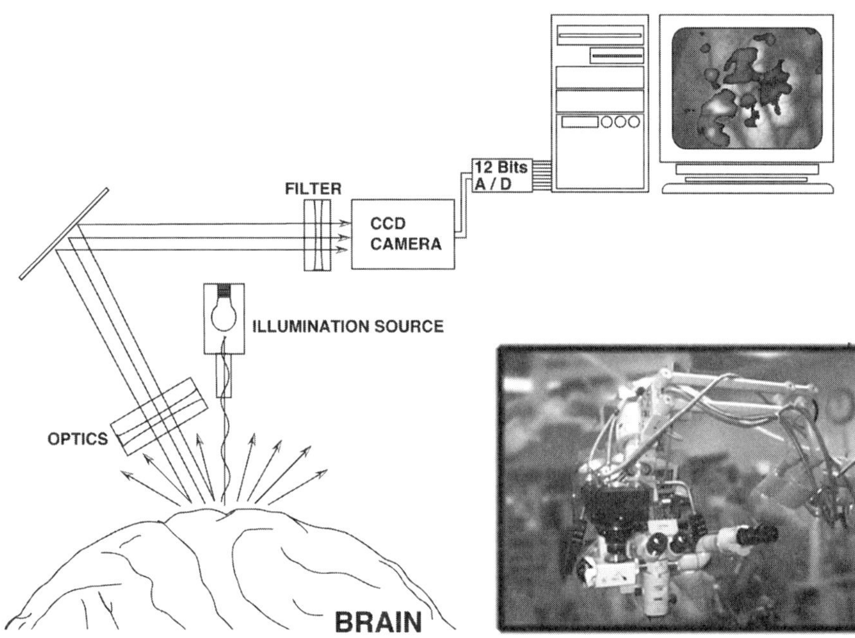

FIGURE 7.1 Schematic diagram of iOIS setup. The cortex is illuminated with white light and the reflected light is filtered at a particular wavelength of interest. Images are captured by a CCD camera during rest and stimulation. The two conditions are compared to determine which pixels demonstrate significant changes across conditions. The window demonstrates how the CCD camera is mounted atop the operating microscope via a custom adaptor.

Although we have opted to use a CCD camera for imaging, photodiode arrays and video cameras are also options for image acquisition. Using a photodiode array offers the opportunity to image with much greater temporal resolution (~1 msec) at the expense of spatial resolution. The increased temporal resolution, however, seems unnecessary when imaging perfusion-related signals, whose time course is on the order of seconds. If, however, optical imaging in humans advances to use of voltage-sensitive dyes (currently not possible due to toxicity), an argument may exist for using photodiode arrays since voltage changes occur on a much faster time scale. Alternatively, video cameras may be used for imaging, but the sensitivity of such devices may not be as great as with CCDs. Sensitivity is an important consideration since image exposure time needs to be minimized in order to avoid excessive intra-image brain movement. The greater sensitivity of modern CCDs allows image exposure times on the order of microseconds.

The human intraoperative optical imaging system (Figure 7.1) has been described in previous publications.[2-5,16,17,24] We utilize a slow-scan CCD camera (TE/CCD-576EFT, Princeton Instruments, Trenton, NJ) mounted via a custom adapter onto the video monitor port of a Zeiss operating microscope. Images are acquired through a transmission filter at 610 (605 to 615) nm (Corion Corp., Holliston, MA). Circular polarizing and heat filters are placed under the main objective of the operating microscope to reduce glare artifacts from the cortical surface. White light illumination is provided by the Zeiss operating microscope light source through a fiber optic illuminator. Since the microscope view is determined by the surgeon's choice of magnifications and microscope placement, the experimental field of view (FOV) varies, depending on lens distance from cortex and specific orientation. FOV varied from 2.75 to 7.50 cm, resulting in spatial resolution of 110 to 425 μm, respectively. The spatial resolution could be increased by asking the surgeon to zoom in further on a cortical area of interest.

It should be noted, however, that with increased magnification, movement of the brain due to respiration and the cardiac cycle become an increasing concern. The temporal resolution of acquisitions ranges from 200 to 500 msec per frame (sufficient to image the slowly evolving perfusion-related response) with image exposure times from 50 to 200 msec. The image exposure time should be kept short in order to minimize intra-image brain movement.

iOIS acquisitions consist of CCD exposure of the cortex during a control state (no stimulation) followed by a subsequent stimulated state (see below). All imaging sessions are controlled via a personal computer running Labview. Control trials and experimental trials are interleaved. During sensorimotor mapping trials, baseline CCD images are acquired, and then physiologically synchronized PC software triggers the stimulations (2-sec duration). Experiments contain 10 to 20 stimulation trials and include an equal number of nonstimulated interleaved controls. During language mapping, patients are asked to perform verbal and nonverbal tasks in 20-sec block trials with a 10-sec rest interval. Paradigm blocks are repeated four to eight times. All intraoperative imaging sessions are planned and timed in advance in order to minimize the duration of intraoperative imaging but attain sufficient signal-to-noise ratio (SNR). We have determined that the number of trials noted seems sufficient to produce reasonable SNR (~10:1) while keeping imaging sessions as short as possible. In most cases, imaging lasts approximately 20 to 40 min. Intraoperative imaging time must be strictly controlled and surgeries should not be unnecessarily lengthened because risk of infection increases with operative duration.

A significant obstacle in detecting the small activity-related reflectance changes in the intraoperative setting is the movement of the cortex due to cardiac and respiratory motion. This becomes an even greater challenge during awake procedures in which respiration rate is not under the direct control of the anesthesiologists. Several strategies have emerged to minimize the effect of cortical movement during imaging, including imaging through a sterile glass plate that lies atop the cortex,[1] synchronizing image acquisition with respiration and heart rate,[2] and using post-acquisition image registration.[1,4,5]

Using a glass plate atop the cortex immobilizes the cortex by applying pressure to the cortical surface to prevent its movement. While this method may achieve immobilization, it is not ideal because it physically interferes with the cortex and may therefore alter normal functional blood volume and cellular swelling patterns. In this respect, the latter two methods (synchronizing with respiration and heart rate and using postacquisition image registration) are preferable since they eliminate contact with the cortex being imaged.

Synchronizing image acquisition with respiration and heart rate requires monitoring of pneumographic and electrocardiographic waveforms. All trials (control and experimental) should begin at the same point in time during the respiration cycle, after which acquisition is controlled from synchronization to the cardiac cycle (500-msec post-R-wave). Experimental and control trials should be collected alternately on sequential respiratory expirations. Each experimental image will have a separate control image taken from either the preceding or subsequent expiration cycle. Since data acquisition occurs at similar time points during every respiration cycle, all images are collected with the brain in a similar position, minimizing the effect of periodic brain motion. While synchronizing image acquisition with respiration and heart rate effectively minimizes movement across images, it is complicated by the need to interface with operating room anesthesiology monitors that may not be easily accessible to investigators. Furthermore, it can potentially reduce temporal resolution in those experiments when the sampling frequency is limited only by heart rate.

Postacquisition image registration utilizes automated image registration (AIR) algorithms.[25] When acquiring images independent of pneumographic or electrocardiographic waveforms, the precise location of the cortex within the field of view of the CCD varies from image to image. AIR can be used to realign all images in a series to a reference image either at the beginning or in the middle of the acquisition series. The realignment of images is intensity based, using the extremely large difference in light intensities emanating from sulci and gyri to place the images into a common space. Functional changes in cortical reflectance do not interfere with this realignment algorithm since functional reflectance changes, which are <1% in magnitude, are minimal compared to the difference in reflectance from sulci and gyri (gyral light intensity levels can be up to 2000 times those of sulci). Furthermore, Woods et al.[25] have demonstrated (using PET data) that simulated cortical activation sites do not interfere with realignment. Postacquisition image registration effectively compensates for movement between images and minimizes the need to interface with operating room monitors.

In order to minimize vascular artifacts, the CCD camera can be focused below the cortical surface.[23] This strategy is especially useful when the area imaged is under or near large draining veins.[23] Changing the depth of focus allows the investigator to focus on optical changes that occur only in deeper layers of cortex and to de-emphasize optical signals that occur at high spatial frequencies (such as from the vasculature). Another strategy recently suggested to reduce vascular artifacts is to average only images collected shortly after stimulus onset (within 1.5 sec), a period of rapid oxygen consumption prior to significant reflectance changes over large vessels.[26] Vascular artifacts may also be wavelength dependent and therefore may be minimized by altering the wavelength at which iOIS is performed.

Once images have been captured, reflectance changes are estimated by performing a pixel-by-pixel ratio for each image ([image–control]/control], using the first image in each trial as a control. In order to increase the SNR, multiple trials of the same task are averaged.

7.3.1 Methods for Intermodality Comparisons

Comparing functional magnetic resonance imaging (fMRI) signals, which are tomographic, with optical signals, which are surface signals, presents a challenge. In order to compare mapping signals from multiple modalities within a single subject, the different modalities must be warped into a common space to facilitate comparisons (Figure 7.2). The common space used is the three-dimensional (3D) cortical model (i.e., cortical extraction) of each subject's brain (generated from high-resolution T1-weighted MR scans; see middle of Figure 7.2, for example). In order to align the fMRI activations with this cortical model, one must determine the transformation necessary to align high-resolution structural scans (co-planar with the fMRI scans of interest) with the T1-weighted MR scans used to create the cortical model.

This transformation should be a rigid-body (i.e., six-parameter: three rotations and three translations) transformation, since two structural scans of the same subject are being aligned. The resulting transformation is then applied to the original fMRI data to line up the functional data with the 3D cortical model (see Pouratian et al.[45] and Cannestra et al.[17] for implemented examples of this approach). OIS maps are projected onto the cortical model by matching sulcal landmarks on the cortical model with identical sulcal landmarks on the raw optical images. The accuracy of these projections is demonstrated in Color Figure 7.7* in the column labeled "region of interest." It is clear from Color Figure 7.7 that sulci in the optical images are continuous with sulci on cortical extractions, thus confirming the success of these projections.

The Ojemann group has adopted a similar methodology;[27] however, instead of relying on sulcal landmarks for registration, they use the position of the cortical blood vessels. Preoperatively, an MR angiogram (an MRI scan sequence is used which highlights surface vasculature) is performed and superimposed onto the subject's cortical extraction. After reflection of the dura and exposure of the brain

* Color figures follow page 112.

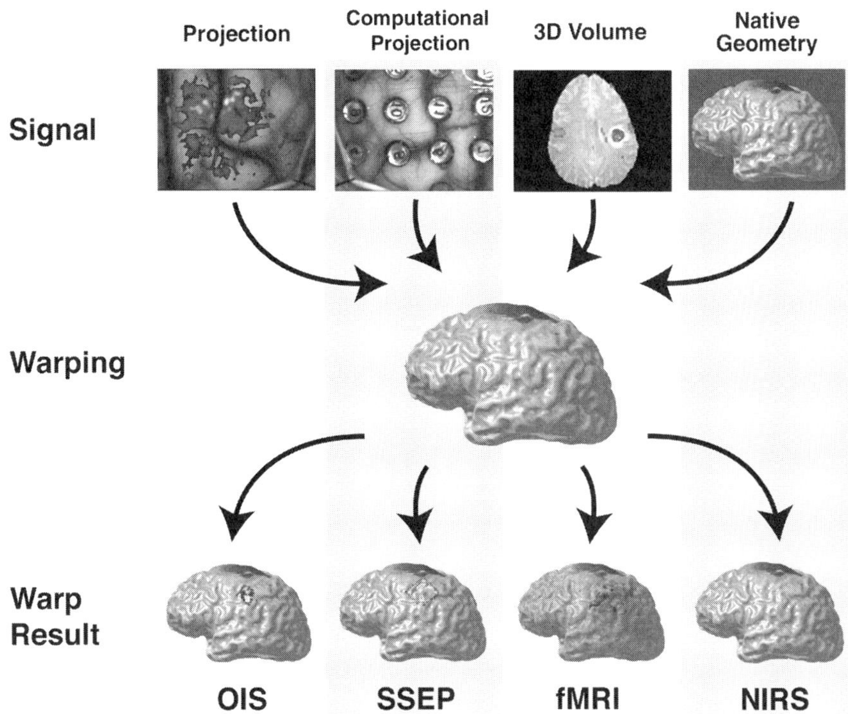

FIGURE 7.2 Intermodality comparisons. In order to compare mapping signals across modalities, we have devised a strategy in which we warp multiple modalities into a common space. This illustration demonstrates how each modality must be warped from its native space into a common space so that the mapping signals from each modality can be compared to each other. The common space is the cortical model of each subject's brain. Depending on the modality, different warping strategies are employed to warp the different mapping signals into a common space. While optical imaging is a surface-imaging technique and requires projection of the data onto the cortical surface, fMRI is a tomographic technique and is warped into the common space using rigid-body transformations to align the fMRI volume with the high-resolution MRI used to extract the cortical surface. The darkened area at the top of each cortical extraction demarcates tumor localization.

surface, the arteries and branch points that have been superimposed onto the cortical extraction are identified on the cortical surface and subsequently used to register intraoperative maps with preoperative scans.

7.4 THE TIMECOURSE OF OPTICAL SIGNALS IN HUMANS

The temporal profile of optical responses in humans is similar to that observed in animal models: appearing within 1 sec, peaking between 3 and 4 sec, and disappearing by 9 sec (Figure 7.3).[2,16] This similarity suggests that analogous phenomena are being imaged in animals and humans. This also suggests that studies elucidating physiology

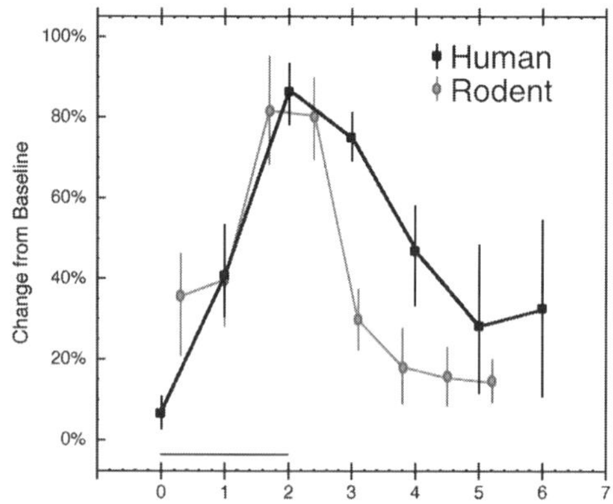

FIGURE 7.3 Rodent and human optical time courses. Comparison of the principal component timecourses from optical response data in rodents and humans (Cannestra, F. et al., *NeuroImage,* 3, 202, 1996). Optical responses were recorded after peripheral stimulation (human: 2 sec transcutaneous median or ulnar nerve stimulation, pulse interval 0.2 sec, pulse duration 200 msec, 15.5 to 17 mA current; rodent: 0.3 sec electromechanical motorized nudger, C1 whisker, 10 Hz) that began at time 0 sec. Signals appeared within 1 sec, peaked between 2 and 3 sec, and decayed by 6 sec in both humans and rodents. Error bars indicate standard errors.

and pathophysiology of the brain in animal models may be generalized to humans and perhaps further investigated in humans. Supporting this assertion, Cannestra et al. demonstrated that hemodynamic refractory periods, or reduced vascular response capacities with temporally close stimuli, originally observed in rodents using optical imaging were also observable in humans.[24] They found, at first in rodents, that the magnitude of the optical response to a stimulus presented immediately after another temporally close stimulus was significantly reduced. Subsequent investigations in humans found that delivering two sensory stimuli within seconds of one another produced a significantly diminished response to the second stimulus (Figure 7.4).

7.5 CORRELATION WITH ELECTROPHYSIOLOGICAL TECHNIQUES

The spatial correlation of perfusion-related responses and underlying electrophysiological activity is not well defined,[28,29] primarily due to differing etiologies of mapping techniques and resolution differences. For example, the blood oxygenation level-dependent (BOLD) fMRI signal is thought to emphasize changes in venous capillary oximetry and therefore represents a venous blood-dependent indicator of brain activation.[5,17,30] In contrast, O^{15} PET studies often measure changes in blood flow in order to produce functional maps.[31] Optical imaging, on the other hand, produces activation maps[6,7] derived from perfusion- and metabolism-related signals. Spatial maps, therefore, may differ across techniques.

Consistent with observations in animal studies, all the human optical studies to date indicate that the observed reflectance changes are spatially correlated with maps

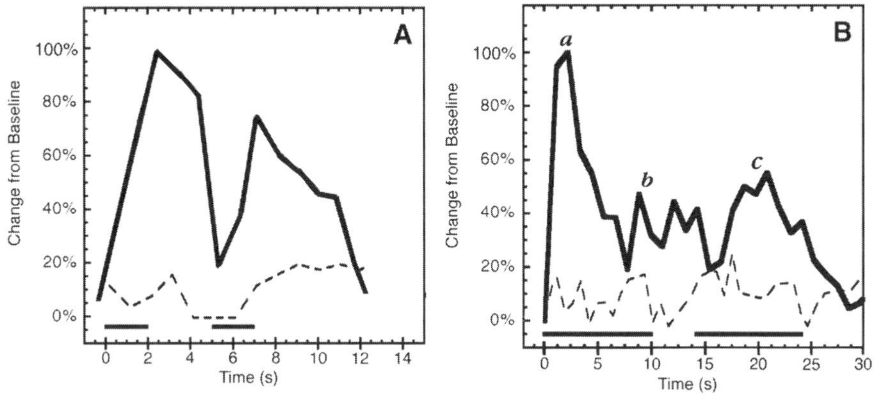

FIGURE 7.4 Hemodynamic refractory periods observed in the human cortex. Human vascular refractory period. Evidence for temporal dynamic behavior and vascular refractory periods in humans. (A) Time course of optical response to two stimuli, each 2 sec in duration, separated by 3 sec. The response to the second stimulus is significantly diminished. (B) Time course of optical response to two stimuli, each 10 sec in duration, separated by 5 sec. Once again, the response to the second stimulus appears much smaller, missing the initial peak in the optical response evident in the response to the first stimulus. Optical responses were recorded at 610 nm. Time courses are individual stimulation trials (solid line) and interleaved nonstimulated control trials (dashed line) from a single subject. (Adapted from Cannestra et al., *J. Neurophysiol.,* 41, 1137, 1998.)

derived using electrophysiological techniques (somatosensory-evoked potentials [SSEPs] and electrocortical stimulation maps [ESMs]).

SSEPs are recorded intraoperatively to help identify the central sulcus. SSEPs are usually recorded using either a 20-electrode grid or an 8-electrode strip placed on the cortical surface. The median or ulnar nerve is stimulated using a bipolar electrode and Grass Instruments (S-12, Quincy, MA) stimulator, using a 2-sec 15.5- to 17.0-mA pulsed current. A maximal amplitude negative peak identifies motor cortex. Somatosensory cortex is identified by phase reversal of the negative peak across the central sulcus.[32] For example, see the top left of Color Figure 7.5, which illustrates a pseudocolored SSEP map, with green areas indicating areas indicating motor cortex and blue areas identifying somatosensory cortex. Stimulus-related cortical optical reflectance changes during median and ulnar nerve stimulation have been shown to co-localize with the largest SSEPs in both sensory and motor cortices.[2]

Perhaps more importantly, optical maps also co-localize with the current gold standard of intraoperative cortical mapping, ESM (Color Figure 7.5).[1,4,5] ESM was the first technique for intraoperative mapping of brain function and was first applied within the neurosurgical setting by Penfield and Boldrey.[33] During ESM, bipolar electrodes are placed on the surface of the brain and direct current applied (4 to 20 mA for 2 to 4 sec). The technique causes a depolarization of the tissue between and around the site where the electrode is placed, which can result in the stimulation of an area resulting in a movement if over motor cortex, a perception (over sensory or limbic structures), or the disruption in task performance in an awake procedure.

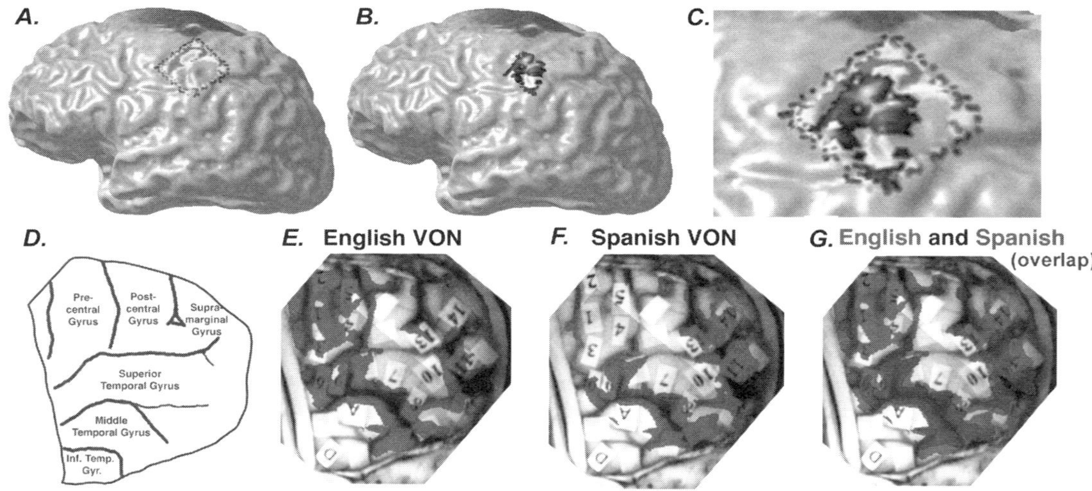

FIGURE 7.5 (See Color Figure 7.5.) Colocalization of optical intrinsic signals (OIS) and electrophysiological maps. (A, B, C) Colocalization of OIS and somatosensory evoked potentials (SSEPs). (A) Pseudocolor representation of SSEPs over left sensorimotor cortex is illustrated. The dark blue color indicates maximum amplitude negative peaking evoked potentials, indicating the location of the somatosensory cortex. The green areas, on the other hand, highlight cortical areas which demonstrate maximal phase reversal of the evoked potential response, corresponding to the motor cortex. Stimulus used was a transcutaneous right median and ulnar nerve stimulation (2-sec, 15-mA pulsed current). (B) Optical response to identical stimulation paradigm. Dark brown areas indicate areas of low intensity optical changes, while whiter areas in the middle of the activation region indicate higher intensity reflectance changes. (C) Superposition of optical response and SSEP, demonstrating a high degree of correlation. Regions of highest measured electrical change correspond to those areas where optical signals were largest in size and intensity. Both sensory and motor strips were activated. The stimulus intensity was sufficient to cause involuntary hand and finger movements. (D, E, F, G) Colocalization of optical signals and electrocortical stimulation maps. Tags indicate critical motor and language areas (numbers) and areas in which after-discharges were observed with stimulation (letters). Blank tags demarcate areas in which no effect of cortical stimulation was observed (ESM). (D) Schematic with labels identifying gyri observed in subsequent images. (E) Optical response to an English visual object-naming (VON) task (in which the patient is asked to name line drawings). Note that optical activity was observed over the same gyri in which critical ESM language sites were identified and no optical activity was observed over blank tags (which indicate areas not deemed essential for language processing by ESM). (F) Optical response to a Spanish VON task. (G) Comparison of English and Spanish VON. Optical maps from images (E) and (F) are thresholded (at one standard deviation above the mean optical response intensity) and displayed. Brown color indicates areas of activation overlap between the two languages tested. iOIS maps during English and Spanish visual object naming in a single subject revealed cortical areas active during both tasks and areas specifically activated by a single language. (Adapted from Pouratian, N. et al., *J. Neurosurg.*, 93, 686, 2000.)

In a recent survey of ten human subjects, we found that 98% of sites deemed active by ESM demonstrate optical changes at 610 nm.[34] In addition, ~25% of areas deemed inactive by ESM are also optically active, indicating that optical maps extend beyond the regions indicated by ESM. This phenomenon of "spread" has also been observed in the rodent somatosensory "barrel" cortex in response to whisker stimulation.[35-37] Narayan et al.[7] reported optical and intravascular fluorescent-dye maps overspilled regions of electrophysiologic activity (using single unit recordings) by about 20%. Optical maps encompass not only the principal barrel but also adjacent barrels. This is believed to be due to low-level neuronal activity that occurs in adjacent barrels in response to stimulation of adjacent, nonprincipal whiskers. In humans, the spread phenomenon may be due, in part, to the fact that optical imaging detects essential and secondary (i.e., active but not necessary for completion of task) cortices, while ESM detects only essential areas. Alternatively, this spread may represent a nonexact co-localization of optical signals and electrophysiological activity due to imprecise physiological coupling of neuronal activity, metabolism, and perfusion.

7.6 iOIS OF SENSORIMOTOR CORTEX

iOIS results demonstrate a spatial correspondence between electrophysiological and optical signal response in the human sensorimotor cortex (Color Figure 7.5). Where the optical signal was diffuse over sensory cortex and more focused over motor areas, the SSEP maps demonstrated similar spatial patterns. On the other hand, a larger contiguous optical signal focus over motor compared to sensory area was less obvious in the EP map. EPs represent neuronal firing and optical signals represent vascular and metabolic factors.[6,7] Nevertheless, the centers of highest response for EPs and optical signals do co-localize on sensory and motor strips.

iOIS studies indicate that optical maps co-localize with electrically active cortex and are also task specific. Haglund et al. originally demonstrated task specificity across cortices, showing that tongue movement and naming tasks activated distinct gyri.[1] Subsequently, we demonstrated that median and ulnar nerve stimulation produced distinct maps within the same gyrus.[2] iOIS has been used to obtain distinct maps of thumb, index finger, and middle finger within the same gyrus.[3] During iOIS imaging, thumb, index, and middle fingers were stimulated individually to obtain separate cortical activation maps, mapping a subset of the homunculus. Although the borders of these maps overlapped, the areas of maximum optical changes were distinct for each activation paradigm. Furthermore, although a large extent of cortex is active during thumb and index finger stimulation, other areas can be identified that are only responsive to one or the other finger. The orientation of individual digit iOIS responses is consistent with the classical homunculus schematic. Peak iOIS responses for each finger reveal a superior-to-inferior orientation for the middle finger, index finger, and thumb.

These studies begin to describe the morphology of perfusion-related responses and cortical activity. Optical reflectance imaging, combined with SSEP mapping of the sensorimotor cortex, illustrate the relationship between neuronal firing and vascular and metabolic activity. Additionally, these experiments suggest that the development and implementation of a noncontact, repeatable, intraoperative tool for functional mapping is of practical significance and may be used in conjunction with SSEPs to avoid eloquent brain during resection.

7.7 INTRAOPERATIVE MAPS OF LANGUAGE

Optical imaging of intrinsic signals (OIS) combined with ESM mapping provides a multimodality approach for the characterization of language cortices (Color Figure 7.5D to G).[1] Building on previous human studies,[2,3,16,24] we investigated the topographical and temporal activation of language cortices within the surgical setting utilizing iOIS and ESM.[4] We provide spatial comparison between disruption (ESM) and high-resolution activation (iOIS) maps, as well as paradigm-specific changes observed with iOIS. This study differs from previous reports[38-40] by characterizing spatial and temporal responses between and within Broca's and Wernicke's areas, once again demonstrating task specificity of optical responses. iOIS was used to identify subregions of Broca's and Wernicke's by identifying different patterns of activation in each region during different language tasks.[4] These observations demonstrate response profile differences dependent upon cortical region and language task.

We have also investigated language representations in a proficient bilingual patient using a novel neuroimaging technique, iOIS, and a visual object-naming task (Color Figure 7.5D to G). Our results indicate cortical areas activated by both English and Spanish (superior temporal sulcus, superior and middle temporal gyri, and parts of supramarginal gyrus). In addition, language-specific areas were identified in the supramarginal (Spanish) and precentral gyrus (English). These results suggest that cortical language representations in bilingual speakers may consist of both overlapping and distinct components. Furthermore, this study demonstrates the utility of iOIS in detecting topographical segregation of cognitively distinct cortices.

7.8 COMPARISON WITH BOLD FMRI

In a recent study we examined the relationship between different hemodynamic mapping signals measured by iOIS and blood oxygen level-dependent (BOLD) fMRI to better elucidate the temporal and spatial characteristics of perfusion-related cortical signals in humans. Current theory holds that the first functional or metabolic change to occur following neuronal activation is likely a brief burst of oxidative metabolism at the site of neuronal activity (Figure 7.6A).[6,8,41] This burst of oxidative metabolism is believed by some to produce a local increase in deoxyhemoglobin concentrations, which are very tightly spatially co-localized with the site of neuronal activity and responsible for the early dip in the 610-nm optical signal and the "initial dip" in the BOLD fMRI response (Figure 7.6A).

Note that this initial increase in deoxyhemoglobin concentrations remains extremely controversial, since not all groups have observed it (see Reference 42 vs. Reference 8, for example). Following this brief burst in oxidative metabolism, a

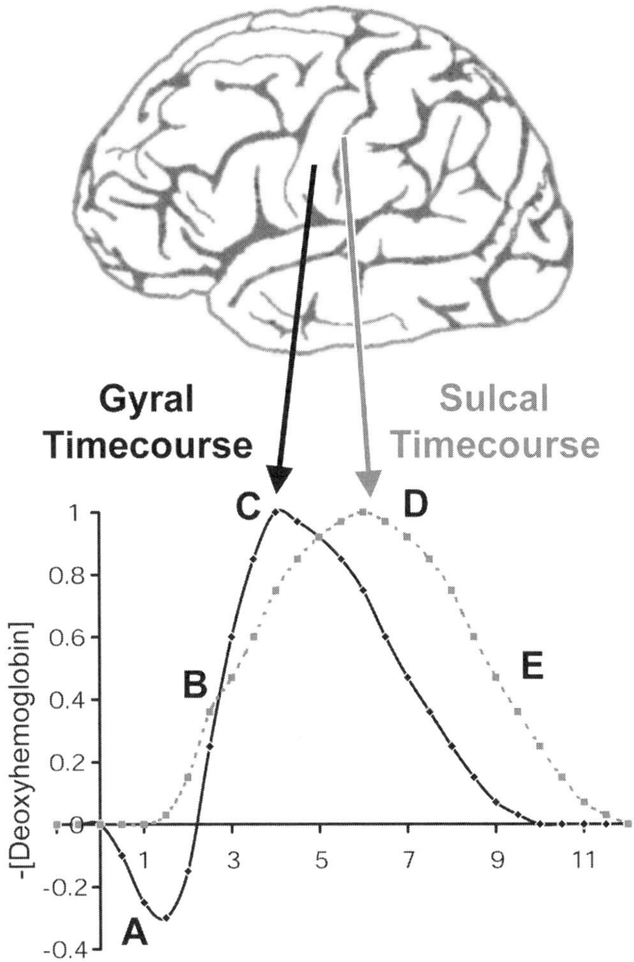

FIGURE 7.6 Schematic model of hemoglobin oxymetry changes that occur following functional activation. The time course of hemoglobin changes may be different in gyri, which contain the capillary beds, and sulci, which contain large venous structures. (A) Following functional activation, there may be a brief increase in deoxyhemoglobin concentrations in the cortex, corresponding to an immediate burst in oxidative metabolism preceding functional perfusion changes. This phase is often referred to as the "initial dip" in the blood oxygen level–dependent fMRI signal or the first phase of the 610-nm optical response. (B) Within seconds, functional recruitment of blood volume and flow reverse the initial increase in deoxyhemoglobin, "washing out" and eventually decreasing deoxyhemoglobin concentrations below baseline. (C, D) The functional blood volume and flow changes may peak in the capillary beds and cortex before peaking in the sulci or in the large venous structures. This phase is referred to as the positive BOLD signal and phase II of the 610-nm optical response. (E) Eventually, functional changes subside and the brain returns to baseline within 10 to 12 sec. *Note:* Time courses displayed are hypothetical and are illustrated to explain a possible model of hemoglobin oxymetry changes with functional activation.

slower onset (approximately 1 sec poststimulation) increase in functional perfusion begins, recruiting increased oxyhemoglobin and "washing out" increases in deoxyhemoglobin concentrations (Figure 7.6B, C, and D).[6,7,43] This increase in oxyhemoglobin and decrease in deoxyhemoglobin probably mediates the positive 610-nm OIS and positive BOLD signals. The increased functional perfusion most likely peaks earlier in the gyral tissue, which contains arterial and capillary components (Figure 7.6C), and peaks 2 to 3 sec later in the sulcal venous structures (Figure 7.6D).

To test this theory, sensorimotor cortex was studied during 110-Hz index finger vibration in eight human subjects undergoing neurosurgical procedures for removal of intracranial pathology (tumors or arteriovenous malformations, which are pathological shunts between arterial and venous systems). The same experimental parameters were used across modalities; all three modalities were co-registered in 3D.

Different spatial patterns and temporal response profiles were observed across modalities (Color Figure 7.7). Although overall fMRI, optical, and SSEP activities were observed with very similar spatial distributions, fMRI and the negative 610-nm optical response were separated by statistically significant distances. Temporally, the total fMRI response was delayed by 2 to 3 sec relative to the optical signals. However, the temporal profile of the responses became more similar when only pixels in the same location from each modality were examined. All modalities showed responses near the level of the superior genu of the central sulcus. SSEPs and optical activity were observed over the surface of pre- and postcentral gyri with very similar spatial distributions. The optical map co-localized with the SSEP map corresponding to a 13-mv potential change (within the 0.5-cm interpolation distance required for SSEPs). This co-localization pattern is consistent with close SSEP and optical spatial coupling observed in previous experiments.[2,3]

Although both BOLD fMRI and iOIS at 610-nm image signals arising from changes in local deoxyhemoglobin concentrations, the two modalities image distinct consequences of this physiological process. BOLD fMRI signals are believed to arise from changes in local magnetic susceptibility due to local changes in deoxyhemoglobin concentrations, while iOIS at 610 nm detects changes in cortical light reflectance related to changes in deoxyhemoglobin concentrations. Consequently, although the signals may be related to perfusion, the two modalities emphasize different components of the response. Based on the signal localization, fMRI may be detecting venous changes occurring in sulci and iOIS may emphasize changes occurring in the capillary beds within the gyri. fMRI's measuring signals from venous structures may effectively reduce the spatial resolution of fMRI, since a single venule may drain blood from multiple cortical areas. On the other hand, studies indicate that distinct capillary plexuses serve functionally discrete areas of cortex.[44] If iOIS maps are in fact derived from capillary signals, iOIS may offer the best resolution possible for a perfusion-dependent imaging modality. This is highlighted by the fact that iOIS may detect areas of activity that fMRI cannot discern.[5]

In another study comparing BOLD fMRI and iOIS, we investigated the temporal and spatial correlation of the positive BOLD signal with the positive 610-nm OIS response.[45] Previously, we compared the early negative 610-nm OIS response with fMRI maps using short (2-sec) somatosensory stimuli. Most studies of fMRI, however, utilize a block design (30 to 60 sec) in order to obtain a steady BOLD signal.

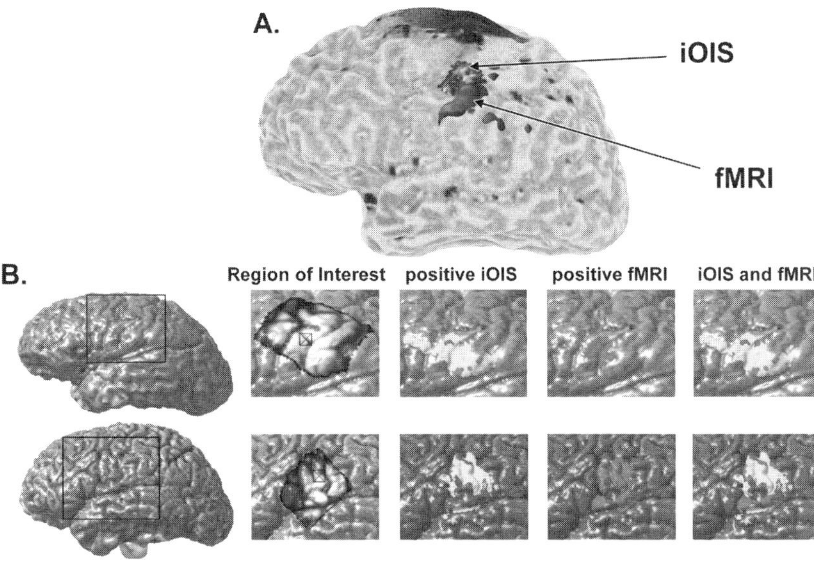

FIGURE 7.7 (See Color Figure 7.7.) Comparison of negative 610-nm OIS and BOLD fMRI signals in response to 2-sec somatosensory stimulus. Sensorimotor cortex was studied using identical stimulation parameters for fMRI and OIS: 2-sec 110-Hz index finger vibration in eight human subjects. The signals occurred adjacent to one another: the center of mass of the signals was separated by ~17 mm. While OIS signals were gyral, fMRI signals centered more superiorly and within sulci, consistent with a venous origin of BOLD fMRI signals. The gyral localization of the iOIS signal is consistent with optical signals emanating from capillary plexuses within the cortex. (The greener areas of the optical signal represent the most intense signal, –.3% reflectance change, while the purple areas of the optical signal represent the lowest-intensity optical reflectance changes observed.) The BOLD fMRI signal is displayed in red as a three-dimensional activation both at the surface of the brain and deep within the brain. The fMRI signal has been thresholded for display: all voxels with $p < .01$ for the correlation of the signal with a predetermined hemodynamic function are displayed. (B) Comparison of positive 610-nm response with BOLD fMRI using a block paradigm tongue motor task (20-sec blocks). Using both modalities, signals were identified near the inferior-lateral aspect of the central sulcus. The positive 610-nm response and the BOLD fMRI signal demonstrate high correlation (last column). These data support the notion that the two signals share similar etiologies, associated with the washout of deoxyhemoglobin by functional recruitment of cerebral blood flow and excess oxyhemoglobin. (Pouratian, N. et al., *Magn. Reson. Med.*, in press, 2001.)

We examined the same motor tongue task in both modalities. This task was chosen because 1) it activates primary motor and sensory cortices often associated with more robust and less complex hemodynamic responses[30] and 2) motor tongue activates the inferior-lateral aspect of the pre- and postcentral gyri and the central sulcus, which are easily accessible neurosurgically for optical imaging experiments.

In all subjects, robust fMRI activations ($p < .001$, correlation with a box-car reference function convolved with a hemodynamic function with a 6-sec lag to peak, 84 timepoints per subject, 5 subjects) were identified that centered on the lateral inferior aspect of the central sulcus and extended into adjacent pre- and postcentral gyri, adjacent to the gyrus containing the ESM tongue localization. OIS and fMRI maps co-localized, although the spatial extent of optical responses was larger ($p < .05$) and contained more gyral components (Color Figure 7.7). In most cases, the fMRI signal was a subset (48%) of the OIS signal. The time courses of the fMRI and OIS signals were similar, appearing within 2.5 sec and peaking within 6 sec of task onset.

Although many physiological processes may contribute to the positive 610-nm response, most of this response is due to a washout of deoxyhemoglobin by oxyhemoglobin. This study therefore supports the theory that BOLD signals represent a similar phenomenon. The inconsistencies between OIS and fMRI maps may be attributed to cell swelling and scattering that also contribute to OIS and fMRI sensitivity. While OIS spatial resolution is on the order of microns, fMRI spatial resolution is on the order of millimeters. Consequently, OIS may be able to detect small changes occurring within the gyri while similar changes occurring within the gyri of the larger fMRI voxel may be "drowned out" and undetectable.[30]

By comparing positive BOLD signals and positive 610-nm OIS signals within human subjects, we have demonstrated the high likelihood that these signals share a common etiology. These findings support the common theory that positive BOLD signals are due to a washout of deoxyhemoglobin by oxyhemoglobin following functional activation. This study also demonstrates that fMRI signals may not precisely co-localize with electrophysiological maps, instead emphasizing changes in venous structures in adjacent sulci.

7.9 CLINICAL UTILITY

iOIS offers several potential advantages over the current gold standard, ESM. The spatial resolution of iOIS (~200 μm) is much greater than that of ESM. Resection within 1 cm of essential areas identified by ESM increases the likelihood of postoperative deficits,[46] suggesting that the limit of resolution of ESM is on the order of 1 cm. This resolution is limited and does not ensure resection of as much pathological tissue as possible. iOIS can potentially provide a more detailed functional map of the exposed cortex.

iOIS also allows a more rapid assessment of large areas of cortex than ESM, which may be particularly important for mapping multiple functions intraoperatively for surgical guidance. Another advantage of iOIS over preoperative mapping techniques, such as fMRI, is that the mapping is done intraoperatively. While preoperative fMRI may be helpful, its application may be limited by the phenomenon of brain shift, which is the herniation or collapse of the brain following craniotomy and dural reflection.[47] The geometric changes that occur intraoperatively prevent the use of preoperative fMRI maps to localize function precisely. Since iOIS is done intraoperatively, brain shift is intrinsically accounted for.

Functional imaging in the operating room environment poses additional challenges. Ideally, functional imaging equipment should interfere with the normal operating environment as little as possible and should be portable so that it can be moved into and out of the sterile field of the operating room easily and rapidly.

The iOIS setup requires neither modification of the traditional operating room configuration nor any major modification to neurosurgical techniques, as has been necessary for implementation of intraoperative MR (iMR). Furthermore, the camera used for optical imaging can be mounted atop a second operating microscope so that it can be moved into and out of the sterile operating field as necessary, minimizing surgical interference.

iOIS offers the additional advantage that it does not perturb the brain during surgery. Since iOIS detects only changes in light reflectance, it does not require any contact with potentially normal tissue that the surgeon may not want to disturb. Moreover, optical imaging equipment is affordable since it requires only three major components: camera, light source, and computer. (Note that the same equipment can be used for both animal and human imaging.) Other intraoperative functional imaging techniques like iMR may, on the other hand, be cost-prohibitive since the investigator must incur the expense not only of the MRI scanner but also of modifying the entire surgical environment to be magnet compatible.

Although several studies have attempted to delineate the precise physiological processes contributing signal to optical responses at each wavelength,[7,8,48,49] it may not be necessary to tease out such details for clinical applications of such a tool. The exact etiology of the signal should not be critical as long as the signals at a particular wavelength are reproducible, consistent, and in good spatial agreement with underlying neuronal activity and other measures of functional activation.

7.10 CONCLUSIONS

Intraoperative optical intrinsic signal imaging (iOIS) offers a unique opportunity to image human perfusion- and metabolism-related responses with high spatial and temporal resolution. Several studies have already demonstrated the feasibility of the technique and its compatibility with the operating room environment. Comparison of iOIS signals with other modalities using the techniques described can potentially contribute great insight into the physiology and the functional organization of the brain.

ACKNOWLEDGMENTS

Thanks go to the staff of the Laboratory of Neuro Imaging, without whom this work would not have been possible. This work was supported, in part, by the National Institute of Mental Health (MH 52083) and the Training Program in Neuroimaging (MH19950). Nader Pouratian is supported, in part, by the Medical Scientist Training Program (GM08042) and National Research Service Award (MH12773).

REFERENCES

1. Haglund, M.M., Ojemann, G.A., and Hochman, D.W., Optical imaging of epileptiform and functional activity in human cerebral cortex, *Nature*, 358, 668, 1992.
2. Toga, A.W., Cannestra, A.F., and Black, K.L., The temporal/spatial evolution of optical signals in human cortex, *Cerebr. Cortex*, 5, 561, 1995.

3. Cannestra, A.F. et al., Topographical and temporal specificity of human intraoperative optical intrinsic signals, *Neuroreport*, 9, 2557, 1998.
4. Cannestra, A.F. et al., Temporal and topographical characterization of language cortices using intraoperative optical intrinsinc signals, *NeuroImage*, 12, 41, 2000.
5. Pouratian, N. et al., Optical imaging of bilingual cortical representations: case report, *J. Neurosurg.*, 93, 686, 2000.
6. Frostig, R.D. et al., Cortical functional architecture and local coupling between neuronal activity and the microcirculation revealed by *in vivo* high-resolution optical imaging of intrinsic signals, *Proc. Natl. Acad. Sci. U.S.A.*, 87, 6082, 1990.
7. Narayan, S.M. et al., Functional increases in cerebral blood volume over somatosensory cortex, *J. Cerebr. Blood Flow Metab.*, 15, 754, 1995.
8. Vanzetta, I. and Grinvald, A., Increased cortical oxidative metabolism due to sensory stimulation: implications for functional brain imaging, *Science*, 286, 1555, 1999.
9. Daghighian, F. et al., Intraoperative beta probe: a device for detecting tissue labeled with positron or electron emitting isotopes during surgery, *Med. Phys.*, 21, 153, 1994.
10. Hill, D.K. and Keynes, R.D., Opacity changes in stimulated nerve, *J. Physiol. (Lond.)*, 108, 278, 1949.
11. Orbach, H.S. and Cohen, L.B., Optical monitoring of activity from many areas of the *in vitro* and *in vivo* salamander olfactory bulb: a new method for studying functional organization in the vertebrate central nervous system, *J. Neurosci.*, 3, 2251, 1983.
12. Grinvald, A. et al., Real-time optical mapping of neuronal activity in vertebrate CNS *in vitro* and *in vivo, Soc. Gen. Physiol. Ser.*, 40, 165, 1986.
13. Masino, S.A. et al., Characterization of functional organization within rat barrel cortex using intrinsic signal optical imaging through a thinned skull, *Proc. Natl. Acad. Sci. U.S.A.*, 90, 9998, 1993.
14. Narayan, S.M. et al., Imaging optical reflectance in rodent barrel and forelimb sensory cortex, *NeuroImage*, 1, 181, 1994.
15. Bonhoeffer, T. and Grinvald, A., Iso-orientation domains in cat visual cortex are arranged in pinwheel-like patterns, *Nature*, 353, 429, 1991.
16. Cannestra, A.F. et al., The evolution of optical signals in human and rodent cortex, *NeuroImage*, 3, 202, 1996.
17. Cannestra, A.F. et al., Temporal spatial differences observed by functional MRI and human intraoperative optical imaging, *Cerebr. Cortex*, 11, 773, 2001.
18. Cohen, L. and Keynes, R.D., Changes in light scattering associated with the action potential in crab nerves, *J. Physiol. (Lond.)*, 212, 259, 1971.
19. Holthoff, K. and Witte, O.W., Intrinsic optical signals in rat neocortical slices measured with near-infrared dark-field microscopy reveal changes in extracellular space, *J. Neurosci.*, 16, 2740, 1996.
20. Malonek, D. and Grinvald, A., Interactions between electrical activity and cortical microcirculation revealed by imaging spectroscopy: implications for functional brain mapping, *Science*, 272, 551, 1996.
21. LeManna, J.C. et al., Detection of an oxidizable fraction of cytochrome oxidase in intact rat brain, *Am. Physiol. Soc.*, 253, C477, 1987.
22. Grinvald, A. et al., High-resolution optical imaging of functional brain architecture in the awake monkey, *Proc. Natl. Acad. Sci. U.S.A.*, 88, 11559, 1991.
23. Hodge, C.J., Jr. et al., Identification of functioning cortex using cortical optical imaging, *Neurosurgery*, 41, 1137, 1997.
24. Cannestra, A.F. et al., Refractory periods observed by intrinsic signal and fluorescent dye imaging, *J. Neurophysiol.*, 80, 1522, 1998.

25. Woods, R.P., Cherry, S.R., and Mazziotta, J.C., Rapid automated algorithm for aligning and reslicing pet images, *J. Comput. Assisted Tomogr.*, 16, 620, 1992.
26. Chen–Bee, C.H. et al., Visualizing and quantifying evoked cortical activity assessed with intrinsic signal imaging, *J. Neurosci. Meth.*, 97, 157, 2000.
27. Corina, D.P. et al., Correspondences between language cortex identified by cortical stimulation mapping and fMRI, *NeuroImage*, 11, S295, 2000.
28. Lou, H.C., Edvinsson, L., and MacKenzie, E.T., The concept of coupling blood flow to brain function: revision required? *Ann. Neurol.*, 22, 289, 1987.
29. Lindauer, U., Villringer, A., and Dirnagl, U., Characterization of CBF response to somatosensory stimulation: model and influence of anesthetics, *Am. J. Physiol.*, 264, H1223, 1993.
30. Cohen, M.S. and Bookheimer, S.Y., Localization of brain function using magnetic resonance imaging, *Trends Neurosci.*, 17, 268, 1994.
31. Phelps, M.E. and Mazziotta, J.C., Positron emission tomography: human brain function and biochemistry, *Science*, 228, 799, 1985.
32. Nuwer, M.R. et al., Topographic mapping of somatosensory evoked potentials helps identify motor cortex more quickly in the operating room, *Brain Topogr.*, 5, 53, 1992.
33. Penfield, W. and Boldrey, E., Somatic motor and sensory representation in the cerebral cortex of man as studied by electrical stimulation, *Brain*, 60, 389, 1937.
34. Pouratian, N. et al., Intraoperative sensorimotor and language mapping using optical intrinsic signal imaging: comparison with electrophysiologic techniques and fMRI in 40 patients, 2000 Am. Assoc. Neurolog. Surgeons Ann. Meet., 2000.
35. Chen–Bee, C.H. and Frostig, R.D., Variability and interhemispheric asymmetry of single-whisker functional representations in rat barrel cortex, *J. Neurophysiol.*, 76, 884, 1996.
36. Masino, S.A. and Frostig, R.D., Quantitative long-term imaging of the functional representation of a whisker in rat barrel cortex, *Proc. Natl. Acad. Sci. U.S.A.*, 93, 4942, 1996.
37. Godde, B. et al., Optical imaging of rat somatosensory cortex reveals representational overlap as topographic principle, *Neuroreport*, 7, 24, 1995.
38. Malow, B.A. et al., Cortical stimulation elicits regional distinctions in auditory and visual naming, *Epilepsia*, 37, 245, 1996.
39. Ojemann, G. et al., Cortical language localization in left, dominant hemisphere. An electrical stimulation mapping investigation in 117 patients, *J. Neurosurg.*, 71, 316, 1989.
40. Buckner, R.L. et al., Functional anatomical studies of explicit and implicit memory retrieval tasks, *J. Neurosci.*, 15, 12, 1995.
41. Fox, P.T. and Raichle, M.E., Focal physiological uncoupling of cerebral blood flow and oxidative metabolism during somatosensory stimulation in human subjects, *Proc. Natl. Acad. Sci. U.S.A.*, 83, 1140, 1986.
42. Kohl, M. et al., Physical model for the spectroscopic analysis of cortical intrinsic optical signals, *Phys. Med. Biol.*, 45, 3749, 2000.
43. Ngai, A.C. et al., Effect of sciatic nerve stimulation on pial arterioles in rats, *Am. J. Physiol.*, 254, H133, 1988.
44. Cox, S.B., Woolsey, T.A., and Rovainen, C.M., Localized dynamic changes in cortical blood flow with whisker stimulation corresponds to matched vascular and neuronal architecture of rat barrels, *J. Cerebr. Blood Flow Metab.*, 13, 899, 1993.
45. Pouratian, N. et al., Spatial/temporal correlation of optical intrinsic signals and BOLD, *Magn. Reson. Med.*, in press, 2002.
46. Haglund, M.M. et al., Cortical localization of temporal lobe language sites in patients with gliomas, *Neurosurgery*, 34, 567, 1994.

47. Maurer, C.R., Jr. et al., Investigation of intraoperative brain deformation using a 1.5-t interventional MR system: preliminary results, *IEEE Trans. Med. Imaging*, 17, 817, 1998.
48. Mayhew, J. et al., An evaluation of linear model analysis techniques for processing images of microcirculation activity, *NeuroImage*, 7, 49, 1998.
49. Nemoto, M. et al., Analysis of optical signals evoked by peripheral nerve stimulation in rat somatosensory cortex: dynamic changes in hemoglobin concentration and oxygenation, *J. Cerebr. Blood Flow Metab.*, 19, 246, 1999.
50. Cohen, L.B., Changes in neuron structure during action potential propagation and synaptic transmission, *Physiol. Rev.*, 53, 373, 1973.

8 Noninvasive Imaging of Cerebral Activation with Diffuse Optical Tomography

David A. Boas, Maria Angela Franceschini, Andy K. Dunn, and Gary Strangman

CONTENTS

0-8493-2389-4/02/$0.00+$1.50
© 2002 by CRC Press LLC

8.1 INTRODUCTION

Optical methods have been used to measure physiological variables for decades, as is apparent from other chapters in this book. However, the use of *diffuse* optical light for physiological monitoring — and for noninvasive monitoring in particular — has a somewhat shorter history. This delayed development resulted from two primary obstacles. The first obstacle was sensitivity — a need to discover a sufficiently nonabsorbing wavelength range and to develop sufficiently sensitive detectors to monitor through relatively large distances (centimeters) of tissue. As it turns out, the near-infrared wavelength range between roughly 600 and 950 nm is relatively poorly absorbed by biological tissue. Furthermore, the primary absorbers in this wavelength range are oxyhemoglobin and deoxyhemoglobin (HbO and Hb, respectively), which are important biological indicators. In the late 1970s, Jöbsis[1] was the first to use near-infrared light to measure hemodynamic parameters of the brain noninvasively. This work employed what is now called near-infrared spectroscopy (NIRS), a point-source measurement of species concentrations within tissue.

It was to be some years before actual imaging with diffuse light (diffuse optical tomography, or DOT) would appear, however, because of the second obstacle: a need for better understanding of how light propagates through highly scattering (diffusive) tissue. Considerable work has appeared on the theory of light propagation through scattering media in recent years (for a review, see Reference 2), and this work, along with phantom studies, has made imaging with diffuse light a reality.

Prior to these developments, and since Jöbsis' initial work, many groups have examined a variety of biologically relevant parameters in tissue, including brain.[3,4] Early research, during the 1980s and early 1990s, focused on the use of diffuse near-infrared light for measuring brain hemoglobin oxygen saturation within neonate and adult humans.[5-9] In the early 1990s, functional magnetic resonance imaging (fMRI) came of age, providing whole-brain imaging of the blood oxygen level–dependent (BOLD) signal associated with altered blood flow and oxygen metabolism within the brain.[10-14] Shortly after this development, NIRS was shown to reveal sensitivity similar to that of fMRI.[15-18] While sensitive to similar species, an important characteristic of the optical measure of brain activity was its ability to separately distinguish changes in oxy- and deoxyhemoglobin concentrations,[19] whereas fMRI only indirectly measures relative changes in deoxyhemoglobin.[20] NIRS has the further advantage of being inexpensive and portable, thus allowing studies of more subjects under a wider range of conditions. These advantages must, of course, be weighed against the principal limitations of the optical techniques: the current lack of anatomical imaging for localization purposes and the relatively shallow penetration depth (particularly during brain monitoring).

To date, many research groups worldwide have used NIRS in functional brain studies as a noninvasive tool to monitor local changes in cerebral oxygenation and hemodynamics.[21-33] In addition, several groups have investigated the possibility of using NIRS to measure the cytochrome oxidase (cyt-ox) redox state.[34,35] Cyt-ox oxidation is a marker of intracellular energy metabolism and has been suggested as a possible parameter to assess the functional state of the brain.[36-40] These optical measurements are usually conducted over a wide spectral range (extending up to 500 nm), because the absorption contributions of cyt-ox are an order of magnitude (or more) smaller than those of hemoglobin.

A few groups have also worked on using diffuse light to measure neuronal activity directly rather than indirectly via hemoglobin. For example, it has been shown *in vitro* that neuronal activity is associated with an increase in light scattering, induced by a change in the index of refraction of the neuronal membranes.[41-43] Following up on this work, near-infrared optical methods have been successfully used to detect such light-scattering changes *in vivo* in adult humans.[44-46] This fast optical signal appears to show the same time course as the electrophysiological response measured with EEG or electrophysiological techniques.

Although still in its infancy, the use of diffuse light for noninvasive imaging (diffuse optical tomography, DOT) is progressing quickly.[2,47] The accumulating literature clearly demonstrates the unique ability of near-infrared techniques to detect hemodynamic, metabolic, and neuronal signals associated with brain activity. Moreover, these techniques hold promise of eventually providing absolute quantitative values for at least hemodynamic and metabolic parameters. No other single imaging modality has this ability; fMRI can measure relative hemodynamic changes, EEG and MEG can measure neuronal signals, and PET can measure blood flow, volume, and glucose metabolism.

The unique abilities of optical methods are further supplemented by the applicability of diffuse optical techniques to a variety of studies that are highly problematic for other brain-imaging techniques. Such studies include neurobehavioral development in infants and young children and functional reorganization during stroke or brain-injury rehabilitation, among others. This wide applicability motivates further development of the technology toward whole-head absolute quantitative imaging.

In the remainder of this chapter, we review the basic principles of NIRS and DOT, discuss example instrumentation implementations, and provide illustrative brain activation data. We then summarize with a discussion of the future directions for functional optical brain imaging.

8.2 THEORY

It is a relatively simple step, conceptually speaking at least, to move from the simple NIRS point measurement to imaging of these same variables. Before going into the details of imaging *per se*, however, we will first briefly review the physical model of photon migration through tissue that is needed to obtain quantitatively accurate estimates of the chromophore concentration changes within the brain during functional activation. We continue to make the distinction between NIRS and DOT, in which the former indicates nonimaging point measurements while the latter is reserved for imaging measurements. The distinction in instrumentation depends only on the number and arrangement of sources and detectors used for measurements. The rest of this section reviews the theoretical concepts of NIRS and DOT.

8.2.1 NIRS Theory*

The theory of the modified Beer–Lambert Law (MBLL) has been explained previously.[5,48] This technique is based on the absorption of near-infrared light by oxy- and deoxyhemoglobin. Changes in the concentrations of these chromophores are quantified using a modified Beer–Lambert law, which is an empirical description of optical attenuation in a highly scattering medium.[5,48] The modified Beer–Lambert law is

$$OD = -\log\frac{I}{I_o} = \varepsilon CLB + G \tag{8.1}$$

where OD is the optical density, I_o is the incident light intensity, I is the detected light intensity, ε is the extinction coefficient of the chromophore, C is the concentration of the chromophore, L is the distance between where the light enters the tissue and where the detected light exits the tissue, B is a pathlength factor that accounts for increases in the photon pathlength caused by tissue scattering, and G is a factor that accounts for the measurement geometry. We use the convention of log base e.

A change in the chromophore concentration causes the detected intensity to change. When the concentration changes, the extinction coefficient ε and distance L remain constant and it is assumed that B and G remain constant. Thus, Equation 8.1 can be rewritten as

$$\Delta OD = -\ln\frac{I_{Final}}{I_{Initial}} = \varepsilon \Delta CLB \tag{8.2}$$

where $\Delta OD = OD_{Final} - OD_{Initial}$ is the change in optical density (the logarithm is base e), I_{final} and $I_{Initial}$ are the measured intensities before and after the concentration change, and ΔC is the change in concentration. L is specified by the probe geometry, ε is an intrinsic property of the chromophore, and B is often referred to as the differential pathlength factor (DPF), which can be determined from independent measurements with ultra-short pulses of light[49] and has been tabulated for various tissues. Thus, given the extinction coefficient, it is possible to quantify the change in chromophore concentration.

Figure 8.1 plots the extinction coefficients for oxy- and deoxyhemoglobin vs. wavelength as measured by Wray et al., Cope, and Matcher et al.[6,48,50] The wavelength range of 700 to 950 nm is the only region in which light is able to penetrate several centimeters through tissue. The other chromophores of significance in tissue in this wavelength range are water, lipids, and cytochrome aa3. We do not consider these chromophores in this chapter, as their contribution in general is an order of

* The generalization of this formula for more than two wavelengths can be found in Cope, M., *The Development of a Near-Infrared Spectroscopy System and Its Application for Noninvasive Monitoring of Cerebral Blood and Tissue Oxygenation in the Newborn Infant,* University College, London, 1991.

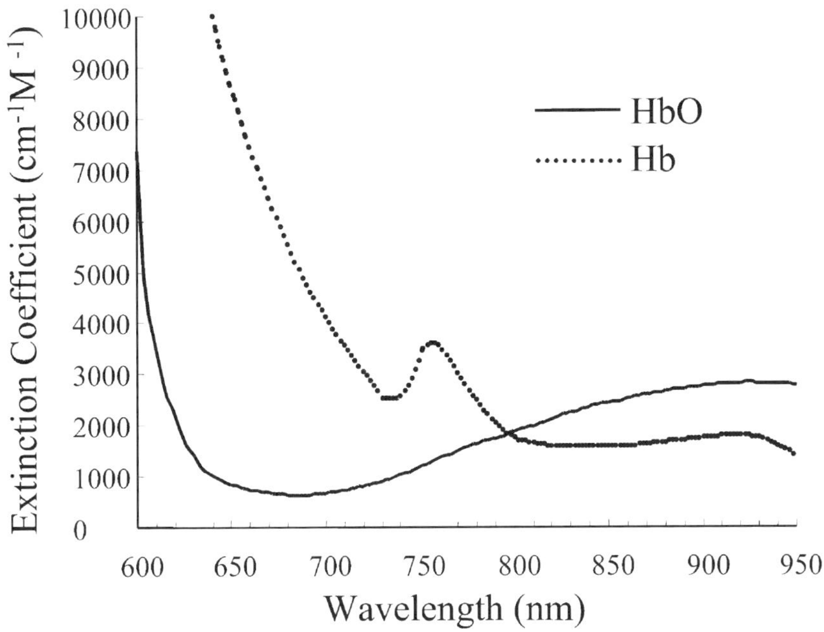

FIGURE 8.1 Absorption spectrum of oxyhemoglobin (HbO) and deoxyhemoglobin (Hb). Below 600 nm the absorption increases by more than a factor of 10, while above 950 nm water absorption increases significantly.

magnitude less significant than hemoglobin and is not easily measured without use of six or more wavelengths. In order to consider the contribution of two chromophores, we must rewrite Equation 8.2 as

$$\Delta OD^{\lambda} = (\varepsilon^{\lambda}_{\text{HbO}}\, \Delta[\text{HbO}] + \varepsilon^{\lambda}_{\text{Hb}}\, \Delta[\text{Hb}])B^{\lambda} L \tag{8.3}$$

where λ indicates a particular wavelength. Equation 8.3 explicitly accounts for independent concentration changes in oxyhemoglobin (Δ[HbO]) and deoxyhemoglobin (Δ[Hb]).

By measuring ΔOD at two wavelengths (λ_1 and λ_2) and using the known extinction coefficients of oxyhemoglobin (ε_{HbO}) and deoxyhemoglobin (ε_{Hb}) at those wavelengths, we can then determine their concentration changes:

$$\begin{aligned}\Delta[\text{Hb}] &= \frac{\varepsilon^{\lambda_2}_{HbO}\dfrac{\Delta OD^{\lambda_1}}{B^{\lambda_1}} - \varepsilon^{\lambda_1}_{HbO}\dfrac{\Delta OD^{\lambda_2}}{B^{\lambda_2}}}{(\varepsilon^{\lambda_1}_{Hb}\varepsilon^{\lambda_2}_{HbO} - \varepsilon^{\lambda_2}_{Hb}\varepsilon^{\lambda_1}_{HbO})L} \\ \Delta[\text{HbO}] &= \frac{\varepsilon^{\lambda_1}_{Hb}\dfrac{\Delta OD^{\lambda_2}}{B^{\lambda_2}} - \varepsilon^{\lambda_2}_{Hb}\dfrac{\Delta OD^{\lambda_1}}{B^{\lambda_1}}}{(\varepsilon^{\lambda_1}_{Hb}\varepsilon^{\lambda_2}_{HbO} - \varepsilon^{\lambda_2}_{Hb}\varepsilon^{\lambda_1}_{HbO})L}\end{aligned} \tag{8.4}$$

The generalization of this formula for more than two wavelengths can be found in Reference 48.

8.2.2 Photon Diffusion Equation

While the MBLL has been shown to work well in highly scattering media with uniform properties, it has known deficiencies in media with more complex structures (e.g., the layered structure of the scalp, skull, and brain), and it does not provide a framework for reconstructing images. The diffusion approximation to the radiative transport equation provides the formal framework required for understanding photon migration in highly scattering medium. The photon diffusion approximation is[51-53]

$$-D\nabla^2\Phi(\mathbf{r},t)+v\mu_a\Phi(\mathbf{r},t)+\frac{\partial\Phi(\mathbf{r},t)}{\partial t}=vS(\mathbf{r},t) \tag{8.5}$$

$\Phi(\mathbf{r}, t)$ is the photon fluence at position **r** and time t (the photon fluence is proportional to the intensity). $S(\mathbf{r}, t)$ is the source distribution of photons. $D = v/(3\mu_s')$ is the photon diffusion coefficient,[54,55] μ_s' is the reduced scattering coefficient, μ_a is the absorption coefficient, and v is the speed of light in the medium. Note that the absorption coefficient is related to the extinction coefficient and the concentration as $\mu_a = \varepsilon C$. For a combination of the hemoglobin chromophores,

$$\mu_a=\varepsilon_{HbO}[HbO]+\varepsilon_{Hb}[Hb] \tag{8.6}$$

Equation 8.5 accurately models the migration of light through highly scattering media, provided that the probability of scattering is much greater than the absorption probability. Note that all factors in Equation 8.5 are wavelength dependent.

Solutions of the photon diffusion equation can be used to predict the photon fluence (or intensity) detected for typical diffuse measurements.[53] Assuming that concentration changes are both global and small, the solution of the photon diffusion equation for a semi-infinite medium is

$$\begin{aligned}\Delta OD &= -\log\frac{\Phi_{Final}}{\Phi_{Initial}}\\ &=\frac{1}{2}\left(\frac{3\mu_s'}{\mu_a^{Initial}}\right)^{1/2}\left[1-\frac{1}{(1+L(3\mu_s'^{\,Initial}\mu_a^{Initial})^{1/2})}\right](\varepsilon_{HbO}\Delta[\mathrm{HbO}]+\varepsilon_{Hb}\Delta[\mathrm{Hb}])L\end{aligned} \tag{8.7}$$

The solution of the photon diffusion equation for representative tissue geometry (Equation 8.7) tells us that the modified Beer–Lambert law is reasonable for tissues with spatially uniform optical properties when the chromophore concentration does not change significantly (i.e., $\Delta[X]/[X] \ll 1$). Equation 8.7 shows that the pathlength factor B in Equation 8.3 is given by

$$B = \frac{1}{2}\left(\frac{3\mu_s'}{\mu_a^{Initial}}\right)^{1/2}\left[1 - \frac{1}{(1 + L(3\mu_s'^{Initial}\mu_a^{Initial})^{1/2})}\right] \tag{8.8}$$

for a semi-infinite medium. This shows that B depends on tissue scattering, initial chromophore concentration, extinction coefficient (thus, B is wavelength dependent), and optode separation. In practice, the validity of the assumption that B is independent of μ_a and L has often been ignored since B is, in general, empirically determined and the changes in μ_a are typically small.

8.2.3 Diffuse Optical Tomography

The preceding equations apply to every individual measurement that might be made from a NIRS or a DOT system. In order to generate an image of spatial variations in Hb and HbO, all that we need to do is to measure the photon fluence at multiple source and detector positions (typically with overlapping sensitivities) and essentially backproject the image. The general solution of the photon diffusion equation at the detector position $\mathbf{r}_d$ for a medium with spatially varying absorption is

$$\Phi(\mathbf{r}_d) = \Phi_{incident}(\mathbf{r}_d) - \int \Phi(\mathbf{r})\frac{v}{D}(\varepsilon_{HbO}\Delta[\mathrm{HbO}] + \varepsilon_{Hb}\Delta[\mathrm{Hb}])G(\mathbf{r}, \mathbf{r}_d)d\mathbf{r} \tag{8.9}$$

For reflectance in a semi-infinite medium, the solutions for $\Phi_{incident}$ and $G(\mathbf{r}, \mathbf{r}_d)$ can be found in Farrell et al.[56] Equation 8.9 is an implicit equation for the measured fluence as $\Phi(\mathbf{r})$ appears in the integral on the right-hand side. One way to solve this equation uses the first Born approximation, which assumes that $\Phi(\mathbf{r}) = \Phi_{incident} + \Phi_{sc}$ where[57,58]

$$\Phi_{sc}(\mathbf{r}_d) = -\int \Phi_{incident}(\mathbf{r})\frac{v}{D}(\varepsilon_{HbO}\Delta[\mathrm{HbO}] + \varepsilon_{Hb}\Delta[\mathrm{Hb}])G(\mathbf{r}, \mathbf{r}_d)d\mathbf{r} \tag{8.10}$$

By measuring $\Phi_{sc}(\mathbf{r}_d)$ using multiple source and detector positions and multiple wavelengths, we can invert the integral equation to obtain images of Δ[HbO] and Δ[Hb]. There are numerous methods for solving Equation 8.10.[59] Most rely on reducing Equation 8.10 to a matrix equation by rewriting the integral as a sum over voxels; i.e., y = Ax where y is the vector of measurements (i.e., $\Phi_{sc}(\mathbf{r}_d)$), x is the vector of image voxels, and A is the transformation matrix obtained from the integrand of Equation 8.10. For the case of fewer measurements than unknowns, the linear inverse problem is underdetermined and is given by the (regularized) Moore–Penrose generalized inverse

$$\hat{x} = -A^T(AA^T + \lambda I)^{-1}y \tag{8.11}$$

where I is the identity matrix and $\lambda = \alpha\max(AA^T)$ is the Tikhonov regularization parameter.

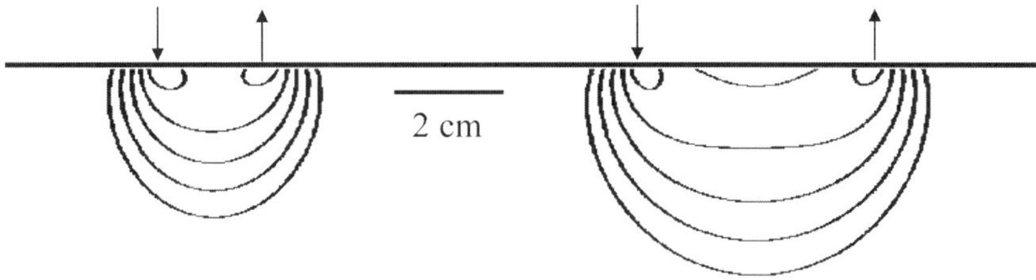

FIGURE 8.2 Photon migration sensitivity profile shown for a semi-infinite medium for 2- and 4-cm source detector separation. Contour lines are shown every half-order of magnitude.

8.2.4 DOT Simulations

Equation 8.10 describes the spatial sensitivity profile of light diffusing through a highly scattering medium. In Figure 8.2 we show the sensitivity profile for a simple reflectance geometry with a continuous-wave (CW) light source using typical tissue optical properties of $\mu_s' = 10\ cm^{-1}$ and $\mu_a = 0.1\ cm^{-1}$. The most important point is that the spatial sensitivity profile is not localized but covers a large volume and therefore intrinsically has relatively poor spatial resolution. Second, notice that as the separation between the source and detector increases, the penetration depth also increases. This fact can be used to probe deeper into the brain and possibly even to distinguish superficial scalp from deeper brain signals.

Figure 8.3 illustrates the process of reconstructing an image with diffuse light. The process is conceptually similar to that of x-ray computed tomography, which

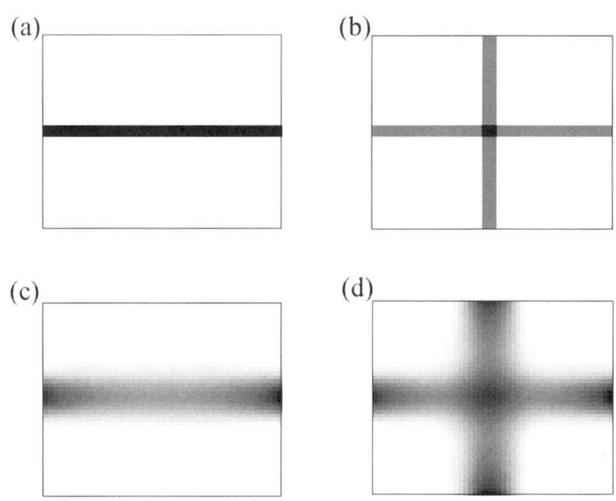

FIGURE 8.3 Comparison of x-ray computed tomography (a, b) and diffuse optical tomography (c, d) of a square absorbing object in the center of the outer square. X-ray tomography has greater resolution, as indicated by the sharper edges of the reconstructed image. Diffuse optical tomography reconstructs the object but with reduced resolution because of the diffusive nature of near-infrared photon migration in tissue.

is included for comparison. In the case of x-rays, a collimated beam transilluminates the tissue, maintaining high spatial resolution, and the attenuation of the beam is measured. The measured attenuation is then backprojected into the tissue along the axis of the x-ray. For a simple object positioned in the center of the plane, we see that a single x-ray projection sharply defines the lateral boundaries of the object but does not localize the object along the x-ray propagation axis. Through the combination of additional measurements at different angles with respect to the object, it is possible to obtain a high-resolution image of the x-ray attenuating object within the tissue. Figure 8.3b shows the result from two orthogonal projections. The artifacts near the boundaries are reduced with additional projections.

Figures 8.3c and d show the results obtained with diffuse light by following the same backprojection procedure. The striking difference obtained with a single projection is that the boundaries of the object are not sharply defined, and the object appears to be near the sources and detectors where the sensitivity of the measurement is greatest. The first observation is an intrinsic limitation of imaging with diffuse light, while the second is an artifact of the backprojection procedure. Utilizing an additional orthogonal projection (Figure 8.3d), we see better localization of the absorbing object. We note that while the backprojection method provides quantitative images with x-ray CT, it is only an approximation for DOT. This difference results from the fact that the measured diffuse light has sampled a volume of tissue, as opposed to the measured x-ray that samples only along a line through the sample. The backprojection method, therefore, does not produce an image that optimally fits the experimental data. More sophisticated optimization is needed to obtain quantitative image reconstruction with diffuse light.[2]

8.2.5 Theoretical Optical Sensitivity to the Brain

Photon migration theory models the propagation of light through tissue and enables the extraction of information about the tissue from measurements of remitted diffuse light. By using Monte Carlo techniques we can simulate the propagation of photons through tissue and obtain a spatial sensitivity map.[60-62] In such simulations, individual photon trajectories are traced by sampling appropriate probability distributions for scattering events. After 10^6 to 10^8 photons are traced, quantities such as the fraction of photons reaching a particular detector or the spatial sampling of the tissue can be determined.

In our Monte Carlo model, the tissue is a three-dimensional volume with optical properties assigned to each voxel; source and detectors can be placed at any voxel. Photons are propagated until they exit the tissue, and the total pathlength in each tissue type is recorded for all photons reaching a detector. Only recently has the computational power become widely available for the practical use of Monte Carlo simulations. (One simulation typically takes 10 to 20 h on current high-end desktop systems.)

Figure 8.4 shows an anatomical MRI of a human head, segmented into five tissue types (air, scalp, skull, cerebral spinal fluid, and gray/white matter; see Reference 63), with a contour overlay indicating the photon migration spatial sensitivity profile for (a) continuous-wave, (b) 200-MHz modulation, and (c,d) pulsed measurements. One contour line is shown for each half-order of magnitude

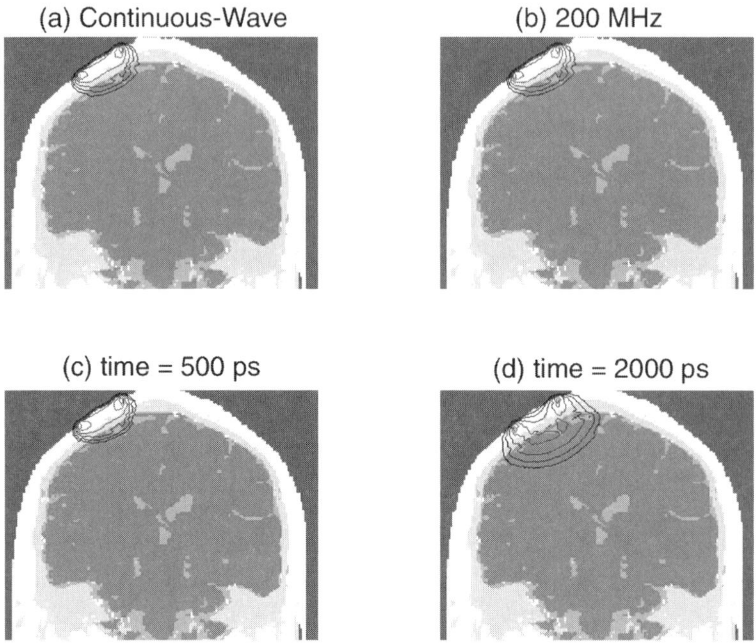

FIGURE 8.4 Contour plots of the photon migration sensitivity profile in a 3D human head as determined from Monte Carlo simulations. Details provided in the text.

(10 dB) signal loss, and the contours end after three orders of magnitude in loss (60 dB). For the 3D Monte Carlo simulation, we assumed that $\mu_s' = 10\ \text{cm}^{-1}$ and $\mu_a = 0.4\ \text{cm}^{-1}$ for the scalp and skull, $\mu_s' = 0.1\ \text{cm}^{-1}$ and $\mu_a = 0.01\ \text{cm}^{-1}$ for the CSF, and $\mu_s' = 12.5\ \text{cm}^{-1}$ and $\mu_a = 0.25\ \text{cm}^{-1}$ for the gray/white matter. Note that the contours extend several millimeters into the brain tissue, indicating sensitivity to changes in cortical optical properties. The depth-penetration difference between the continuous-wave and 200-MHz measurements is difficult to discern. A ratio of the two sensitivity profiles (not shown) shows that the 200-MHz profile is shifted slightly towards the surface. The time domain sensitivity profiles suggest the possibility of obtaining greater penetration depths in the head from measurements made at longer delay times.

8.3 INSTRUMENTATION

Several technical solutions exist for implementation of NIRS and DOT, including time domain (TD), frequency domain (FD), and continuous-wave (CW) systems. TD systems[47,64-67] introduce into tissue extremely short (picosecond) incident pulses of light that are broadened and attenuated (i.e., scattered and absorbed) by the various tissue layers such as skin, skull, cerebrospinal fluid (CSF), and brain. A TD system detects the temporal distribution of photons as they leave the tissue; the shape of this distribution provides information about tissue scattering and absorption. In FD systems,[68-71] the light source is on continuously but is amplitude modulated at frequencies on the order of tens to hundreds of megahertz. Information about the absorption and scattering properties of tissue is obtained by recording amplitude decay and phase shift (delay) of the detected signal with respect to the incident one.[72] In CW systems,[73-76] light sources emit light continuously as FD systems do but at constant amplitude or modulated at frequencies not higher than a few tens of kilohertz. These systems measure only the amplitude decay of the incident light and therefore do not provide sufficient information to quantify baseline absorption and scattering separately.

Each of these techniques has intrinsic advantages and drawbacks. TD systems can theoretically obtain the highest spatial resolution and can accurately determine absorption and scattering. Drawbacks include long acquisition times to achieve reasonable signal-to-noise ratios, a need to mechanically stabilize the instrument, and the large dimensions and high cost of the necessary ultrafast lasers.[77] These drawbacks have limited the use of TD systems, but technological advances are beginning to make them more attractive.[78]

State-of-the-art FD systems cover a wide range of applications for clinical use and can achieve considerably higher temporal resolution than TD systems. FD instruments are typically less expensive to build than TD systems but are still more expensive than CW systems, and they require more careful engineering and optimization to eliminate noise, ground loops, and RF-transmitter effects. While they are able to measure scattering effects, the noise in such measurements remains a significant issue.

CW technology can be engineered with relatively inexpensive and widely available components; a hospital pulse oximeter is an example of the small dimensions achievable for a CW NIRS instrument. The CW's primary shortcoming compared to TD and FD systems is the inability to uniquely quantify the effects of light scattering and absorption.[79] Despite their relatively minimal design requirements, commercially available systems until recently featured only a single detector (NIRO 500; see Reference 5) or had a restricted bandwidth (2 Hz: OXYMON, University of Nijmegen, see Reference 80). We are now seeing considerable development of CW technology to greater bandwidths and greater numbers of source and detectors.[81-84]

It is likely that CW technology will quickly find widespread use among brain researchers because of its relatively low cost, portability, and ease of implementation and use compared to FD and TD systems while still displaying sensitivity to cerebral hemodynamic features. Arguably, a CW imaging system will also achieve the best signal-to-noise ratio at image frame rates faster than 1 Hz. For these reasons, we focus below on discussing basic instrument design of CW NIRS and DOT systems. While CW imaging techniques have the potential to provide quantitative images of hemodynamics changes during brain activation, quantitative imaging of baseline brain states is likely to be possible only with FD and TD methods because they provide time-delay and amplitude information.

8.3.1 NIRS Instrument

A basic NIRS system delivers light of preferably two or more wavelengths to a single position on the tissue and then collects and detects the diffusely re-emitted light at one or more locations. Because the light must travel several centimeters through the scalp and skull to sample the adult human brain, it is necessary that light sources be chosen with sufficient power and in the wavelength range between 600 and 950 nm where light absorption is minimized. This typically requires using laser diodes, although filtered white light sources have also been successfully used in adult humans.[39] The optical power incident on the tissue must be no more than 4 mW per 1 mm^2 for safe, long-term tissue exposure to laser light.[85]

To gain sensitivity to the brain in the adult human, detectors must be placed at least 2.5 cm from the source (less for babies, due to their thinner scalps and skulls), as the depth of sensitivity is roughly proportional to the source–detector separation. With such separations, the amount of light reaching the detector is on the order of 10 pW (an attenuation of 9 decades). Therefore, to obtain a decent signal-to-noise ratio at desired bandwidths of 1 Hz or greater, it is necessary to use high-sensitivity detectors: photo-multiplier tubes (PMTs), avalanche photodiodes (APDs), or CCD cameras.

Figure 8.5 shows a photograph of our CW NIRS system. The sources in this system are two low-power laser diodes emitting light at discrete wavelengths, typically 785 nm (Sanyo, DL7140-201) and 830 nm (Hitachi, HL8325G). These lasers are powered by a stabilized current, intensity modulated by an approximately 5-kHz square wave at a 50% duty cycle. Both diodes are driven at the same frequency but phase shifted by 90° with respect to one another. This phase encoding, known as an in-phase/quadrature-phase (IQ) circuit, allows simultaneous laser operation as well as separation of the contributions of each source to a given detector's signal.

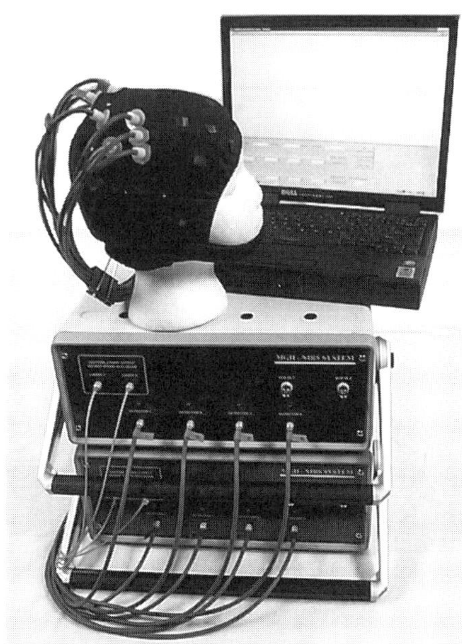

FIGURE 8.5 Photograph of the NIRS system and cap used to hold fiber optics on the head.

The system has four separate detector modules (Hamamatsu C5460-01), which are optically and electrically isolated from each other. Each module consists of a silicon avalanche photodiode (APD) with a built-in, high-speed, current-to-voltage amplifier and temperature compensation. The modules achieve a gain of typically 10^8 V/W with a noise equivalent power of 0.04 pW/(Hz)$^{-0.5}$ resulting from the high detector sensitivity. The incoming signal at a given detector is composed of both source colors, which are separated by synchronous (lock-in) detection using the two source signals as a reference. The output of the decoding portion of the IQ circuit includes two signal components, corresponding to the two laser wavelengths. These two components are low-pass filtered at 20 Hz and digitized by a computer.

8.3.2 Imaging Instrumentation

Conceptually, an imaging instrument is a simple extension of a NIRS system to include more sources and detectors. Adding more detectors is straightforward because each detector operates independently, without interaction with the other detectors. The sources, on the other hand, introduce light into the tissue and detectors can only detect the total amount of light present at a given time. Thus, in order to assign light received at a detector properly to each individual source, the source light must be encoded.

Several encoding strategies exist: time sharing, time encoding, and frequency encoding. Time sharing is the easiest to implement. In this scheme, each source is turned on, one at a time, long enough for the detectors to acquire a decent signal (typically 10 to 100 msec). Large numbers of sources give rise to slow image frame rates and temporal skew between sources. Time encoding overcomes the temporal skew problem by switching between sources at a much faster rate (100 μsec or faster) but integrating the signal from each source over several switch cycles (e.g., the same 10 to 100 msec) and resulting in roughly the same image frame rate as with time sharing. Finally, frequency encoding turns all sources on at the same time but modulates the intensity of each one at a slightly different frequency. The individual source signals are then discriminated at the detector by employing simple analog or digital band-pass filters at the frequencies of source modulation. While the duty cycle in this scheme is maximized, the effective dynamic range of the system is limited by the fact that all sources are on at the same time. That is, it can be difficult to detect a distant source several orders of magnitude weaker than a source close by, given that each detector has a fixed dynamic range (typically three to four orders of magnitude).

Figure 8.6 shows a block diagram and photograph of our DOT imaging system with 18 lasers and 16 detectors. This system is being extended to 32 lasers (intensities driven at 32 different frequencies) and 32 detectors. At present, the 18 lasers are divided into 9 lasers at 785 nm (Sanyo, DL7140-201) and 9 lasers at 830 nm (Hitachi, HL8325G), although they can be divided among as many different wavelengths as desired. The detectors are avalanche photodiodes (APDs, Hamamatsu C5460-01). A master clock generates the 18 distinct frequencies between 4.0 and 7.4 kHz in approximately 200-Hz steps. These frequencies are then used to drive the individual lasers with current stabilized square-wave modulation. Following each APD module

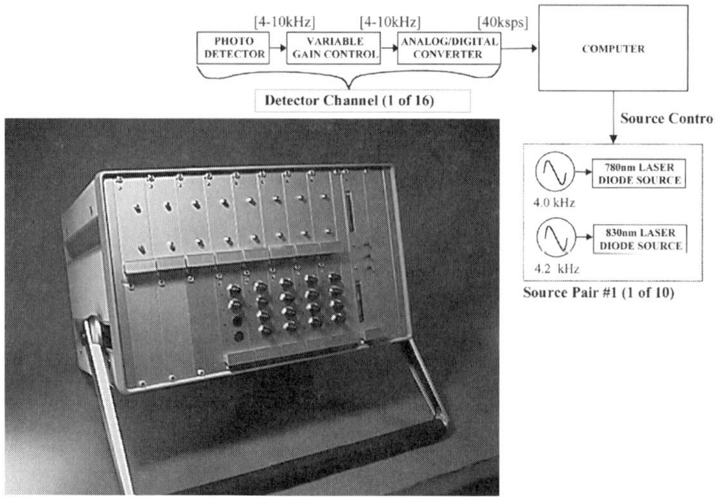

FIGURE 8.6 Photograph and block diagram of the frequency-encoded, continuous-wave diffuse optical tomography system. Details provided in the text.

is a bandpass filter with a cut-on frequency of ~500 Hz to reduce 1/f noise and the 60-Hz room light signal and a cut-off frequency of ~10 kHz to reduce the third harmonics of the square-wave signals. After the bandpass filter is a programmable gain stage to match the signal levels with the acquisition level on the analog-to-digital converter within the computer. Each detector is digitized at ~40 kHz and the individual source signals are then obtained by use of a digital bandpass filter — for example, a discrete Fourier transform or an infinite impulse–response filter.

8.4 NIRS AND DOT EXPERIMENTS

8.4.1 NIRS DOT Paradigm

To demonstrate that such an instrument has the ability to measure cerebral hemodynamic signals in humans noninvasively, we conducted simple experiments with young, healthy volunteers. A basic motor protocol was chosen to investigate the recordability of optical signals from the brain and to allow comparison with validated findings of deoxyhemoglobin and volume changes found in fMRI studies. Either two NIRS systems were used in parallel to record time-locked signals from two separate but broad regions of the brain or our newer DOT imaging system was used. In the NIRS experiments, the multiple detectors were used primarily to simplify the localization process. All subjects consented to the experimental procedure, which was approved by the Massachusetts General Hospital institutional review board.

The positions of the optodes were separately determined for each paradigm so as to cover an area over the cortex activated by similar fMRI studies.[11,86-89] Figure 8.7 depicts the optode placement relative to the international 10-20 system.[90,91] To couple the fibers to the head, we used a flexible plastic spine as the substrate to which

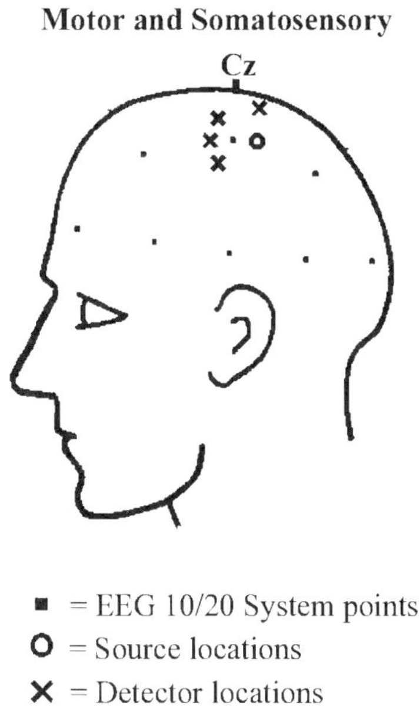

FIGURE 8.7 Diagram showing the placement of the light source (o) and four photo detectors (x) for the NIRS measurements.

we attached the fibers. Side-firing fiber bundles (3-mm core diameter) were developed to maintain a low profile on the scalp and to minimize motion artifact. Simple SMA connectors achieved coupling of the sources and detectors to the instrument. Figure 8.5 shows an image of the setup with the cap attached to a head-phantom.

At the start of the experiment, the subject remained seated in a chair while the cap was put on. No gel or special hair treatment was required except to wiggle the fibers through the hairs to improve the contact with the skin. The first subject was simply asked to close his eyes and relax during two 40-sec recording periods. For the first period, the subject sat upright in a chair; for the second, the subject was asked to lie supine. The purpose of these two recordings was to see what physiological variables could be recorded by the instrument in a simple resting baseline condition. Another subject in the supine position was asked to perform a four-finger flexion and extension task. The task was to flex and extend the four fingers of a designated hand for 15-sec blocks, alternated with 15-sec periods of rest, paced at 3 Hz by visual stimuli on a computer monitor. Signals were recorded continuously for 315-sec runs (10 active periods, 11 resting periods) from the region surrounding C3 and C4 (international 10-20 system designations).

Data acquisition for the DOT experiments was similar. The subject was first fitted with the optical cap (fibers again attached to a plastic substrate, anchored to a well-secured headband), and positioned supine. Again, visual stimuli paced the four-finger flexion and extension task, with optical recordings beginning 30 sec prior to the onset of stimuli.

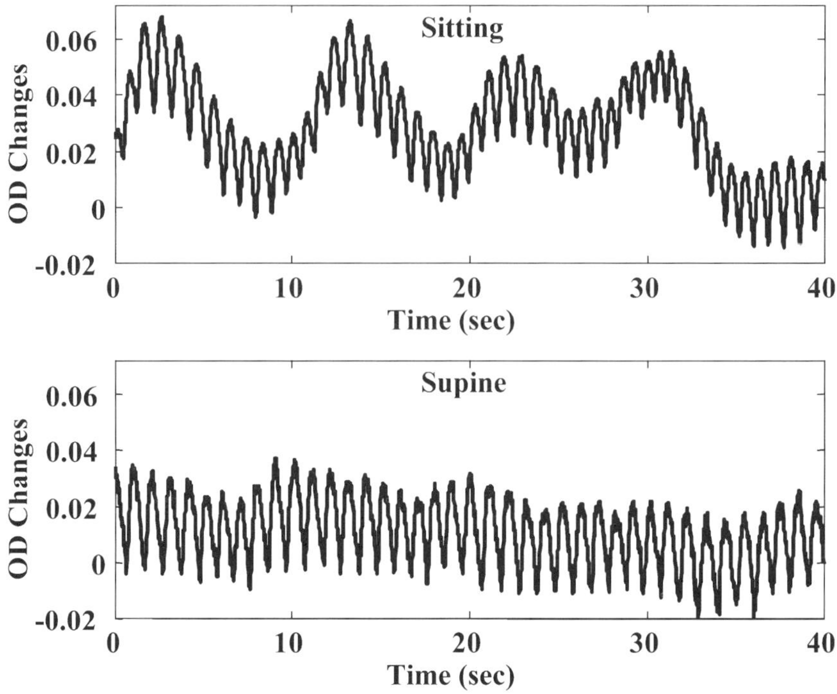

FIGURE 8.8 Results from 40 sec of recording during an eyes-closed resting condition in a subject while in a sitting position (top) and supine (bottom). Optical density changes observed from the 830-nm laser are shown. In both traces, heart pulsations are clearly evident. In the sitting case, a second, high-amplitude periodicity of ~0.1 Hz is also evident, while it is essentially absent in the supine position (see text).

8.4.2 NIRS Results

8.4.2.1 Brain Recording Experiments

Baseline Recording: Figure 8.8 shows the observed amplitude modulations from a single subject during two different baseline conditions (in optical density units). The top trace — with data gathered at 830 nm from approximately location C3 in the international 10-20 system — shows 40 sec of eyes-closed baseline while the subject was seated upright. The bottom trace is from the same subject, again eyes-closed baseline, but while lying supine. Heart pulsations are clear in both records, while in the upright case an additional high-amplitude oscillation of ~0.1 Hz appears. The frequency of this periodicity corresponds to the Mayer wave — a systemic blood pressure oscillation more prominent when standing or sitting than when lying down.[92,93] Further investigation is needed to fully characterize this signal and verify its source.

Motor Task: The data gathered from a single detector (again sampling from approximately location C3) for the motor task appear in Figure 8.9 for a supine subject. The horizontal bar indicates the period of motor activity, while the upper

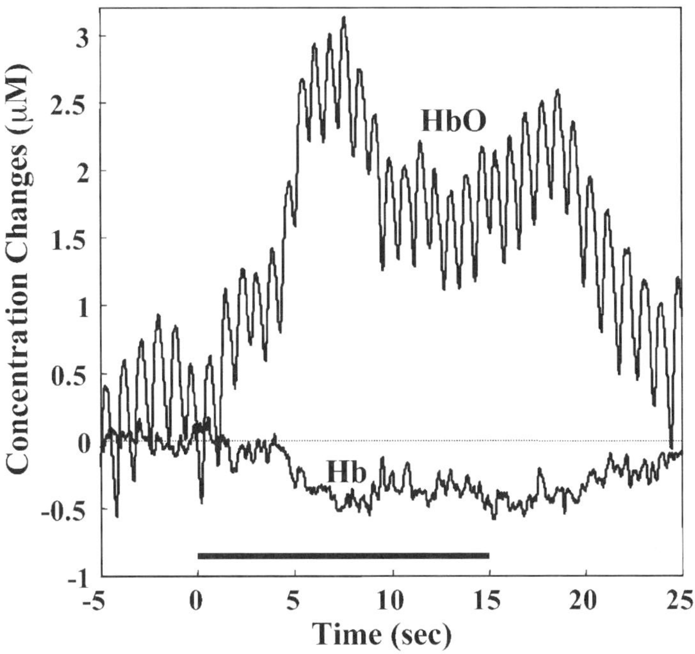

FIGURE 8.9 Changes in oxyhemoglobin (HbO) and deoxyhemoglobin (Hb) beneath C3 in a single subject while performing right-hand, four-finger flexions and extensions at approximately 3 Hz (horizontal bar). Concentration changes were calculated from raw optical data via the MBLL. No averaging was required due to the large activation amplitude typically associated with motor tasks.

and lower traces show the change in HbO and Hb (respectively) from the rest period. These traces involve no averaging, only a boxcar smoothing function with a 2-Hz bandwidth. The expected hemodynamic response for this task, based on previous fMRI experiments, would be an increase in HbO (and concomitant decrease in Hb) starting ~3 sec after motor activity onset, peaking around 6 to 9 sec post-onset, and decaying to baseline some 7 to 11 sec following the cessation of motor activity.[94] Both optical time courses match this profile. Also noteworthy is that the MBLL, Equation 8.4, has segregated the pulsatile changes in the optical signal more or less exclusively to the HbO signal,[93] and the activity-induced change in this signal is approximately a factor of 4 larger in magnitude than the observed change in Hb. For the MBLL analysis we assumed that the differential pathlength factor was equal to 6 at each wavelength.

8.4.3 Imaging Result

For the DOT experiment, the arrangement of the 9 sources and 12 detectors spanning the subject's left and right hemispheres is shown in Figure 8.10a. Fifteen seconds of rest data were collected, followed by 8 blocks of 15 sec of right-finger extension and flexion and 15 sec of rest. This 255-sec run was repeated twice: once with visual

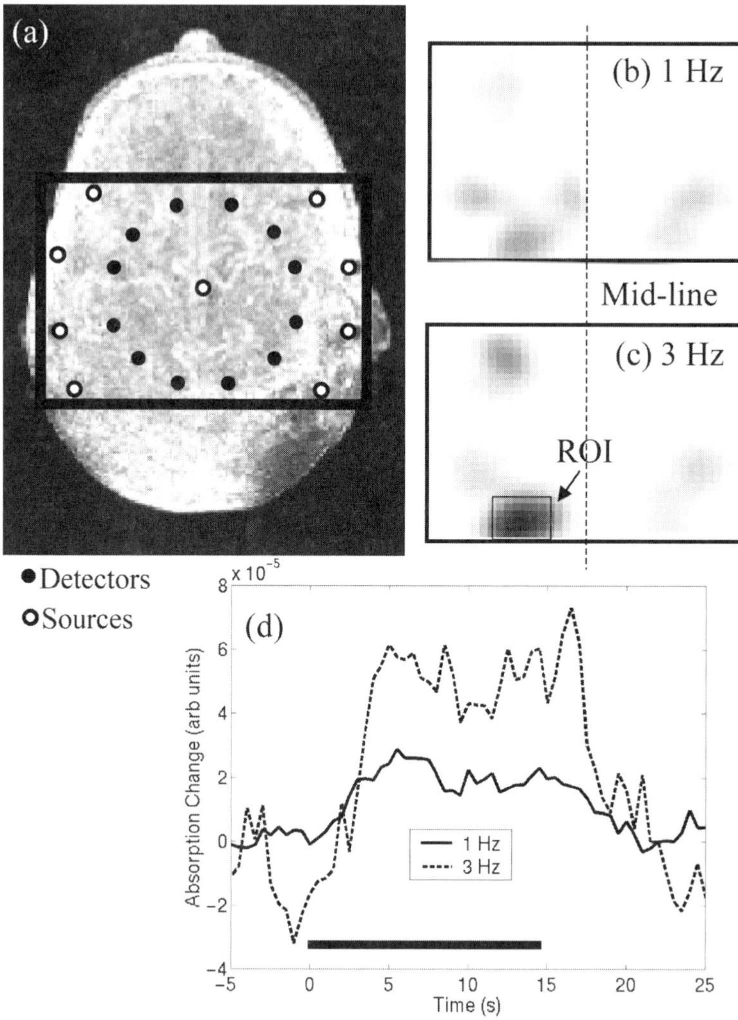

FIGURE 8.10 Images of motor activation obtained with a continuous-wave diffuse optical tomography system with 9 sources and 12 detectors (a). The images reconstructed at 785 nm for a 1- and 3-Hz finger flexion and extension are shown in (b) and (c). (d) shows the time course of the absorption change in the region-of-interest (ROI) shown in (c).

cues guiding a 1-Hz extension and flexion, and once with 3 Hz. Images were reconstructed in a single plane with Tikhonov regularization, using a regularization parameter of $\alpha = 0.8$, assuming that the head was flat and spatially uniform (i.e., semi-infinite homogeneous space) and that the activation occurred in a single plane 2 cm deep in the brain. The assumption of activation in a single plane will project all activity above and below it into the plane, causing image artifacts. Likewise, the assumption of a flat head will introduce model errors, resulting in image artifacts.

For these reasons, the images are considered qualitative and preliminary, as the magnitude of the errors introduced by the assumptions is as yet unexplored.

The images reconstructed from the block averaged data from 1- and 3-Hz stimulation are shown in Figure 8.10b and c, respectively. These images show absorption changes at 785 nm at the end of the stimulation block prior to signal recovery to baseline. The observed spatial patterns are qualitatively similar, with stronger amplitude modulation in the 3-Hz image. Based on the anatomical MRI, the posterior activations roughly correspond to primary sensorimotor cortex and show a typical pattern of stronger activation on the left (contralateral to movement) than on the right (ipsilateral) side. The left anterior activation corresponds roughly to premotor cortex (Brodmann area 6) and shows less lateralization, in keeping with the more bilateral interconnectivity of premotor cortex. In Figure 8.10d we show the time course of the absorption change at 785 nm in response to activation in the region of interest indicated in Figure 8.10c. As is expected for functional activation, the onset of the absorption change is delayed by 2 to 3 sec and time-to-peak is between 5 and 9 sec. Furthermore, the amplitude is observed to be a factor of 3 greater at 3 Hz than at 1 Hz.

At this stage the data are too preliminary to extract oxy- and deoxyhemoglobin concentration changes from the signal. Confidence in the imaged concentrations will come with further development and validation of the methodology, in particular with a better understanding of the image reconstruction accuracy. Our initial exploration of the image accuracy indicates that the presented measurement geometry is too sparse for obtaining accurate image magnitudes (and thus concentrations) with the present reconstruction algorithms. Accuracy can be improved with a denser measurement geometry or improved reconstruction algorithms.

8.5 FUTURE DIRECTIONS

Given the relative immaturity of DOT as an imaging modality, a broad spectrum of future directions remains open to inquiry. At one end of the spectrum lies the important theoretical work on light diffusion through inhomogeneous media and improved methods for 3D-image reconstruction. Such work will help improve the spatial and quantitative nature of the data obtainable from DOT. At the other end of the spectrum lie the clinical applications such as monitoring patients for signs of stroke or low oxygen saturation, especially in a bedside setting. In between the theoretical and clinical realms lies the ground of our most recent efforts. This work involves validation and comparision of DOT with other imaging modalities (e.g., EEG and fMRI) and, more importantly, the synergy generated by simultaneous application of two or more brain-monitoring technologies.

8.5.1 Quantitative Accuracy

On the theoretical side, the Modified Beer–Lambert Law (MBLL), discussed earlier, is a common method for calculating Hb and HbO concentrations from optical signals, with known pitfalls. Implementation of this model requires a value for the differential pathlength factor (DPF) for each wavelength (Equation 8.3). The DPF is simply a scaling factor transforming the source–detector separation into a measure of the average path length that each wavelength of light takes through the tissue from source to detector. While the DPF can be measured by using FD and TD optical systems,[95-98] the complexity and cost of such systems often prohibits DPF measurement for each subject prior to functional recording. When using a CW system, one has to rely on previously tabulated DPF values.[95,99-101] This approach suffers from uncertainties in the DPF, which varies between people and tissue types[95] and with age.[99] Alternatively, one can calibrate the DPF against the cardiac-induced arterial pulsation,[93] but this requires an accurate measurement of the cardiac pulsation within the optical signal, which is not always achievable. The greatest concern with using the MBLL is that it assumes that changes in the hemoglobin concentrations are spatially uniform over the measurement sampling volume.

During functional brain imaging, however, changes in optical properties result from localized changes in the blood oxygenation and blood volume within the brain; that is, the changes are not uniform over the measurement sampling volume. This is most clearly understood by considering the geometry involved in noninvasive measurements. In an adult human, the first layer of tissue is 0.5 to 1 cm of scalp, followed by 0.5 to 1 cm of skull, followed by a thin layer (0 to 2 mm) of cerebral spinal fluid and, finally, the gray and white matter of the brain. Functional activation changes occur primarily in the cortex, the outermost layer of the brain, and are therefore up to 2 cm below the source and detector. Thus, light traveling from the source to the detector will at best encounter the functional change over only a very small portion of the optical sampling volume, even for a spatially large activation. Such local changes violate the MBLL assumption of a global change and thereby introduce an error reminiscent of partial volume averaging.

With fMRI, for example, the signal from a given voxel is a weighted average of all the changes occurring within that voxel. Because of this, a 50% signal change restricted, for example, to half of the voxel will produce a 25% change in fMRI signal for that voxel. It is clear that this type of error will always produce an underestimate of the signal change when the change is assumed to be global to the sampling volume (as is done with the MBLL analysis). Similarly, a functional change in Hb and HbO concentrations within the optical sampling volume will be underestimated when the change is assumed to be global but is actually focal in nature. Furthermore, if the partial volume effect is different for each wavelength, it could lead to cross-talk in the determination of oxy- and deoxyhemoglobin concentration changes.[32]

The cross-talk produced by the differential wavelength partial volume effect needs further investigation to better understand the significance of the error in interpretation of optical signals. For instance, anecdotal evidence suggests that cross-talk can be significant when using 785 and 830 nm but less significant when using 760 and 830 nm. The difference may arise from the differing sensitivity that each wavelength pair has to the change in oxygen saturation of the hemoglobin. In addition, better models can be employed for analyzing the experimental data. These models would attempt to consider the focal nature of the hemodynamic change when analyzing the data, as is done by DOT.

8.5.2 MRI–DOT Synergy

As we have discussed, DOT has the potential to provide quantitative measures of deoxyhemoglobin (Hb) as well as oxyhemoglobin (HbO), provided partial volume errors can be minimized. If the tissue anatomy is known, then DOT will have an easier job determining the optical properties of the different structures and characterizing any regional activation. The anatomical data obtainable via MRI can provide just such data, resulting in a synergetic relationship between MRI and DOT.[101-103]

To see how this synergy develops, consider the following sequence of events. First, we acquire anatomical MRI scans that (1) help localize our optical probes with respect to the underlying brain anatomy and (2) can be used to segment the head into a variety of tissue types (typically air, scalp, skull, cerebral spinal fluid, and white and gray matter). The segmented head can then be used for Monte Carlo studies in photon diffusion to determine the DOT instrument's sensitivity to various tissue types and depths. The functional MRI data, once processed to provide functional maps, can also provide an estimate of the spatial extent and magnitude of the hemodynamic changes due to functional challenge.

These estimates can, of course, be used to compare the functional sensitivity of DOT with fMRI. More importantly, however, these estimates can also be used as a model of optical changes that would allow more accurate quantitation of the Hb and HbO changes during functional challenge. This way, simultaneous MRI and CW-DOT recordings have the potential to provide more accurate spatial localization of an activation and absolute quantitation of the Hb and HbO changes found there — something impossible with either technique alone.

A first step toward merging MRI and DOT is to examine simultaneously acquired fMRI and NIRS time courses from spatially co-registered locations. Figure 8.11 shows an example of this comparison for a single subject performing the unimanual motor task described earlier. The fMRI signal is showing a percent change in the deoxyhemoglobin concentration (positive percent represents a decrease in concentration of unknown absolute magnitude). The NIRS signal, on the other hand, shows the relative concentration changes in oxy- and deoxyhemoglobin. Qualitatively, the signals are similar.

The next step toward combining MRI and DOT is to examine spatial correspondences in the diffuse optical image and the fMRI, completing an analysis of the spatial–temporal correlation of the two methods. One can then proceed to use the anatomical information provided by the MRI to obtain more accurate estimates of where the light has sampled within the head, using the Monte Carlo method described earlier. With these better estimates of the photon migration sensitivity profile, one can expect to obtain a more accurate characterization of the spatially varying optical properties within the medium.

This synergy work is still at a very early stage. Although further developments and computational refinements are definitely necessary, this approach promises to provide accurate, quantitative measures of Hb and HbO changes in brain tissue — something heretofore lacking from both imaging modalities. Combining this quantitative information with the spatial detail of simultaneously acquired fMRI will

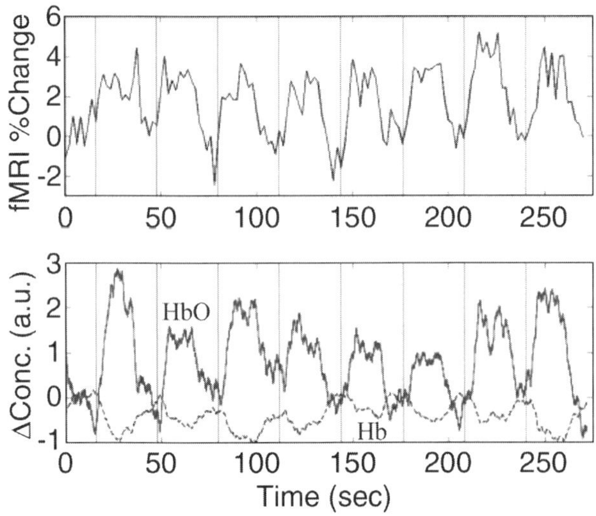

FIGURE 8.11 Time series plots for a subject performing the 3-Hz, right-hand, four-finger flexion and extension. Top: the fMRI signal averaged across all significantly activated pixels within contralateral (left) primary motor cortex. Vertical bars indicate the onset of motor activity; offset occurred 15 sec later. Bottom: NIRS signals from the optode sampling the region closest to the significant fMRI activation region, time-locked with the fMRI signal. The top curve shows changes in [HbO], in arbitrary units (a.u.), and the bottom curve shows changes in [Hb].

result in a more complete and accurate picture of the hemodynamic changes and a better understanding of the brain's response to functional challenge.

8.5.3 Imaging Newborn Babies

In a more clinical domain, Hintz et al.[104] have demonstrated that DOT can be used bedside in the neonatal intensive care unit to measure brain activation in response to passive motor stimulus. An illustrative result of their work with an earlier version of our DOT imaging system is reproduced in Figure 8.12. For the study presented in this figure, the patient was a 33-week estimated gestational age infant. The infant was quiet and asleep but not pharmacologically sedated or paralyzed. The patient had normal heart rate and O_2 saturation by pulse oximetry during the approximately 40-min study period. The probe used for this study was 3×6 cm and was placed in true midline position in the area that corresponded to the motor cortex. Baseline data were collected for 20 sec, at which point the patient's right arm was flexed and extended at the elbow. The data for each image were collected over a 5-sec interval. Data were collected at 780 and 830 nm.

As seen on the second row of images, there are clear increases in absorbance seen at 830 nm, indicated by the increase in yellow to red color, but also, as seen in the first row of images, much greater absorbance at 780 nm. Taken together, these data indicate an increase in blood flow to the imaged area but an overall increase

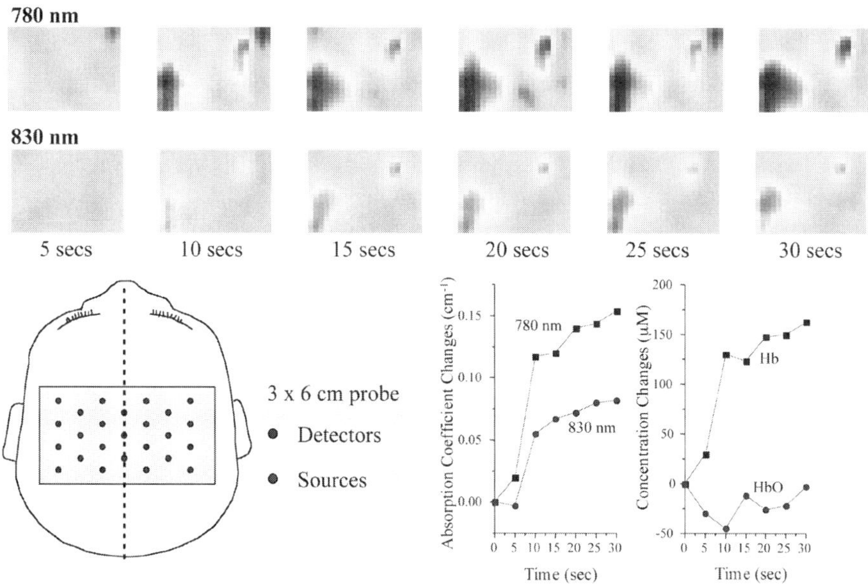

FIGURE 8.12 Time-series images at 780 and 830 nm presented every 5 sec following onset of a passive motor-sensory task whereby the baby's right arm is flexed ~30° once a second for 30 sec. Darker area indicates increased absorption. A greater absorption increase is seen at 780 than at 830 nm. The arrangement of the nine sources at each of the two wavelengths and the 16 detectors is shown as the overlay on a drawing on the baby's head. The probe spans the left and right hemispheres. The time course of the absorption change in focal activation in the contralateral hemisphere is shown, along with the corresponding changes in hemoglobin concentrations.

in deoxyhemoglobin concentration, as well. A graphic representation of the greater absorbance at 780 vs. 830 nm is seen in Figure 8.12, along with the indicated increase in deoxyhemoglobin.

This work demonstrates the feasibility of using DOT for imaging brain activation in the infant brain. This same technology can be equally applied to young children. Both of these populations have traditionally been difficult to study with any other neuroimaging methods because of extreme sensitivity to motion artifacts. DOT is less sensitive to motion artifacts for two main reasons: 1) as opposed to EEG, DOT is not as sensitive to muscular activity and 2) the DOT imaging probe can be made lightweight and moves with the subject's head.

8.6 SUMMARY

The rapidly growing literature from the past ten years clearly demonstrates the ability of near-infrared spectroscopy to noninvasively measure cerebral hemodynamic changes in response to stimuli. Furthermore, the literature is growing in support of the unique ability of NIRS to measure metabolic (cytochrome oxidase) and neuronal (fast optical scattering changes) signals. NIRS is unique as a neuromonitoring and neuroimaging technique because it is the only method that can potentially measure hemodynamic, metabolism, and neuronal signals simultaneously. Furthermore, the technology is relatively inexpensive and portable, allowing studies of brain activation in populations not easily studied in the past, such as babies and children and freely behaving paradigms.

The application to brain imaging of diffuse optical tomography, the imaging extension of NIRS, is in its infancy. Its sensitivity to cerebral signals is identical to NIRS (because they are the same measurements), but the algorithms used for reconstructing images need further investigation as do the technological advancements of instrumentation and probes for securing optical fibers to the head. Early results are nonetheless encouraging, as they demonstrate that spatial–temporal signals of brain activation can be imaged with a subsecond temporal resolution. While the images presented here are of the intrinsic absorption changes at 785 nm, with further algorithm development and careful selection of optical wavelengths, it will become possible to reconstruct accurate images of the concentration changes in oxy- and deoxyhemoglobin.

This is an exciting time to be working on the theoretical, instrumental, and experimental aspects of diffuse optical tomography.

REFERENCES

1. Jobsis, F.F., Noninvasive infrared monitoring of cerebral and myocardial sufficiency and circulatory parameters, *Science*, 198, 1264, 1977.
2. Arridge, S.R., Optical tomography in medical imaging, *Inverse Probl.*, 15, R41, 1999.
3. Chance, B. et al., Comparison of time-resolved and unresolved measurements of deoxyhemoglobin in brain, *Proc. Natl. Acad. Sci. U.S.A.*, 85, 4971, 1988.
4. Chance, B., Optical method, *Annu. Rev. Biophys. Chem.*, 20, 1, 1991.
5. Cope, M. and Delpy, D.T., System for long-term measurement of cerebral blood flow and tissue oxygenation on newborn infants by infrared transillumination, *Med. Biol. Eng. Comput.*, 26, 289, 1988.
6. Wray, S., Cope, M., and Delpy, D.T., Characteristics of the near infrared absorption spectra of cytochrome aa3 and hemoglobin for the noninvasive monitoring of cerebral oxygenation, *Biochim. Biophys Acta*, 933, 184, 1988.
7. McCormick, P.W. et al., Noninvasive cerebral optical spectroscopy for monitoring cerebral oxygen delivery and hemodynamics, *Clin. Care Med.*, 19, 89, 1991.
8. Benaron, D.W. et al., Noninvasive methods for estimating *in vivo* oxigenation, *Clin. Pediatr.*, 31, 258, 1992.
9. Elwell, C.E. et al., Quantification of adult cerebral hemodynamics by near-infrared spectroscopy, *J. Appl. Physiol.*, 77, 2753, 1994.
10. Belliveau, J.W. et al., Functional mapping of the human visual cortex by magnetic resonance imaging, *Science*, 254, 716, 1991.
11. Kwong, K.K. et al., Dynamic magnetic resonance imaging of human brain activity during primary sensory stimulation, *Proc. Natl. Acad. Sci. U.S.A.*, 89, 5675, 1992.
12. Ogawa, S. et al., Intrinsic signal changes accompanying sensory stimulation: functional brain mapping with magnetic resonance imaging, *Proc. Natl. Acad. Sci. U.S.A.*, 89, 5951, 1992.
13. Frahm, J. et al., Dynamic MR imaging of human brain oxygenation during rest and photonic stimulation, *J. Magn. Reson. Imaging*, 2, 501, 1992.

14. Bandettini, P.A. et al., Time course EPI of human brain function during task activation, *Magn. Reson. Med.*, 25, 390, 1992.
15. Hoshi, Y. and Tamura, M., Detection of dynamic changes in cerebral oxygenation coupled to neuronal function during mental work in man, *Neurosci. Lett.*, 150, 5, 1993.
16. Villringer, A. et al., Near infrared spectroscopy (NIRS): a new tool to study hemodynamic changes during activation of brain function in human adults, *Neurosci. Lett.*, 154, 101, 1993.
17. Okada, F. et al., Gender- and handedness-related differences of forebrain oxygenation and hemodynamics, *Brain Res.*, 601, 337, 1993.
18. Chance, B. et al., Cognitive activated low frequency modulation of light absorption in human brain, *Proc. Natl. Acad. Sci. U.S.A.*, 90, 2660, 1993.
19. Villringer, A. and Chance, B., Non-invasive optical spectroscopy and imaging of human brain function, *Trends Neurosci.*, 20, 435, 1997.
20. Prichard, J.W. and Rosen, B.R., Functional study of the brain by NMR, *J. Cerebr. Blood Flow Metab.*, 14, 365, 1994.
21. Meek, J.H. et al., Regional changes in cerebral haemodynamics as a result of a visual stimulus measured by near infrared spectroscopy, *Proc. R. Soc. London B*, 261, 351, 1995.
22. Hock, C. et al., Age dependency of changes in cerebral hemoglobin oxygenation during brain activation: a near infrared spectroscopy study, *J. Cerebr. Blood Flow Metab.*, 15, 1103, 1995.
23. Gratton, G. et al., Rapid changes of optical parameters in the human brain during a tapping task, *Neuroscience*, 7, 446, 1995.
24. Maki, A. et al., Spatial and temporal analysis of human motor activity using noninvasive NIR topography, *Med. Phys.*, 22, 1997, 1995.
25. Heekeren, H.R. et al., Cerebral haemoglobin oxygenation during sustained visual stimulation — a near-infrared spectroscopy study, *Philos. Trans. R. Soc. Lond. B Biol. Sci.*, 352, 743, 1997.
26. Obrig, H. and Villringer, A., Near-infrared spectroscopy in functional activation studies. Can NIRS demonstrate cortical activation? *Adv. Exp. Med. Biol.*, 413, 113, 1997.
27. Hoshi, Y. and Tamura, M., Near-infrared optical detection of sequential brain activation in the prefrontal cortex during mental tasks, *NeuroImage*, 5, 292, 1997.
28. Colier, W.N. et al., Human motor-cortex oxygenation changes induced by cyclic coupled movements of hand and foot, *Exp. Brain. Res.*, 129, 457, 1999.
29. Toronov, V. et al., Near infrared study of fluctuation in cerebral hemodynamics during rest and motor stimulation: temporal analysis and spatial mapping, *Med. Phys.*, 27, 801, 2000.
30. Franceschini, M.A. et al., On-line optical imaging of the human brain with 160-ms temporal resolution, *Optics Express*, 6, 49, 2000.
31. Benaron, D.A. et al., Noninvasive functional imaging of human brain using light, *J. Cerebr. Blood Flow Metab.*, 20, 469, 2000.
32. Boas, D.A. et al., The accuracy of near infrared spectroscopy and imaging during focal changes in cerebral hemodynamics, *NeuroImage*, 13, 76, 2001.
33. Colier, W.N. et al., Simultaneous near-infrared spectroscopy monitoring of left and right occipital areas reveals contra-lateral hemodynamic changes upon hemi-field paradigm, *Vision Res.*, 41, 97, 2001.
34. Jobsis, F.F., Noninvasive, infrared monitoring of cerebral and myocardial oxygen sufficiency and circulatory parameters, *Science*, 198, 1264, 1977.

35. Wyatt, J.S. et al., Quantification of cerebral oxygenation and haemodynamics in sick newborn infants by near infrared spectrophotometry, *Lancet*, ii, 1063, 1986.
36. Chance, B. and Williams, G.R., The respiratory chain and oxidative phosphorylation, *Adv. Enzimol.*, 17, 65, 1956.
37. Lockwood, A.H. et al., Effects of acetazolamide and electrical stimulation on cerebral oxidative metabolism as indicated by cytochrome oxidase redox state, *Brain Res.*, 308, 9, 1984.
38. Wong Riley, M.T., Cytochrome oxidase: an endogenous metabolic marker for neuronal activity, *Trends Neurosci.*, 12, 94, 1989.
39. Kohl, M. et al., Changes in cytochrome-oxidase oxidation in the occipital cortex during visual stimulation: Improvement in sensitivity by the determination of the wavelength dependence of the differental pathlength factor, *Proc. SPIE*, 3194, 18, 1998.
40. Wobst, P. et al., Linear aspects of changes in deoxygenated hemoglobin concentration and cytochrome oxidase oxidation during brain activation, *NeuroImage*, 13, 520, 2001.
41. Salzberg, B.M. and Obaid, A.L., Optical studies of the secretory event at vertebrate nerve terminals, *Exp. Biol.*, 139, 195, 1988.
42. Stepnoski, R.A. et al., Noninvasive detection of changes in membrane potential in cultured neurons by light scattering, *Proc. Natl. Acad. Sci. U.S.A.*, 88, 9382, 1991.
43. Rector, D.M., Rogers, R.F., and George, J.S., A focusing image probe for assessing neural activity *in vivo*, *J. Neurosci. Methods*, 91, 135, 1999.
44. Gratton, G. et al., Shades of gray matter: noninvasive optical images of human brain responses during visual stimulation, *Psychophysiology*, 32, 505, 1995.
45. Gratton, G. et al., Toward non-invasive 3-D imaging of the time course of cortical activity: investigation of the depth of the event-related optical signal (EROS), *NeuroImage*, 11, 491, 2000.
46. Steinbrink, J. et al., Somatosensory evoked fast optical intensity changes detected non-invasively in the adult human head, *Neurosci. Lett.*, 291, 105, 2000.
47. Hebden, J.C., Arridge, S.R., and Delpy, D.T., Optical imaging in medicine: I. Experimental techniques, *Phys. Med. Biol.*, 42, 825, 1997.
48. Cope, M., *The Development of a Near-Infrared Spectroscopy System and Its Application for Noninvasive Monitoring of Cerebral Blood and Tissue Oxygenation in the Newborn Infant*, University College London, 1991.
49. Delpy, D.T. et al., Estimation of optical pathlength through tissue from direct time of flight measurement, *Phys. Med. Biol.*, 33, 1433, 1988.
50. Matcher, S.J. et al., Performance comparison of several published tissue near-infrared spectroscopy algorithms, *Anal. Biochem.*, 227, 54, 1995.
51. Ishimaru, A., *Wave Propagation and Scattering in Random Media*, Academic Press, San Diego, 1978.
52. Patterson, M.S., Chance, B., and Wilson, B.C., Time resolved reflectance and transmittance for the non-invasive measurement of tissue optical properties, *Appl. Optics*, 28, 2331, 1989.
53. Haskell, R.C. et al., Boundary conditions for the diffusion equation in radiative transfer, *J. Opt. Soc. Am. A*, 11, 2727, 1994.
54. Furutsu, K. and Yamada, Y., Diffusion approximation for a dissipative random medium and the applications, *Phys. Rev. E*, 50, 3634, 1994.
55. Durduran, T. et al., Does the photon diffusion coefficient depend on absorption? *J. Opt. Soc. Am. A*, 14, 3358, 1997.

56. Farrell, T.J., Patterson, M.S., and Wilson, B., A diffusion theory model of spatially resolved, steady state diffuse reflectance for the noninvasive determination of tissue optical properties *in vivo, Med. Phys.*, 19, 879, 1992.
57. Jackson, J.D., *Classical Electrodynamics*, 2nd ed., John Wiley & Sons, New York, 1975.
58. Kak, A.C. and Slaney, M., *Principles of Computerized Tomographic Imaging*, IEEE Press, New York, 1988.
59. Arridge, S.R. and Hebden, J.C., Optical imaging in medicine: II. Modelling and reconstruction, *Phys. Med. Biol.*, 42, 841, 1997.
60. Wang, L., Jacques, S.L., and Zheng, L., MCML — Monte Carlo modeling of light transport in multi-layered tissues, *Comput. Methods Programs Biomed.*, 47, 131, 1995.
61. Firbank, M. et al., An investigation of light transport through scattering bodies with non-scattering regions, *Phys. Med. Biol.*, 41, 767, 1996.
62. Okada, E. et al., Theoretical and experimental investigation of near-infrared light propagation in a model of the adult head, *Appl. Optics*, 36, 21, 1997.
63. Dale, A.M., Fischl, B., and Sereno, M.I., Cortical surface-based analysis. I. Segmentation and surface reconstruction, *NeuroImage*, 9, 179, 1999.
64. Benaron, D.A. and Stevenson, D.K., Optical time-of-flight and absorbance imaging of biologic media, *Science*, 259, 1463, 1993.
65. Chance, B. et al., Comparison of time-resolved and unresolved measurements of deoxyhemoglobin in brain, *Proc. Natl. Acad. Sci. U.S.A.*, 85, 4971, 1988.
66. Cubeddu, R. et al., Time-resolved imaging on a realistic tissue phantom: μs' and μa images vs. time-integrated images, *Appl. Opt.*, 35, 4533, 1996.
67. Grosenick, D., Wabnitz, H., and Rinneberg, H., Time-resolved imaging of solid phantoms for optical mammography, *Appl. Optics*, 36, 221, 1997.
68. Gratton, E. et al., Measurements of scattering and absorption changes in muscle and brain, *Philos. Trans. R. Soc. Lond. B Biol. Sci.*, 352, 727, 1997.
69. Jiang, H. et al., Simultaneous reconstruction of optical absorption and scattering maps in turbid media from near-infrared frequency-domain data, *Opt. Lett.*, 20, 2128, 1995.
70. Pogue, B.W. and Patterson, M.S., Frequency-domain optical-absorption spectroscopy of finite tissue volumes using diffusion-theory, *Phys. Med. Biol.*, 39, 1157, 1994.
71. Pogue, B.W. et al., Instrumentation and design of a frequency-domain diffuse optical tomography imager for breast cancer detection, *Optics Express*, 1, 391, 1997.
72. Chance, B. et al., Phase measurement of light absorption and scattering in human tissues, *Rev. Sci. Instru.*, 689, 3457, 1998.
73. Nioka, S., Luo, Q., and Chance, B., Human brain functional imaging with reflectance CWS, *Adv. Exp. Med. Biol.*, 428, 237, 1997.
74. Siegel, A.M., Marota, J.J.A., and Boas, D.A., Design and evaluation of a continuous-wave diffuse optical tomography system, *Optics Express*, 4, 287, 1999.
75. Maki, A. et al., Visualizing human motor activity by using non-invasive optical topography, *Front Med. Biol. Eng.*, 7, 285, 1996.
76. Colier, W. et al., A new and highly sensitive optical brain imager with 50 Hz sample rate, *NeuroImage*, 11, 542, 2000.
77. Hebden, J.C., Evaluating the spatial resolution performance of a time-resolved optical imaging system, *Med. Phys.*, 19, 1081, 1992.
78. Ntziachristos, V. et al., Multichannel photon counting instrument for spatially resolved near infrared spectroscopy, *Rev. Sci. Instrum.*, 70, 193, 1999.

79. Arridge, S.R. and Lionheart, W.R.B., Nonuniqueness in diffusion-based optical tomography, *Optics Lett.*, 23, 882, 1998.
80. Colier, W.N.J.M. et al., Age dependency of cerebral oxygenation assessed with near infrared spectroscopy, *J. Biomed. Optics*, 2, 162, 1997.
81. Colier, W.N. et al., Human motor-cortex oxygenation changes induced by cyclic coupled movements of hand and foot, *Exp. Brain Res.*, 129, 457, 1999.
82. Takahashi, K. et al., Activation of the visual cortex imaged by 24-channel near-infrared spectroscopy, *J. Biomed. Opt.*, 5, 93, 2000.
83. Schmitz, C.H. et al., Instrumentation and calibration protocol for imaging dynamic features in dense-scattering media by optical tomography, *Appl. Optics*, 39, 6466, 2000.
84. Quaresima, V. et al., Noninvasive measurement of cerebral hemoglobin oxygen saturation using two near infrared spectroscopy approaches, *J. Biomed. Opt.*, 5, 201, 2000.
85. A.N.S., *American National Standard for Safe Use of Lasers*, ANSI Z136.1-1993, American National Standards Institute, New York, New York, 1993.
86. Bandettini, P.A. et al., Processing strategies for time-course data sets in functional MRI of the human brain, *Magn. Reson. Med.*, 30, 161, 1993.
87. Belliveau, J.W. et al., Functional mapping of the human visual cortex by magnetic resonance imaging, *Science*, 254, 716, 1991.
88. Halgren, E. et al., Location of human face-selective cortex with respect to retinotopic areas, *Hum. Brain Map.*, 7, 29, 1999.
89. Drummond, S.P. et al., Sleep deprivation-induced reduction in cortical functional response to serial subtraction (in process citation), *Neuroreport*, 10, 3745, 1999.
90. Jasper, H.H., The ten-twenty electrode system of the International Federation, *EEG Clin. Neurophysiol.*, 10, 371, 1958.
91. Klem, G.H. et al., The ten-twenty electrode system of the International Federation. The International Federation of Clinical Neurophysiology, *Electroencephalogr. Clin. Neurophysiol. Suppl.*, 52, 3, 1999.
92. Taylor, J.A. et al., Low-frequency arterial pressure fluctuations do not reflect sympathetic outflow: gender and age differences, *Am. J. Physiol.*, 274, H1194, 1998.
93. Kohl, M. et al., Determination of the wavelength dependence of the differential pathlength factor from near-infrared pulse signals, *Phys. Med. Biol.*, 43, 1771, 1998.
94. Bandettini, P.A. et al., Time course EPI of human brain function during task activation, *Magn. Reson. Med.*, 25, 390, 1992.
95. Duncan, A. et al., Optical pathlength measurements on adult head, calf and forearm and the head of the newborn infant using phase resolved optical spectroscopy, *Phys. Med. Biol.*, 40, 295, 1995.
96. van der Zee, P. et al., Experimentally measured optical pathlengths for the adult's head, calf and forearm and the head of the newborn infant as a function of interoptode spacing, *Adv. Exp. Med. Biol.*, 316, 143, 1992.
97. Fantini, S. et al., Non-invasive optical imaging of the piglet brain in real time, *Optics Express*, 4, 308, 1999.
98. Franceschini, M.A., Gratton, E., and Fantini, S., Non-invasive optical method to measure tissue and arterial saturation: an application to absolute pulse oximetry of the brain, *Opt. Lett.*, 24, 829, 1999.
99. Duncan, A. et al., Measurement of cranial optical path length as a function of age using phase resolved near infrared spectroscopy, *Pediatr. Res.*, 39, 889, 1996.

100. Essenpreis, M. et al., Wavelength dependence of the differential pathlength factor and the log slope in time-resolved tissue spectroscopy, *Adv. Exp. Med. Biol.*, 333, 9, 1993.
101. Barbour, R.L. et al., MRI-guided optical tomography:prospects and computation for a new imaging method, *IEEE Computation Sci. Eng.*, 2, 63, 1995.
102. Pogue, B.W. and Paulsen, K.D., High-resolution near-infrared tomographic imaging simulations of the rat cranium by use of a priori magnetic resonance imaging structural information, *Optics Lett.*, 23, 1716, 1998.
103. Ntziachristos, V. et al., Concurrent MRI and diffuse optical tomography of breast after indocyanine green enhancement, *Proc. Natl. Acad. Sci. U.S.A.*, 97, 2767, 2000.
104. Hintz, S.R. et al., Bedside functional imaging of a premature infant during passive motor activation, *J. Perinat. Med.*, 29, 335, 2001.

9 Fast Optical Signals: Principles, Methods, and Experimental Results

Gabriele Gratton and Monica Fabiani

CONTENTS

0-8493-2389-4/02/$0.00+$1.50
© 2002 by CRC Press LLC

9.1 INTRODUCTION

The purpose of this chapter is to describe the use of optical methods to record in a noninvasive fashion intrinsic signals related to neuronal activity (i.e., fast signals). Fast signals can be used to study the time course of activity in selected cortical areas noninvasively. We will first present a general introduction to the problem of noninvasive functional imaging and describe different approaches that can be used in this area of research. We will then describe the rationale and procedures related to the use of optical methods to create *functional* images of brain activity. We will then discuss examples of application of the technique and conclude with a discussion of some advantages and limitations of the technique.

9.1.1 "Fast" Optical Signals and the Event-Related Optical Signal (EROS)

"Fast" optical signals are newly developed brain-imaging methods whose purpose is to provide spatiotemporal maps of the transmission of near-infrared (NIR) light through the head. With these methods, NIR light is shone through the scalp and skull, and changes in the time taken by the photons to move through the head tissues[1] and/or in the amount of light reaching the detector[2] are observed . These two parameters (i.e., the photons' "time of flight" and the light "intensity") are influenced by the scattering and absorption properties of the tissue. Of particular importance here is the observation that the scattering properties of brain tissue vary concurrently with neural activity in the tissue.[3-5] Because optical measures are relatively localized (to volumes of just a few mm in diameter), it is possible to use these methods to derive estimates of the time course of activity in specific brain areas. The recording of fast optical signals can be time-locked to particular events (such as stimuli or responses). In this case we use the term event-related optical signal, or EROS. The EROS technique's advantages and limitations will be described later in this chapter.

9.1.2 Why Noninvasive Functional Brain Imaging?

The last 20 years have seen a phenomenal expansion of noninvasive functional brain imaging.[6] One of the major advantages of neuroimaging is the possibility of measuring *in vivo* physiological signals from various brain areas in normal human subjects. This possibility is greatly significant for scientists and clinicians alike. Of course, a cost is also attached to the use of these measures: at present, noninvasive imaging techniques allow us only to measure the cumulative physiological response of macroscopic areas of the brain (thus summarizing the activity of thousands of neurons at a time) and cannot provide data about the activity of individual neurons. Further, experimental lesion studies are not possible; it is therefore difficult to ascertain causal links between the activity observed in various brain areas and behavioral and physiological outcomes.

9.1.3 Types of Functional Imaging Methods

Several types of noninvasive functional methods are now available that differ on a number of dimensions, most notably the type of signal imaged to obtain estimates of brain activity. Most of the experimental work in functional neuroimaging is based on either hemodynamic–metabolic or neuronal signals (although other types of functional imaging signals, such as receptor–ligand distribution, have also been used[7]). Optical methods are unique in that they can be used to visualize both hemodynamic and neuronal signals.[4,8] In the next section we will describe some general principles of each of these two types of signals.

9.1.3.1 Hemodynamic–Metabolic Methods

This class of techniques aims at measuring changes in the concentration of metabolically significant substances during the course of an experiment. These changes in concentration can be inferred directly (e.g., by labeling the substance using radioactive, magnetic, or optical markers) or indirectly (e.g., by studying the effects of a target substance, such as deoxyhemoglobin, on the magnetic resonance properties of the tissue where the substance occurs naturally). Techniques that allow us to measure hemodynamic–metabolic effects include positron emission tomography (PET), functional magnetic resonance imaging (fMRI), and near-infrared spectroscopy (NIRS; for reviews see Toga and Mazziotta[7]). In particular, fMRI has become widely used because of its excellent spatial resolution, its easily attainable coupling with anatomical imaging (structural MRI), and the wide diffusion of the required hardware (as MR scanners are available in most hospitals because of their clinical use). However, most fMRI data do not provide absolute estimates of the concentration of oxy- and deoxyhemoglobin but only estimates of their relative concentration (the blood oxygen level-dependent, or BOLD, signal). Quantitative estimates can instead be obtained using optical spectroscopic techniques,[9,10] although this requires solving several practical problems (see the chapter on NIRS in this volume).

These hemodynamic–metabolic changes are interesting to neuroscientists and psychologists because they allow us to study not only metabolism and blood circulation in the brain but also, indirectly, neuronal activity. In fact, it has been known for a long time that blood flow and metabolism increase in brain areas that are active during a particular task (e.g., Grinvald et al.[11]). Thus, hemodynamic–metabolic neuroimaging is based on the assumption that, by studying changes in these parameters, one can infer whether a particular brain area is involved in the neural activity associated with a particular task.

Although this is a reasonable assumption in many cases, it also points to a limitation of these techniques as measures of neuronal activity. Namely, the inference of the occurrence of neuronal activity from hemodynamic–metabolic data is indirect because it is mediated by "neurovascular coupling."[12] This term refers to the set of biochemical and biophysical steps that occur between neuronal activity and the physiological (hemodynamic) events measured. Neurovascular coupling complicates the measures for two reasons: 1) it introduces an inherent delay in the physiological variable measured (e.g., change in the concentration of deoxyhemoglobin) with respect to the phenomenon inferred (i.e., neuronal activity), which, in turn, limits the temporal resolution of the method, and 2) neurovascular coupling may not be constant but may in fact vary as a function of the state of the organism, of the brain area from which the measures are taken, and of individual and group variables.[13]

9.1.3.2 Neuronal Methods

Another type of signal used in functional neuroimaging is related to the movement of ions through and around the cell membranes, which occurs during neuronal activity. Several methods can be used to record ion movement noninvasively. Some methods rely on changes in electromagnetic fields that can be detected from sensors located outside the head. Examples of these techniques are the electroencephalogram (EEG), evoked and event-related potentials (EPs and ERPs), and magnetoencephalography (MEG). However, differently from single- and multiple-units activity measures obtained with implanted microelectrodes, noninvasive electromagnetic measures taken at a distance (e.g., from the surface of the head) cannot provide information on individual neurons but need to rely on the spatial and temporal summation of the fields generated by individual neurons. Spatial summation occurs only if the individual neurons are oriented in a consistent direction (with respect to their dendritic field–axon axis), generating an "open-field" configuration.[14]

Thus, only a portion of the neuronal activity generated in the brain can be studied with these methods. In the case of the cerebral cortex, electromagnetic methods are most sensitive to the activity of pyramidal neurons (the output cells of the cortex). They are also most sensitive to the postsynaptic activity measurable in the dendrites rather than to the all-or-none activity typical of axons, because of the comparatively large size of the dendritic fields and the longer duration of postsynaptic potentials. Therefore, the effects observed with these methods differ from those observed with single-unit methods (which do not have a directional bias and are based mostly on action potential measures in axons). Further, the spatial resolution of general-purpose 3D reconstruction algorithms for electromagnetic measures, allowing investigators to reconstruct the distribution of the activity inside the head from surface measures, is limited[15] and needs to rely on assumptions often difficult to verify.[16] Optical methods, reviewed in the next section, provide a complementary approach for studying neuronal and hemodynamic activity and their coupling.

9.2 PRINCIPLES OF FAST OPTICAL IMAGING

9.2.1 The Scattering Signal

This book reviews techniques that use optical methods for studying brain function. Several optical signals can be used. This chapter focuses on signals temporally concurrent with neuronal changes. Changes of this type have been reported elsewhere in this book and can generally be described in terms of changes of the scattering coefficient of neural tissue when the tissue is active. These changes have been demonstrated in individual axons,[3] brain slices,[4] and depth optical recording in animals.[5]

For instance, Rector et al.[5] developed a methodology for recording changes in the transparency of rat hippocampus or other deep brain structures *in vivo*. They showed that stimulation of the Schaeffer's collateral elicits a change in the transparency of the CA1 field for which latency is consistent with that of the evoked potential from CA1 recorded with electrophysiological methods. They attributed these effects to changes in the scattering properties of this area, presumably associated with the migration of ions through the neuronal membrane. These and other data reported in this book provide strong support to the claim that neuronal activity is associated with changes in optical properties (probably scattering coefficient) of the neural tissue.

9.2.2 Noninvasive Measurement of Optical Properties

9.2.2.1 Continuous vs. Time-Resolved Measures

Although scattering changes accompanying neuronal activity have been demonstrated for quite some time using invasive techniques, only recently have methods become available for their noninvasive measurement. Because of the high absorption by hemoglobin and water at other wavelengths, noninvasive measurements are restricted to the near-infrared range. Further, the scattering-based functional changes are relatively small and require appropriate instrumentation for their measurement. Two types of measures have been used. We have used measures of delay in the flight of NIR photons through active areas (delay measures[1]), whereas others have used measures of the amount of light moving through active areas (intensity measures[2]). In heterogeneous surface-bound media such as the head, delay and intensity measures are sensitive to both absorption and scattering changes, although in a different manner.[17,18]

Intensity measures are conceptually simpler. The simplest recording apparatus (the optode) consists of a source of NIR light and a light measurement instrument, both of which can be connected to the head using an optic fiber. Several types of light sources can be used, including incandescent lamps, light-emitting diodes (LEDs), laser diodes, and regular laser, as well as several types of detectors, including light-sensitive diodes, photomultiplier tubes, CCD cameras, etc. Irrespective of the sources and detectors used, it is important to exclude environmental light sources. This can be achieved by insulating the detector instrument (and the head) from environmental light sources (e.g., Steinbrink et al.[2]) or, preferably, by labeling the instrumental light source in a special manner, e.g., by modulating the light source at a particular frequency (e.g., Wolf et al.[19]).

Photon delay measures are more complex because the delays to be measured (and their changes) are of the order of picoseconds or fractions thereof. The measurement requires that the light source vary over time, which can be achieved by making it pulsate or oscillate in intensity at a very high frequency. The latter measure is more convenient in terms of cost.

Time-resolved measures are used in NIR spectroscopy because they make computation of the concentration of oxy- and deoxyhemoglobin easier.[20] However, if the main interest of the investigation is in detecting effects (such as those related to neuronal events) but not necessarily in providing quantification of substance concentration, the advantage of time-resolved measures over continuous measures may be less important. Another differentiation between measures obtained with continuous light (i.e., measures of tissue transparency, usually labeled intensity measures) and time-resolved measures (i.e., measures of modulation and delay of the light waves or pulses when passing through the head tissue) is their relative sensitivity to environmental light and to superficial vs. deep layers. Continuous measures cannot distinguish between light coming from the instrumental source and light coming from other environmental sources. Modulated or pulsated sources make this distinction possible because they produce light that changes with a special frequency or time course.

9.2.2.2 Penetration and 3D Reconstruction

Noninvasive measures are obtained from instruments placed on the surface of the head. A central issue for this research is to determine which parts of the brain are responsible for the effects observed. Photons injected into a particular point of the head (a scattering medium) propagate randomly through the tissue. However, since a large number of photons is always involved, the propagation of the photon population can, in principle, be described mathematically, using equations that describe the statistical volume explored by each source–detector pair. Any change in the measurement parameters can then be probabilistically ascribed to this volume.

These equations are relatively simple for surface-bound homogenous media. In this case, photons moving between a particular source and detector (both located on the surface of the medium at a certain distance from each other) are most likely to be contained within a volume shaped like a curved spindle whose extremes are located at the source and detector (see Figure 9.1). The width of the spindle (which is important in defining the spatial resolution of the technique) depends on various factors, including the parameter (intensity or photon delay) used for the measurement. In fact, the spindle's width is substantially larger for measures of the amount of light transmitted (continuous measures) than for measures of photon delay. Thus the latter possess a better spatial resolution.

An important consideration is that head tissues vary considerably with respect to their optical properties (scattering and absorption). For instance, bone has a relatively high scattering coefficient but a low absorption coefficient compared to the gray matter. The cerebrospinal fluid is almost transparent to light, but the subarachnoid space in which it is contained is traversed by a dense web of blood vessels. Thus, within this space, NIR light can travel without much resistance in a direction perpendicular to the surface but can travel very little in a direction parallel to the surface because it will be absorbed. For this reason, the subarachnoid space has little influence on the transmission of light through the head.[21]

The opposite occurs with the white matter, which has a very high scattering coefficient and whose effect is to reflect most of the light reaching it. The heterogeneity of the optical properties of the head tissues has the consequence of deforming the volume investigated by a particular source–detector pair.[18,21] The effects of this distortion appear to be relatively moderate (a few mm) at source–detector distances of less than 5 cm[22] but are potentially much greater at longer source–detector distances.

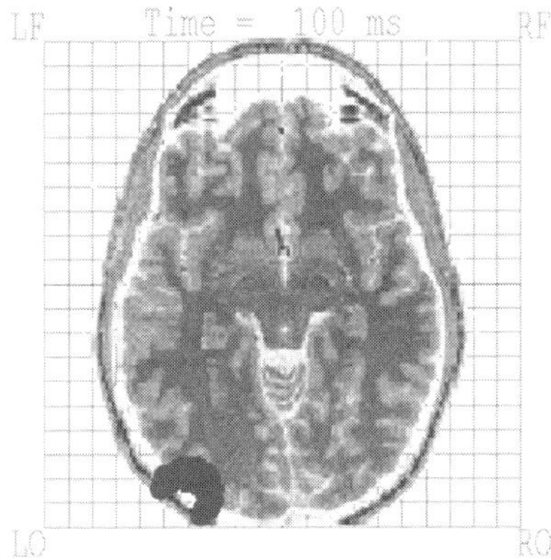

FIGURE 9.1 Schematic representation of the volume in the brain explored by a particular source–detector pair located in left occipital areas. (From Gratton, G. and Fabiani, M., *Int. J. Psychophysiol.*, 2001.)

The maximum depth at which fast (scattering-dominant) optical effects can be studied noninvasively is mostly determined by the source–detector distance. For homogenous media, the maximum depth is between one-half and one-fourth of the source–detector distance.[23] However, the sensitivity to phenomena occurring at different depths varies depending on the type of measure used. Measures of amount of light transmission are more sensitive to superficial phenomena, whereas measures of photon delay are more sensitive to deep phenomena.[17,21,23] This may lead to discrepancies in the results obtained with these two sets of measures.

Various investigators have reported methods for the 3D reconstruction of photon migration data.[24-32] Most methods are based on finite element models of the propagation of photons through tissue and use an iterative process to determine the scattering and absorption coefficients of various compartments of the tissue. These models are therefore computation intensive.

We have recently used a simpler approach[33] consisting of approximating the path between a particular source–detector pair using a volume of arbitrary shape (to be scaled in size, depending on the source–detector distance). This approach introduces approximations in the 3D reconstruction on the order of several mm but may be adequate for many practical applications and is computationally very simple. Gratton and colleagues[33] reported data comparing estimates of the depth of regions of brain activity obtained using this approach and fMRI. The study was based on manipulating the eccentricity of visual stimuli, which influences the depth of the region of medial occipital cortex where activity should be observed (i.e., more eccentric visual stimuli are processed in deeper areas of the calcarine fissure than more central visual stimuli).

Optical imaging data were recorded using several source–detector distances; the distance at which the largest effect was observed was used to estimate the depth of the optical effect (estimated as half of this distance). Functional MRI data were also recorded from the same subjects, using the same paradigm. Results indicate that the two estimates differ, on average, by less than 1 mm. This provides support for the simplified approach to the estimation of the depth of the optical effects (and therefore to 3-D reconstruction) described earlier.

9.2.2.3 Data Collection and Artifact Correction

At present, several companies produce recording instruments suitable for recording fast optical data. These instruments are capable of collecting measures of the intensity and delay of the photons reaching the detector at a very fast pace (up to more than 1 kHz). Multiple-channel systems are available (up to 64 sources and 8 detectors). The use of multiple channels may involve time-multiplexing to distinguish signals associated with different source–detector pairs. Time-multiplexing may reduce the temporal resolution of the instrument. All systems are designed to be movable (and therefore easily portable) and use low-energy, nonionizing radiation, which makes them safe and suitable for extended and/or repeated recordings with the same subject. Also, most systems allow for an external event channel or at least for external triggering, so as to enable synchronization with the system producing the stimulation and/or recording the subject response. This allows for the recording of event-related brain activity (EROS).

In addition to fast neuronal signals, the raw data obtained with optical recording systems (and in particular the intensity data) carry a number of other signals. These are related to 1) arterial and capillary pulse (about 1 Hz), 2) respiration (about 0.25 Hz), and 3) pressure waves (about 0.1 Hz). These signals (which can be of interest in a number of cases) are considerably larger than the neuronal signal and can, in fact, obscure it. Thus, they should be considered artifacts if the research focuses on recording neuronal signals. These artifacts are quite evident for intensity data but less so for time-delay data, which are less sensitive to superficial phenomena (and therefore less sensitive to capillary pulse). Some of these artifacts (as well as slow drifts related to heating or cooling of the apparatus) can be eliminated using appropriate high-pass filters. The pulse artifact is more problematic because its frequency is closer to that of the neuronal signals. However, this artifact (ubiquitous in intensity recordings) generates very characteristic saw-tooth waves; its shape regularity can be used to devise algorithms for its correction. Gratton and Corballis[34] proposed one such method, now used in several laboratories.

Other types of artifacts are due to movements of the subjects or of the optical apparatus. These movements produce large signals that are temporally well localized and can be discarded using automatic procedures. In Figure 9.2 we report an example of the headgear (a modified motorcycle helmet) that we use to hold the source and recording fiber in place.

9.2.2.4 Signal Processing and Statistical Analysis

The recording apparatus described yields time series that can be time-locked to particular events (i.e., to stimuli or to the subject's responses to the stimuli). These time series can be processed with standard methodologies used for analyzing other types of time series, such as those obtained with ERPs or fMRI recordings. A simple way of extracting the fast optical signal is by averaging the time-locked waveforms

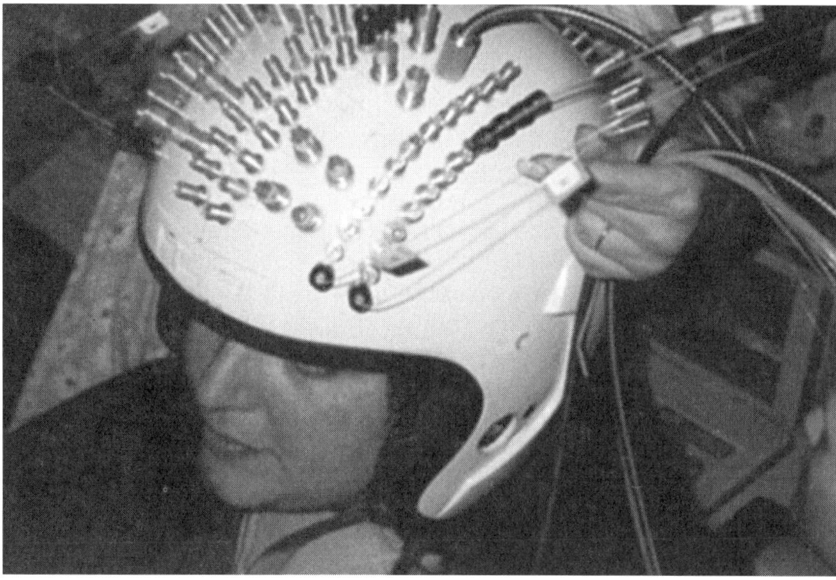

FIGURE 9.2 The headgear (a modified motorcycle helmet) used to hold in place the source and detector fibers.

obtained for particular stimuli or response conditions. More complex analysis procedures can also be used, such as cross-correlation, wavelet analysis, and combined time-frequency analysis. In addition, low- and high-pass filtering can be used to improve signal-to-noise ratio. Statistics about the reliability of the observed effects can then be computed within or across subjects, using standard statistical methods. Fabiani et al.[35] have shown that nonparametric statistics based on bootstrap methods can also be useful to determine the reliability of effects.

As previously mentioned, data from a single source-detector pair (or channel) refer to a curved spindle-shaped volume with vertices at the source and detector. These data, therefore, provide an estimate of the origin of the signal. Given the variability in cortical anatomy across individuals, it is usually necessary to record from multiple surface locations to make sure to "hit" the active area in a particular subject. Whenever multiple recordings are obtained, it is important to establish that effects are not due to chance, as the probability of capitalizing on chance increases with the number of locations. This can be achieved in several ways:

1. Bonferroni-corrected confidence intervals or p-values can be computed.
2. An alternative approach is to first compute omnibus statistics across all locations and then, if the omnibus effect is significant, determine the exact location where the effect is maximum. Bootstrap methods may be useful for this second step.[35]
3. Finally, a multivariate approach can be used to deal with the problem of multiple comparison or at least to minimize the number of locations investigated.

When collapsing data across subjects, it is important to consider that recordings from the same surface location (identified using scalp landmarks) in different subjects may not correspond to the same functional cortical area in all subjects. The first studies using optical imaging data largely ignored this issue, which may, however, lead to a marked reduction of the signal-to-noise ratio because active locations from one subject may be averaged with inactive locations from another. This problem is not unique to optical imaging, as it is typical of any brain-imaging technique with a spatial resolution higher than a few cm. Several investigators have proposed a number of tools for realigning and rescaling brain and cortical areas in different subjects. In principle, all of these tools can be applied to EROS data as well. However, this requires:

1. That the exact locations used for each source and detector be recorded within a fiducial space, which can be achieved by using a magnetic 3D digitizer such as the Polhemus® 3D digitizer system
2. That the same fiducial locations (e.g., nasion and preauricolar points) be marked on structural MR images (e.g., by using vitamin E pills or other markers)
3. That a 3D reconstruction method be used to determine the volume from which data are recorded from, separately for each individual subject (see earlier description)
4. That if the volumes from different source–detector pairs overlap, at least in part, this can be used to further constrain the volume to which the data are referred (an example of this approach is given by the PI detector proposed by Wolf et al.[19])
5. That the 3D reconstructed data from different subjects be aligned and rescaled using one of the procedures described in the literature (the simplest of which is the Talairach transformation)

Once these transformations have been obtained, data from different subjects can be pooled together and across-subjects statistics can be computed. A comparison between data that were or were not aligned using the anatomical landmarks of individual subjects is presented in Color Figure 9.3.* From this figure it is evident that realignment greatly improves the reliability of the results and that apparent differences between young and older adults are most likely due to individual differences in brain anatomy, which are more pronounced in older than in younger subjects.

9.3 EXPERIMENTAL RESULTS

As this chapter is being written, there are fourteen published reports (including conference abstracts) of studies investigating the fast optical signal. All but one[36] reported the observation of a fast optical signal, although the methods used in different studies varied significantly. The majority of these studies have been conducted in our laboratory and have used photon-delay measures as the main dependent

* Color figures follow page 112.

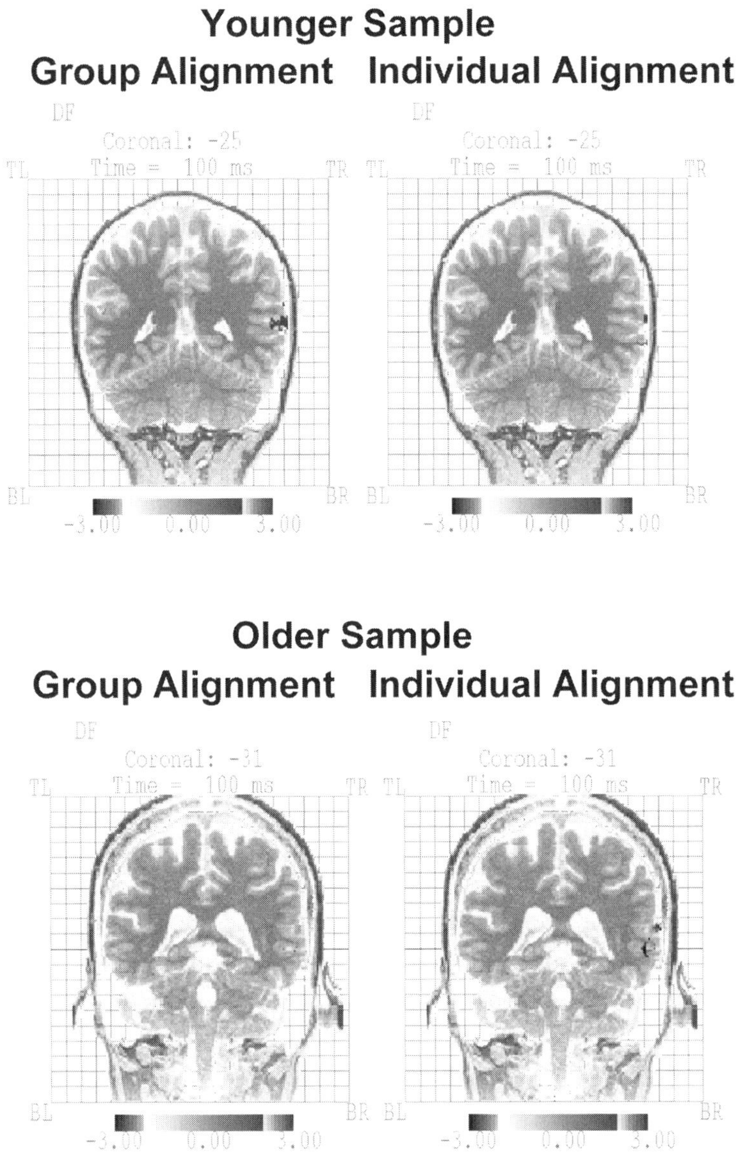

FIGURE 9.3 (See Color Figure 9.3.) Effect of using a procedure to align optical data on individual brain anatomy in young (top row) and old (bottom row) subjects. The left panels refer to 3D reconstructions based on group coordinates; the right panels refer to 3D reconstructions based on individual subjects' coordinates. The data were collected during an auditory stimulation paradigm, at a latency of 100 msec from stimulation.[44] Note the increased similarity of the effects in the two groups of subjects after alignment based on individual subject coordinates.

variable. Our major goal has been to show that the fast optical signal can be used to study the time course of activity in selected cortical areas noninvasively. To support this claim, we have run experiments demonstrating the following:

1. Rapid changes in the optical signal can be recorded immediately after stimulation (within 100 msec or less) or immediately before a motor response. This is essential to demonstrate that a fast optical signal occurs. The short latency of the signal excludes that this signal can be due to hemodynamic effects, which take several hundred msec to develop.[37]
2. These changes are temporally overlapping with other neuronal signals, such as the evoked potential, recorded from the same subjects and conditions. This is important if one intends to use these measures to study the time course of neuronal activity.
3. The fast optical effects are localized to brain areas where hemodynamic measures with high spatial resolution (such as fMRI) indicate that neuronal activity occurs. This is important if these measures are used as indices of local neuronal activity in selected cortical areas.
4. Similar effects are recorded from different areas when different modalities are used. This is important to demonstrate that fast optical signals are not the property of one particular cortical area but are a general property of the cortex. This, in turn, allows us to generalize our studies to different domains.

Table 9.1 shows how different studies provide support for each of these four points. Figures 9.4 and 9.6 and Color Figure 9.5 show examples of the experimental support obtained in these studies. Next, we will describe some of the findings obtained in each modality and then some additional data illustrating how fast optical imaging can be used in the study of brain physiology, perception, and cognition.

9.3.1 Visual Modality

Most studies on the fast signal have been conducted in the visual modality. With one exception,[36] these studies indicate that a fast optical signal can be recorded after the presentation of visual stimuli. Recordings have been made from medial and more lateral occipital areas. The first study[1] was based on the reversal of one of four grids placed in different quadrants of the visual field; in this study, the reversals occurred every 500 msec. The original study was based on a small sample size (N = 3), but quite similar results were obtained in a subsequent study using a larger sample size (N = 8).[38] A comparison of the results of these two studies is presented in Figure 9.4.

Both studies indicated the existence of a fast optical effect consisting of an increase in the photon delay that occurred approximately 100 msec after stimulation (grid reversal). This effect was localized in that it occurred at locations placed on the contralateral hemisphere where the stimulation occurred. Further, stimulation of upper quadrants resulted in larger effects in lower locations in the occipital cortex (and vice versa) in a manner consistent with the well-known representation of the visual field in medial occipital cortex. The location of the response of our original

TABLE 9.1
Papers Providing Support for Fast Optical Signals

Detection of fast optical signals (only first reports from different laboratories)

Gratton, G. et al., *Psychophysiology,* 32, 505, 1995.
Steinbrink, J. et al., *Neurosci. Lett.,* 291, 105, 2000.
Wolf, U. et al., *OSA Biomed. Top. Meeting, Tech. Dig.,* 427, 2000.

Concurrence with electrophysiological signals[a]

Gratton, G. et al., *NeuroImage,* 6, 168, 1997.
Rinne, T. et al., *NeuroImage,* 10, 620, 1999.
Gratton, G. et al., *Hum. Brain Mapping,* 13, 13, 2001.
De Soto, M.C. et al., *J. Cognit. Neurosci.,* 13, 523, 2001.

Co-localization with fMRI signals[a]

Gratton, G. et al., *NeuroImage,* 6, 168, 1997.
Gratton, G. et al., *NeuroImage,* 11, 491, 2000.

Effect in different modalities (first reports for each modality)

Visual: Gratton, G. et al., *Psychophysiology,* 32, 505, 1995.
Auditory: Rinne, T. et al., *NeuroImage,* 10, 620, 1999.
Somatosensory: Steinbrink, J. et al., *Neurosci. Lett.,* 291, 105, 2000.[b]
Motor: Gratton, G. et al., *J. Cognit. Neurosci.,* 7, 446, 1995.

[a] This includes only studies in which both modalities (optical and electrophysiological or fMRI) were recorded with the same subjects and conditions.
[b] The Steinbrink et al. paper was based on a different method from those used in the other studies.

study was found to correspond closely with that of the hemodynamic response measured with fMRI, whereas the time course was similar to that of the visual evoked potential (VEP).[39] Syre et al.[36] reported being unable to replicate these original findings; however, they did not use a systematic set of recording locations and did not report statistics computed across subjects. A fast optical effect elicited by visual stimulation was instead reported by Wolf et al.,[19] who also used a very small sample size but a different type of detector (PI detector; see above).

Our 1995 study, as well as the subsequent replication shown in Figure 9.4, was based on a fixed interstimulus interval (ISI). It is therefore theoretically possible that the observed response could be due to a time-locked response to some previous stimulation. However, two subsequent studies rule out this possibility. In one of these studies,[7] we systematically varied the ISI from 100 to 1000 msec; in each case, a response (characterized by an increase in photon delay) was observed with a latency between 60 and 100 msec from stimulation. In fact, the latency of the response increased slightly with the ISI. The same study also indicated that the fast optical response increased in amplitude with ISI, in a manner consistent with that of the concurrently recorded VEP.

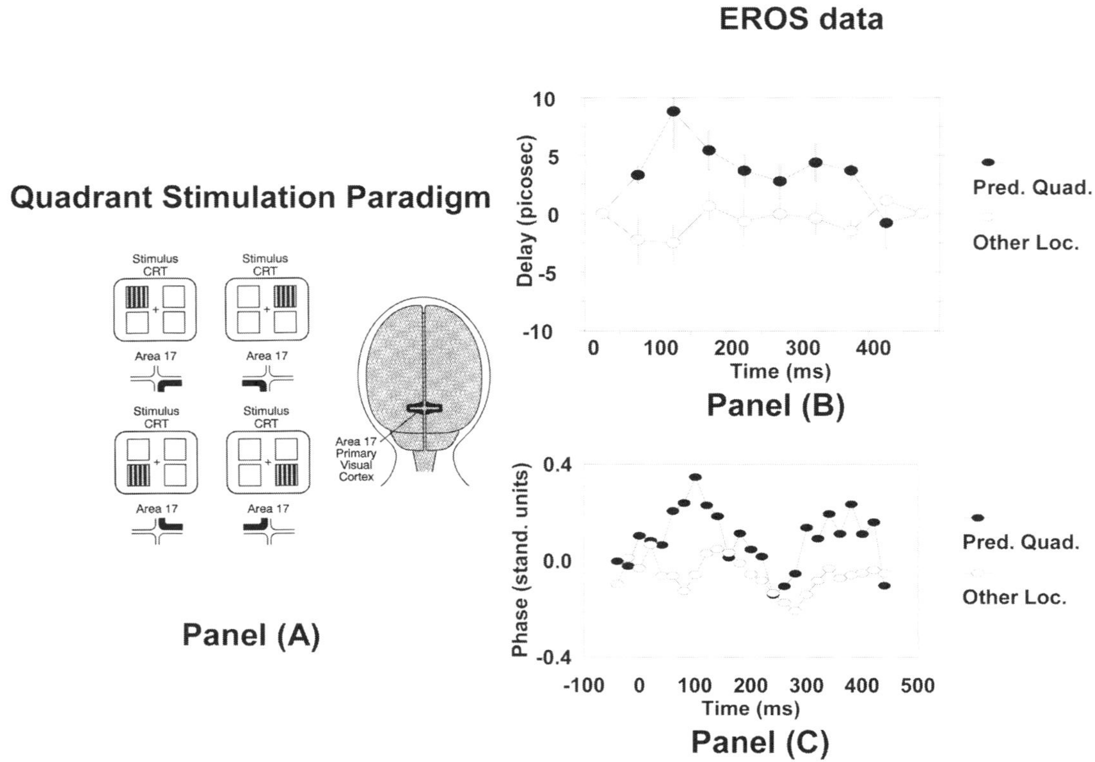

FIGURE 9.4. EROS recorded from medial occipital areas during two visual stimulation experiments using the same paradigm.[1,38] Panel (A): Diagram of the stimulation conditions. The stimulation consisted of reversals of black-and-white grids occurring every 500 msec; only one quadrant of the visual field was stimulated during a trial block, and the stimulated quadrant was rotated across blocks. The head diagram represents the predicted location of the response, based on a contralateral, inverted representation of the visual field in medial occipital areas. Panels (B and C): EROS effects in the two experiments. In both experiments Panel (B): Gratton, G. et al., *Psychophysiology,* 32, 505, 1995; Panel (C): Gratton, G. et al., in preparation), the filled circles refer to the EROS activity recorded from predicted locations for each quadrant stimulation condition, and the open circles refer to the EROS activity recorded from the same locations when the other quadrants were stimulated. The second experiment (Panel C) was based on a larger data sample and had a higher sampling rate than the first experiment (Panel B).

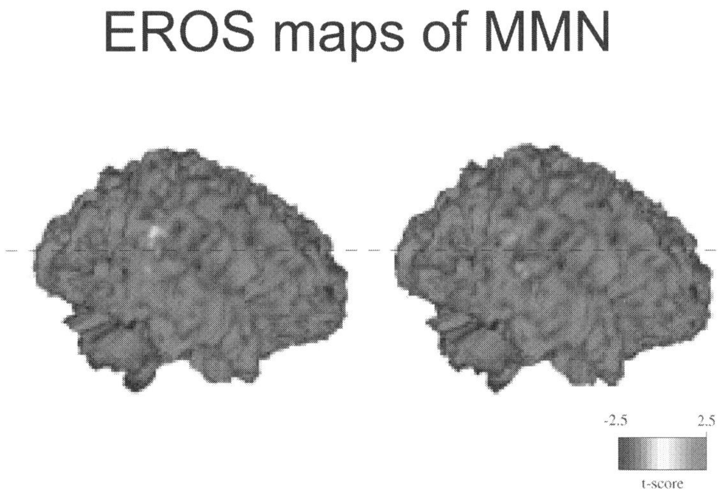

FIGURE 9.5 (See Color Figure 9.5.) Maps of the right hemisphere optical response to stimuli in a passive auditory oddball paradigm. The response to standard stimuli (latency 100 msec) is presented on the left panel; the response to the deviant stimuli (latency 160 msec) is presented in the right panel. (From Rinne, T. et al., *NeuroImage,* 10, 620, 1999, with permission.)

In another study, we used a variable ISI between flashing stimuli and observed an increase in phase delay with a peak latency of approximately 60 to 80 msec.[40] Increases in phase delay with similar peak latency (60 to 100 msec) were also reported in several other studies.[22,41,42] In all cases, when the stimulus was unilateral, this fast response was only observed in the hemisphere contralateral to stimulation, thus confirming the spatial localization of the response. Further, we found that the more lateral the area stimulated was, the larger was the source–detector distance at which the response (i.e., an increase in photon delay) was observed.[22] This is consistent with the idea that lateral stimuli elicit activity in deeper cortical areas than foveal stimuli and consistent with fMRI findings obtained in the same study and with the well-known functional organization of medial occipital cortex. The latency of the optical effects is also generally consistent with what is known about the visual system. For instance, typically stimuli with abrupt onset tend to elicit activity with a shorter latency (about 60 to 80 msec) than isoluminant alternating stimuli (about 80 to 100 msec). This is also generally consistent with what is observed for visual EPs.

In some of our studies, the recording was extended to more lateral occipital areas. Fast optical responses could also be recorded from these areas with a latency similar to or only slightly longer than that in more medial areas.[7,42] Finally, several studies showed optical responses with much longer latencies (200 to 300 msec).[1,38,42] These responses were often ipsilateral or bilateral, perhaps reflecting subsequent feedback processes from other cortical or subcortical areas.

In summary, a number of studies indicate the presence of a fast optical response to visual stimuli in medial occipital areas, characterized by an increase in phase delay with a latency of 60 to 100 msec from stimulation. The localization of this response is consistent with the well-known representation of the visual field in medial

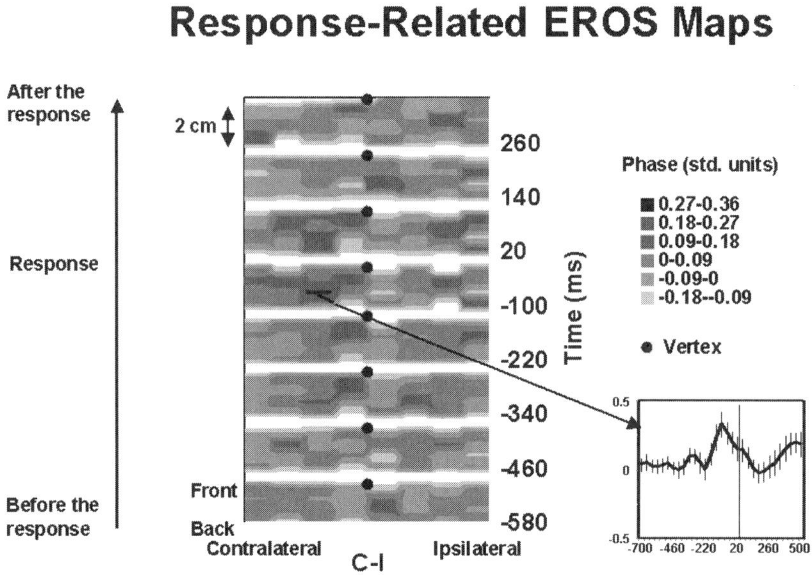

FIGURE 9.6 Maps of the optical activity preceding and following a motor response. The maps are presented as if viewed from the top of the head. The brain area contralateral to the responding hand is presented on the left, and that ipsilateral to the responding hand is presented on the right. The numbers beside the maps refer to the relative time (in msec) with respect to the response (negative values refer to time before the response). The small panel on the right shows the time course of the response (with standard-error bars) for the location of largest activity. (From DeSoto, M. et al., *J. Cognit. Neurosci.*, 13, 523, 2001, with permission.)

occipital cortex, and the temporal properties of the signal are consistent with the time course of neuronal signals such as the VEP.

9.3.2 Auditory Modality

Two studies have been completed in the auditory modality.[43,44] Both are based on a passive auditory oddball paradigm; in fact, the second study is essentially a replication of the first using a larger sample size and both young and older subjects. In these studies, the subjects read books while series of tones were presented to them via headphones. The tone could be either long (75-msec duration) or short (25-msec duration). The long-duration tone occurred 80% of the time and was labeled the standard stimulus. The short-duration tone occurred 20% of the time and was labeled the deviant stimulus. The ERPs elicited in this paradigm have been the subject of extended investigation[45] and are typically characterized by two types of activities. First, some potentials are elicited by all stimuli (although their amplitude is larger for stimuli with higher overall intensity). They are labeled exogenous potentials, the most evident of which is the N1 (probably itself the sum of several components). Second, some potentials are typically generated only by deviant stimuli; among them is the mismatch-negativity (MMN), an ERP component with a peak latency of approximately 120 to 200 msec.

There is some uncertainty about whether the N1 and MMN components are generated in the same or different cortical areas. The Rinne study was set up to show 1) that fast optical effects can be recorded from auditory areas, 2) that different optical responses with characteristics similar to those of the N1 and the MMN could be obtained, and 3) whether these two fast optical responses were generated in the same or different areas. The results confirmed the first two hypotheses and showed that the optical response to deviant stimuli (corresponding to the electrical MMN) was systematically recorded approximately 1 cm below the location where the response to the high-energy standard stimuli (corresponding to the electrical N1) was maximum. The response to the deviant stimuli had a latency of 160 to 180 msec, whereas the response to the standard stimuli had a latency of 100 msec.

These latencies closely correspond to the latencies of the electrical MMN and N1, respectively; both consisted of an increase in the photon delay parameter. The results of the Wee study were very similar to those of the Rinne work and also indicated that responses to the deviant stimuli occurred later (160 to 180 msec) and at lower locations (1 cm) with respect to responses to the standard stimuli (for which peak latency was 100 msec). Further, anatomical 3D reconstruction of the Wee data indicated that the optical responses occurred in superior or medial temporal cortex.

In summary, the data from the auditory modality show that fast optical responses with latency consistent with those of auditory evoked potentials can be recorded from temporal cortex. Several phenomena can be distinguished on the basis of their location, time course, and response to experimental manipulations.

9.3.3 Somatosensory Modality

Fast optical imaging in the somatosensory modality has been conducted so far only by the Berlin group[2]; however, this work was based on a different technique than that used by our group in the other modalities. Steinbrink and colleagues used a continuous light source (a lamp) instead of a radio-frequency-modulated light source. The measures obtained were changes in the amount of light transmitted from the source to the detector in the period immediately following somatosensory stimulation. Stimulations were brief low-intensity electric shocks delivered to either arm (median nerve stimulation). Continuous measures were obtained from parietal locations. Rapid changes (60 to 160 msec from stimulation) were observed when the contralateral side was stimulated but not when the ipsilateral side was stimulated. These data confirm that a fast optical signal can be recorded noninvasively and that this signal is localized (at least to the contralateral hemisphere). However, the authors pointed out that their technique is different from that used in our laboratory. Whether this is an important distinction and, in general, which techniques are more appropriate for recording fast optical signals remain to be ascertained.

9.3.4 Motor Modality

We have conducted two studies investigating fast optical signals in the motor modality. One of them was an early study[17] that involved unilateral hand- or foot-tapping tasks at a frequency of 0.8 Hz. In this study, there was no synchronization between the subjects' movements and the optical recordings, and a single-channel system was used. The analysis was conducted in the frequency domain and showed significant oscillations of the optical signal (photon delay) at the frequency of stimulation and its harmonics. These oscillations were larger on the side contralateral to the tapping limb.

Recently, we recorded fast optical imaging data during a choice reaction time task.[46] This experiment was based on a synchronized multichannel system, which allowed us to generate maps of brain activity at different times with respect to the response. Some of these maps are shown in Figure 9.6. They show that a fast optical signal, consisting of an increase in phase delay and analogous to those observed in the visual and auditory modalities, can be observed in the last 100 msec before the button press on the side contralateral to the responding hand. In our data, the time course of the optical response is very similar to that of the lateralized readiness potential (LRP), an electrical measure used to index response preparation.[47] Similar data have been recently obtained by Wolf et al. (personal communication, March 2001), who showed that a fast optical signal shortly preceding the response can be observed using either photon delay or intensity as a dependent variable.

These data provide further support to the generality of the fast optical signal as a measure of localized cortical activity. They also indicate that different methodologies can be used to record fast optical signals.

9.3.5 A Physiological Study: Relationship between Neuronal and Vascular Signals

The data described in the previous sections support the idea that fast optical signals can be used to study neuronal activity in localized cortical areas. In this and the following section, we will describe examples of studies in which the fast optical signal is used to investigate questions of interest to neurophysiologists and psychologists. The first example is related to the use of optical methods to study the relationship between neuronal and hemodynamic events.[7] This relationship is important for two reasons: 1) the use of hemodynamic methods to map brain function is predicated on the relationship between hemodynamic changes and neuronal activity, and 2) the relationship between neuronal and vascular phenomena may be influenced by variables such as age, cardiovascular fitness, and hypertension and may be the cause of brain dysfunction. Optical methods appear particularly suited for studying this relationship because they can be used to index both neuronal and vascular events. Further, hemodynamic and neuronal optical measures can be related to approximately the same volume of tissue, something difficult to achieve with other methods.

Our study was based on a manipulation similar to that used by Fox and Raichle.[48] This seminal PET study showed that the amplitude of the blood flow change in medial occipital areas is proportional to the frequency of visual stimulation, at least for stimulation frequencies up to 8 Hz. This justified the use of subtraction and parametric methods (such as statistical parametric mapping — SPM) in the analysis of hemodynamic imaging data. Subsequent studies using fMRI also confirmed this proportionality in occipital areas, although other modalities show different effects.[49]

However, none of these studies involved simultaneous measures of neuronal and vascular signals. Therefore, the actual relationship between neuronal and vascular phenomena is not well known. In our study, we recorded simultaneously fast (neuronal) and slow (hemodynamic) optical signals and varied the visual stimulation frequency between 0 (constant stimulus) and 10 Hz. Fast and slow signals were distinguished on the basis of their time courses. Further, the fast signal was assessed using photon delay measures, whereas the slow signal was assessed using intensity measures.

We found that the fast signal amplitude decreased with increasing stimulation frequency; in contrast, the amplitude of the slow signal increased with stimulation frequency up to 5 Hz but then declined slightly at the highest stimulation frequency (10 Hz) (see Figure 9.7). This latter result is consistent with that reported by Fox and Raichle.[48] In addition, when the amplitude of the fast signal was multiplied by the stimulation frequency (thus obtaining an estimate of the fast response integrated over time), the two signals were found to be linearly proportional to each other ($r = 0.98$). This finding provides some support for the use of hemodynamic data in the analysis of brain activity. However, it should be taken with some caution for two reasons: 1) a single-wavelength system was used, thus providing only a summary measure of hemodynamic phenomena, and 2) only one modality was investigated. Therefore, more studies are needed to further elucidate the relationship between neuronal and hemodynamic phenomena. However, this study indicates that optical imaging can be particularly useful for this purpose.

9.3.6 An Attention Study: Visual Selective Attention in Medial Occipital Cortex

The second example is a study investigating the issue of the level, within the visual information processing stream, at which selective attention effects become apparent — a controversial issue in cognitive neuroscience. On the one hand, ERP data indicate that some type of early brain activity following visual stimulation is not affected by selective attention, whereas later activity is.[50] This suggests that some cortical responses are automatic. On the other hand, hemodynamic brain imaging studies show that attention effects are already visible in primary visual cortex (e.g., Worden et al.[51]). The relevance of this finding, however, has been criticized by Hillyard and his group,[50] who suggest that the attention effects visible in primary visual cortex are due to late re-entrant activity and not to the first response observed in this area. They provide some dipole modeling of ERP data in support of this possible alternative explanation.

A procedure that can provide information about the time course of activity in a selected cortical area (such as primary visual cortex) may be very useful in this case. For this reason, we applied optical imaging to study the time course of activity in medial and lateral occipital cortex, in order to determine the effect of spatial selective attention on the early response in these cortical areas. The paradigm used was similar to that used by Hillyard and his group in the study of visual selective attention.

On each short block of trials, subjects were presented with a train of stimuli, some of which appeared to the left and some to the right of fixation. Most of the

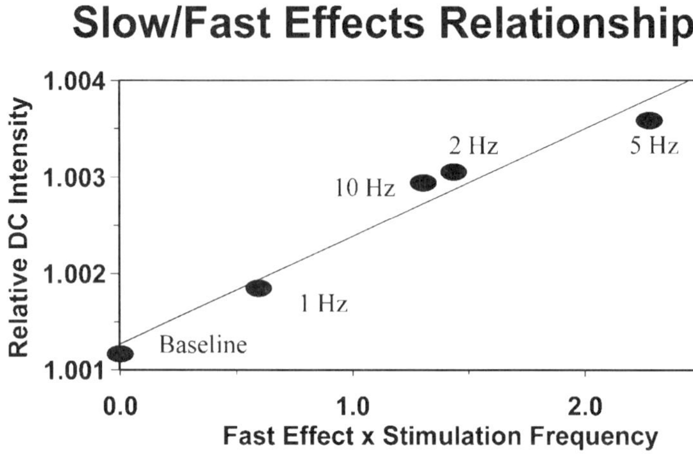

FIGURE 9.7 Scatter plot describing the relationship between fast and slow response as a function of stimulation frequency. For each stimulation frequency condition, the fast response was measured as the amplitude of the EROS response at a latency of 80 msec after each individual stimulation (grid reversal), and the slow response was quantified as the average increase in the slow signal between 4 and 14 sec after onset of the stimulation period. It was expected that the slow signal (related to hemodynamic phenomena) would be related to the fast signal (related to neuronal phenomena) cumulated over time. Therefore, if the relationship between neuronal and vascular phenomena was linear, then the slow signal should be proportional to the amplitude of the fast signal multiplied by the stimulation frequency. The latter value is reported in the abscissa. (From Gratton, G. et al., *Hum. Brain Mapping,* 13, 13, 2001, with permission.)

stimuli were small squares, but rectangles were presented in 20% of the trials. The task of the subjects was to monitor the stimuli on one side of fixation (which side they had to monitor was indicated by an arrow presented at the beginning of each block) and to press a button whenever a rectangle occurred on the attended side. The stimuli on the other side were to be ignored. Since changes of fixation may help to perform the task but confound the results, the position of the eyes was monitored during each block (using horizontal EOG recordings) and blocks in which eye movements occurred were discarded.

The results indicated that stimuli elicited a rapid increase in photon delay (latency = 80 msec) in occipital areas contralateral to the side where the stimulus was presented. In medial occipital areas, this rapid increase was similar whether the stimuli occurred on the attended or on the unattended side. However, in more lateral occipital areas there was a large differentiation between the responses for attended and unattended stimuli. To highlight these different responses, two types of maps were computed. One map represented the stimulation effect and was obtained by averaging all brain responses, whether the stimuli occurred on the attended or unattended side; the other represented the attention effect and was obtained by subtracting brain responses obtained when the stimulus occurred on the unattended side from those obtained when the stimulus occurred on the attended side. Both maps corre-

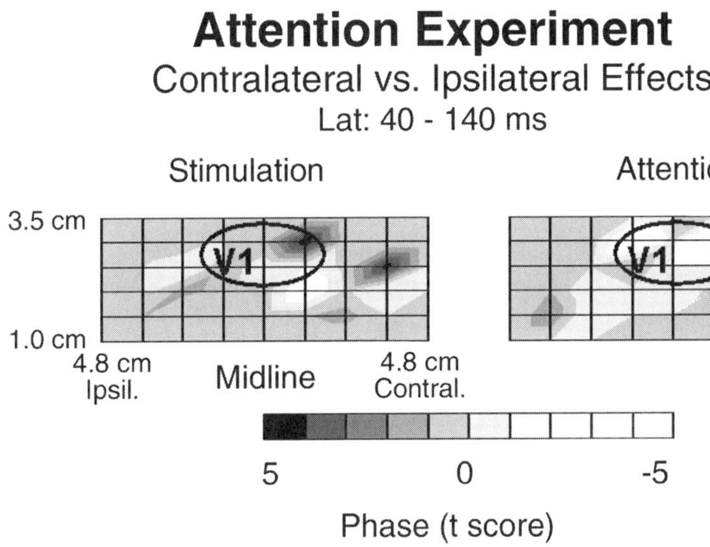

FIGURE 9.8 Maps of the stimulation and attention effects in a spatial selective attention experiment. The maps refer to the optical activity recorded in occipital areas at a latency of 40 to 140 msec after stimulation (an approximate average extension of the V1 area is presented as a reference). In each map, the right side represents the hemisphere contralateral to the stimulation, and the left represents the hemisphere ipsilateral to the stimulation. The maps are based on "t score" values computed across subjects (N = 4).

spond to a latency between 40 and 140 msec after stimulation and therefore can be considered maps of the early cortical response as measured with optical imaging. These maps (shown in Figure 9.8) indicated an early stimulation effect in medial occipital cortex; the attention effect was evident at more lateral locations.

These data suggest that attention effects are not evident in medial occipital cortex but emerge in more lateral occipital areas, thus supporting the thesis of Hillyard and his group. However, this study used a spatial selective attention task. Some of the fMRI studies observing attention effects in medial occipital cortex (e.g., Worden et al.[51]) used other types of features to define target and nontarget stimuli. It is therefore possible that whether attention effects are observed in medial occipital cortex depends on the type of stimuli used. To test this hypothesis, a new optical imaging study should be run using different types of stimulus materials. In any case, the present study shows how a technique combining spatial and temporal resolution can be useful to address issues relevant to cognitive neuroscientists.

9.4 DISCUSSION

9.4.1 Summary of Empirical Data

In this chapter we have presented evidence indicating that it is possible to record fast optical signals using a noninvasive procedure. Fast optical signals have been recorded in different modalities (including visual, auditory, somatosensory, and motor) and are presumably related to scattering changes associated with neuronal activity. In general, the response is characterized by an increase in photon delay in active cortical areas (a reduction in light transmission was observed in the Steinbrink et al.[2] study).

9.4.2 Relationship with Other Techniques

Several studies show that the time course of the fast optical signal is similar to the event-related electrical activity recorded in the same conditions. Also, several studies show that the optical signal is localized in a manner consistent with the known functional anatomy of the cortical areas involved and, in at least two studies, is consistent with fMRI data. These data provide support for the idea that fast optical imaging can be used to study the time course of activity in localized cortical areas.

9.4.3 Unresolved Issues and Future Research Directions

Several issues remain to be addressed. First, it has been hypothesized that the fast optical signal (or EROS) is due to changes in the scattering coefficient in active cortical areas determined by the movement of ions across the neuronal membrane. However, this has not yet been proven directly and more investigation on the physiological basis of the signal is needed. Some of this research will need to be conducted on *in vitro* preparations or with animal models. The research conducted using depth recordings with miniature optical instruments or implanted optic fibers (e.g., Rector et al.[5]) is clearly relevant for this purpose, as is work by various groups on *in vitro* preparations (e.g., Frostig et al.[4]).

Second, more research is needed to determine what is the best method for recording fast optical signals. Both photon delay and intensity signals have been used. A direct and systematic comparison of different methods is needed. It is also possible that a combination of the two methods may be useful to increase signal-to-noise ratio.

Third, although substantial advances have been made in the area of 3D reconstruction, a single method still has to obtain general acceptance. This area of research will gain from the developments of easy-to-use, general-purpose analysis software.

Fourth, so far the technique appears to be useful for relatively shallow areas (3 cm or so from the surface of the head), but its penetration to deeper structures is severely limited. However, the penetration may be increased by improvements in signal-to-noise ratio and by other technological developments (such as the use of the PI detector).

Fifth, raw data are at present quite noisy, often requiring data collection from a large number of trials and subjects. Several tools can be used to increase the signal-to-noise ratio, including instrumental improvements (such as the use of more powerful light sources) and analytical methods (such as the use of appropriate filtering methods such as wavelet analysis, cross-correlation, autoregression, etc.).

9.4.4 Conclusions

Fast optical imaging appears to be a promising new tool for studying brain function and, in particular, the time course of activity in selected cortical areas. Advantages of the technique include noninvasivity, portability, good combination of spatial and temporal resolution, relatively low cost, and easy integration with other techniques such as fMRI, electrophysiological methods, and NIR spectroscopy. The main limitations are its reduced penetration and, at least at present, its low signal-to-noise ratio. The technique appears particularly well suited to provide a bridge between hemodynamic and neuronal methods and can be used profitably in the study of brain physiology and in cognitive neuroscience.

ACKNOWLEDGMENTS

The work presented in this chapter was supported in part by NIMH grant MH57125 to Dr. Gratton and by grant #97-32 from McDonnell–Pew to Dr. Fabiani.

REFERENCES

1. Gratton, G. et al., Shades of gray matter: noninvasive optical images of human brain responses during visual stimulation, *Psychophysiology*, 32, 505, 1995.
2. Steinbrink, J. et al., Somatosensory evoked fast optical intensity changes detected non-invasively in the adult human head, *Neurosci. Let.*, 291,105, 2000.
3. Cohen, L.B., Changes in neuron structure during action potential propagation and synaptic transmission, *Physiol. Rev.*, 53, 373, 1972.
4. Frostig, R.D. et al., Cortical functional architecture and local coupling between neuronal activity and the microcirculation revealed by *in-vivo* high resolution optical imaging of intrinsic signals, *Proc. Natl. Acad. Sci.*, 87, 6082, 1990.
5. Rector, D.M. et al., Light scattering changes follow evoked potentials from hippocampal Schaeffer collateral stimulation, *J. Neurophysiol.*, 78, 1707, 1997.
6. Raichle, M.E., Visualizing the mind, *Sci. Am.*, April, 58, 1994.
7. Toga, A.W. and Mazziotta, J.C., Eds., *Brain Mapping. The Methods*, Academic Press, San Diego, 1996.
8. Gratton, G. et al., Comparison of neuronal and hemodynamic measure of the brain response to visual stimulation: an optical imaging study, *Hum. Brain Mapping*, 13, 13, 2001.
9. Jobsis, F.F., Noninvasive, infrared monitoring of cerebral and myocardial oxygen sufficiency and circulatory parameters, *Science*, 198, 1264, 1977.
10. Villringer, A. and Chance, B., Non-invasive optical spectroscopy and imaging of human brain function, *Trends Neurosci.*, 20, 435, 1997.
11. Grinvald, A. et al., Functional architecture of cortex revealed by optical imaging of intrinsic signals, *Nature*, 324, 361, 1986.
12. Villringer, A. and Dirnagl, U., Coupling of brain activity and cerebral blood flow: basis of functional neuroimaging, *Cerebrovasc. Brain Metab. Rev.*, 7, 240, 1995.
13. Miller, K.L. et al., Non-linear transport dynamics of the cerebral blood flow response, *Hum. Brain Mapping*, 13, 1, 2001.

14. Allison, T. et al., The central nervous system, in *Psychophysiology: Systems, Processes, and Applications*, Coles, M.G.H., Porges, S.W., and Donchin, E., Eds., Guilford, New York, 1986, 5.
15. Pascual–Marqui, R.D. et al., Low resolution electromagnetic tomography: a new method for localizing electrical activity in the brain, *Int. J. Psychophysiol.*, 18, 49, 1994.
16. Dale, A.M. et al., Dynamic statistical parametric mapping: combining fMRI and MEG for high-resolution imaging of cortical activity, *Neuron,* 26, 55, 2000.
17. Gratton, G. et al., Rapid changes of optical parameters in the human brain during a tapping task, *J. Cognit. Neurosci.,* 7, 446, 1995.
18. Okada, E. et al., Theoretical and experimental investigation of near-infrared light propagation in a model of the adult head, *Appl. Opt.*, 36, 21, 1997.
19. Wolf, U. et al., Detecting cerebral functional slow and fast signals by frequency-domain near-infrared spectroscopy using two different sensors, *OSA Biomed. Top. Meeting, Tech. Dig.,* 427, 2000.
20. Gratton, E. et al., The possibility of a near-infrared optical imaging system using frequency-domain methods, *Proc. III Int. Conf. Peace through Mind/Brain Sci.,* 183, 1990, Hamamatsu City, Japan.
21. Firbank, M. et al., A theoretical study of the signal contribution of regions of the adult head to near-infrared spectroscopy studies of visual evoked responses, *NeuroImage*, 8, 69, 1998.
22. Gratton, G. et al., Toward non-invasive 3-D imaging of the time course of cortical activity: investigation of the depth of the event-related optical signal (EROS), *NeuroImage,* 11, 491, 2000.
23. Gratton, G. et al., Feasibility of intracranial near-infrared optical scanning, *Psychophysiology,* 31, 211, 1994.
24. Alfano, R.R. et al., Advances in optical imaging of biomedical media, *Ann. New York Acad. Sci.,* 820, 248, 1997.
25. Arridge, S.R. and Hebden, J.C., Optical imaging in medicine: II. Modeling and reconstruction, *Phys. Med. Biol.*, 42, 841, 1997.
26. Arridge, S.R. and Schweiger, M., Image reconstruction in optical tomography, *Phil. Trans. R. Soc. London — Series B: Biol. Sci.,* 352, 717, 1997.
27. Benaron, D.A. et al., Non-recursive linear algorithms for optical imaging in diffusive media, *Adv. Exper. Med. Biol.,* 361, 215, 1994.
28. Chang, J. et al., Optical imaging of anatomical maps derived from magnetic resonance images using time-independent optical sources, *IEEE Trans. Med. Imaging*, 16, 68, 1997.
29. Franceschini, M.A. et al., On-line optical imaging of the human brain with 160-ms temporal resolution, *Opt. Expr.,* 6, 49, 2000.
30. Jiang, H. et al., Optical image reconstruction using DC data: simulations and experiments, *Phys. Med. Biol.,* 41, 1483, 1996.
31. Paulsen, K.D. and Jiang, H., Spatially varying optical property reconstruction using a finite element diffusion equation approximation, *Med. Phys.,* 22, 691, 1995.
32. Zhu, W. et al., Iterative total least-squares image reconstruction algorithm for optical tomography by the conjugate gradient method, *J. Opt. Soc. Am. A-Opt. Image Sci.,* 14, 799, 1997.
33. Gratton, G., "Opt-cont" and "Opt-3D": a software suite for the analysis and 3D reconstruction of the event-related optical signal (EROS), *Psychophysiology,* 37, S44, 2000.

34. Gratton, G. and Corballis, P.M., Removing the heart from the brain: compensation for the pulse artifact in the photon migration signal, *Psychophysiology,* 32, 292, 1995.
35. Fabiani, M. et al., Bootstrap assessment of the reliability of maxima in surface maps of brain activity of individual subjects derived with electrophysiological and optical methods, *Beh. Res. Meth., Instr., Comp.,* 30, 78, 1998.
36. Syre, F. et al., Are VEP correlated fast optical changes detectable in the adult by non invasive near infrared spectroscopy (NIRS)? *Adv. Exp. Med. Biol.*, in press.
37. Malonek, D. and Grinvald, A., Interactions between electrical activity and cortical microcirculation revealed by imaging spectroscopy: implications for functional brain mapping, *Science,* 272, 551, 1996.
38. Gratton, G. et al., The Event-Related Optical Signal (EROS) elicited by stimulation of different quadrants of the visual field: A replication, article, in preparation.
39. Gratton, G. et al., Fast and localized event-related optical signals (EROS) in the human occipital cortex: comparison with the visual evoked potential and fMRI, *NeuroImage*, 6, 168, 1997.
40. De Labra, C. et al., Optical and electrical recordings of event-related activity in primary visual cortex during binocular rivalry, *Psychophysiology*, 37, S35, 2000.
41. Gratton, G., Attention and probability effects in the human occipital cortex: an optical imaging study, *NeuroReport*, 8, 1749, 1997.
42. Gratton, G. et al., Memory-driven processing in human medial occipital cortex: an event-related optical signal (EROS) study, *Psychophysiology,* 38, 348, 1998.
43. Rinne, T. et al., Scalp-recorded optical signals make sound processing from the auditory cortex visible, *NeuroImage*, 10, 620, 1999.
44. Wee, E. et al., Optical response in auditory cortex to standard and deviant stimuli: a replication, *Psychophysiology*, 37, S103, 2000.
45. Naatanen, R., Mismatch negativity (MMN): perspectives for application, *Int. J. Psychophysiol.,* 37, 3, 2000.
46. DeSoto, M.C. et al. When in doubt, do it both ways: brain evidence of the simultaneous activation of conflicting responses in a spatial Stroop task, *J. Cognit. Neurosci.,* 13, 523, 2001.
47. Gratton, G. et al., Pre- and poststimulus activation of response channels: a psychophysiological analysis, *J. Exp. Psychol.: Hum. Perc. Perf.*, 11, 331, 1988.
48. Fox, P.T. and Raichle, M.E., Stimulus rate determines regional brain blood flow in striate cortex, *Ann. Neurol.*, 17, 303, 1985.
49. Rees, G. et al., Characterizing the relationship between BOLD contrast and regional cerebral blood flow measurements by varying the stimulus presentation rate, *NeuroImage*, 6, 270, 1997.
50. Martinez, A. et al., Involvement of striate and extrastriate visual cortical areas in spatial attention, *Nature Neurosci.,* 2, 364, 1999.
51. Worden, M. et al., Determining the locus of selective attention with function magnetic resonance imaging, *NeuroImage,* 3, S244, 1996.
52. Gratton, G. and Fabiani, M., The event-related optical signal: a new tool for studying brain function, *Int. J. Psychophysiol.*, 42, 109–121, 2001.

Index

A

B

C

D

E

F

G

M

N

O

P

R

S

T

V

W

X

Y